GW00494381

NATURAL GAS DISTRIBUTION
FOCUS ON WESTERN EUROPE

OECD
OCDE
PARIS 1998

INTERNATIONAL ENERGY AGENCY
9, RUE DE LA FÉDÉRATION, 75739 PARIS CEDEX 15, FRANCE

The International Energy Agency (IEA) is an autonomous body which was established in November 1974 within the framework of the Organisation for Economic Co-operation and Development (OECD) to implement an international energy programme.

It carries out a comprehensive programme of energy co-operation among twenty-four* of the OECD's twenty-nine Member countries. The basic aims of the IEA:

- to maintain and improve systems for coping with oil supply disruptions;
- to promote rational energy policies in a global context through co-operative relations with non-Member countries, industry and international organisations;
- to operate a permanent information system on the international oil market;
- to improve the world's energy supply and demand structure by developing alternative energy sources and increasing the efficiency of energy use;
- to assist in the integration of environmental and energy policies.

* IEA Member countries: Australia, Austria, Belgium, Canada, Denmark, Finland, France, Germany, Greece, Hungary, Ireland, Italy, Japan, Luxembourg, the Netherlands, New Zealand, Norway, Portugal, Spain, Sweden, Switzerland, Turkey, the United Kingdom, the United States. The Commission of the European Communities also takes part in the work of the IEA.

ORGANISATION FOR ECONOMIC CO-OPERATION AND DEVELOPMENT

Pursuant to Article 1 of the Convention signed in Paris on 14th December 1960, and which came into force on 30th September 1961, the Organisation for Economic Co-operation and Development (OECD) shall promote policies designed:

- to achieve the highest sustainable economic growth and employment and a rising standard of living in Member countries, while maintaining financial stability, and thus to contribute to the development of the world economy;
- to contribute to sound economic expansion in Member as well as non-member countries in the process of economic development; and
- to contribute to the expansion of world trade on a multilateral, non-discriminatory basis in accordance with international obligations.

The original Member countries of the OECD are Austria, Belgium, Canada, Denmark, France, Germany, Greece, Iceland, Ireland, Italy, Luxembourg, the Netherlands, Norway, Portugal, Spain, Sweden, Switzerland, Turkey, the United Kingdom and the United States. The following countries became Members subsequently through accession at the dates indicated hereafter: Japan (28th April 1964), Finland (28th January 1969), Australia (7th June 1971), New Zealand (29th May 1973), Mexico (18th May 1994), the Czech Republic (21st December 1995), Hungary (7th May 1996), Poland (22nd November 1996) and the Republic of Korea (12th December 1996). The Commission of the European Communities takes part in the work of the OECD (Article 13 of the OECD Convention).

FOREWORD

This study is a companion volume to two earlier IEA publications on natural gas (Natural Gas Transportation and the IEA Natural Gas Security Study). It completes the IEA's analysis of the gas supply chain by focussing on gas distribution, with particular emphasis on the six largest gas countries in Western Europe Belgium, France, Germany, Netherlands, Italy and the UK. It also, more briefly, considers gas distribution in other IEA countries Canada, USA, Japan and Turkey.

The study provides a comprehensive analysis of the market, structure, regulation, costs, pricing and profitability of gas distribution in the six countries. The recent adoption of the EU gas directive sets the framework for the introduction of competition in the natural gas sector. The expectation is that competition will as in other markets generate strong incentives for greater efficiency through pressures to reduce costs, improve productivity and reduce prices to end consumers. The study suggests that there is a potential for increasing efficiency in gas distribution, which is estimated to account for some 50% of value added in the total gas supply chain. Unlocking this potential will, however, require important changes in the way gas distribution is organised and regulated, to make room for competition.

The last chapter of this study explores the different ways in which these changes might be made and competition introduced. The IEA intends to deepen its analysis of natural gas markets through further work. This further work will consider the wider implications of developing competition, including security of supply.

The IEA Secretariat wishes to express its gratitude to Member Countries, in particular Belgium, and to industry contacts for the analytical support they have provided.

The book is published on my authority as Executive Director of the IEA.

Robert Priddle
Executive Director

ACKNOWLEDGEMENTS

The IEA wishes to acknowledge the very helpful cooperation of the relevant ministries and regulatory bodies, as well as natural gas companies in the reviewed member countries, with special thanks to Douglas Paquin of the US DOE, and to Belgium.

The IEA would also like to thank DG XVII of the European Commission, TOTAL and Jonathan Stern for their useful advice, as well as Annelene Decaux for her valuable contribution.

The IEA authors of this study are: Bjørn Saga (project initiator and main researcher), Jochen Hierl, Elizabeth Kuhlenkamp and Caroline Varley.

TABLE OF CONTENTS

LIST OF TABLES

LIST OF FIGURES

I. EXECUTIVE SUMMARY AND RECOMMENDATIONS

INTRODUCTION

The European gas industry has started to undergo fundamental changes as traditional monopoly structures open to competition. Triggered by the EU directive on gas market opening, adopted in May 1998, governments are reviewing the regulatory framework of their gas industry. Their main objective is to improve the industry's economic efficiency, without compromising other important policy objectives such as security of supply, social equity and environmental protection.

The EU gas directive focuses on gas transmission. However, gas distribution - that part of the industry defined by the low pressure grid - is at least equally important in the search for greater efficiency. It accounts for the biggest share of the total cost of bringing gas from the wellhead to the consumer. The absence of competitive pressure in gas distribution means that costs, profits and prices to end users in continental Europe are likely to be higher than necessary.

MARKET AND INDUSTRY STRUCTURE

The six countries studied (Belgium, France, Germany, Italy, the Netherlands and the UK) account for about 90% of gas consumption in OECD Europe and are the most mature gas markets in the region. The share of gas in total primary energy supply varies from 12% in France to 46% in the Netherlands. Household gas use is very high in the Netherlands and the UK, much lower elsewhere. There are 67 million gas customers in the six countries, of which about 96% are households.

With the exception of Great Britain[1] (which has an integrated transmission and distribution infrastructure), the industry in these countries typically consists of a transmission and distribution sector. Local distribution companies (LDCs) generally buy the gas from one transmission company based on long term supply contracts. Transmission companies also sell gas directly to large end users (who account for 30-50% of total consumption). The number of LDCs varies considerably among countries, from France with less than 20 LDCs (GDF accounts for 88% of the market) to Italy with more than 800. LDCs are predominantly in public ownership although the proportion varies among countries. This, however, is gradually changing in favour of private ownership.

---

1. In the following, the UK is referred to as Great Britain in recognition of the fact that Northern Ireland has a distinct gas industry structure and regulatory regime of its own.

Gas is often distributed to end users along with other services - electricity, heat, water and more recently telephone and cable TV services. As LDCs normally enjoy monopoly rights in their supply areas this inevitably reduces competitive pressure and choice in the provision of services.

REGULATION

Gas industry regulation is pervasive in the six countries studied and takes many forms, ranging from government legislation, public ownership, rules applied by regulatory bodies and taxation. It has evolved over time and, with the exception of Great Britain's fairly recent regulatory framework, now often lacks clarity and coherence.

There are two key elements of regulation. The first is the concession or franchise, which typically gives LDCs an exclusive right to distribute gas in specified areas in return for which LDCs have an obligation to supply in those areas. The second key element is price regulation, especially for small, captive consumers (i.e., those who have no choice of supplier). The current practice in Continental Europe is generally to price gas on the basis of its market value in competition with other fuels, i.e., the price at which gas is an attractive alternative to other fuels, notably oil. In some cases, the market-value approach is linked to a cost-plus approach in which the actual costs of supplying the gas are also taken into account. Either way, this is not how prices would be set in a competitive market featuring a choice of gas suppliers, where the incentive would be to minimise underlying costs as well as to price gas in relation to the true costs of supply.

ECONOMICS AND COSTS

Gas distribution is economically significant. Gas sales in the six countries equalled about half a percent of their combined GDP in 1995. It is capital intensive, requiring about twice as much investment per kilometre of pipeline as transmission requires. In a country without gas production, gas distribution investment can account for 70% to 80% of total investment in the supply chain to the end user. Investment is very long term and much of it is sunk, i.e., it cannot be retrieved if a company leaves the market.

A study of distribution costs suggests that about 25% of costs could be reduced by economies of scale, i.e., unit costs should decrease with increases in company size, and about 30% by economies of scope, i.e., cost of supply per unit of gas should fall when done in combination with other services. A sample analysis of LDCs in the Netherlands, Germany and Italy suggests, however, that this potential for economies is not fully exploited. It is likely that LDCs have little incentive to reduce their costs in the absence of competitive pressure.

PRICING AND COSTS

LDCs are generally charged by transmission companies for the gas according to the market value principle. End users generally pay a tariff which is made up of a fixed and a variable element. Industry and power generators generally pay individual tariffs, though on the basis of the same principles.

In terms of economic efficiency, that prices should reflect costs and should not be distorted, the tariff systems fall very short. Cost identification and allocation is difficult in gas distribution, but also generally not well developed. The market value approach to pricing does not make a connection with underlying costs and does not reflect costs explicitly in tariffs. As a result, cross subsidies (allocation of one activity's costs to other activities) are widespread. Only Great Britain and, to some extent Belgium, currently seek to disaggregate the cost components of gas supply. But only the former also applies enough incentives to reduce costs and prices by way of competition.

A comparison of end user prices across the countries studied confirms this. The consistently lowest end-user gas prices in Europe are found in Great Britain, the only country studied with widespread competition within the gas market.

The price comparison also reveals considerable differences in prices among countries in Continental Europe. These are likely to be the result of varying policies on taxation, subsidies and other policy mechanisms which affect prices.

PROFITABILITY

Profitability comparisons across countries are fraught with methodological difficulties, mainly because of different accounting rules. Comparisons must therefore be treated with appropriate caution.

Gas transmission appears to be more profitable in Italy. Gas distribution appears to be more profitable in Belgium, the Netherlands and probably France.

Pure distribution companies, i.e., those which distribute gas as their sole activity, appear to be more profitable than mixed companies. This links to the earlier observation that the potential for economies of scope, which are presumed to exist, does not in fact materialise.

On average, Dutch distribution companies appear to be the most profitable, followed by Italian and Belgian distribution companies. German distribution companies do least well.

In all the continental countries studied, return on equity in both transmission and distribution appears to be particularly high. Profits are distributed among a variety of owners, including public municipalities, other gas distribution or transmission companies, and other private companies.

DEVELOPING COMPETITION: POLICY ANALYSIS

Our analysis of gas distribution economics suggests that economic efficiency is not optimised under the current monopoly conditions. The absence of economies of scale and scope where they might be expected to exist, the very high profitability of the industry, the market value approach to pricing, the relative lack of disaggregated pricing for key cost components such as storage all point in this direction.

There is a need therefore to engage a debate about the best way to introduce incentives into gas distribution that will encourage economic efficiency, whilst sustaining other policy objectives such as security.

Two steps need to be considered. The first is to make LDCs fully eligible for Third Party Access to the high pressure transmission grid. This will give them a choice of suppliers - they will not longer be tied to one transmission company as is the often now the case. The overall effect should be to lower the cost to LDCs of purchasing gas. However, the second step is as important. A greater choice of supplier for LDCs will obviously benefit LDCs, but not necessarily LDC customers, unless there is a further incentive or requirement on LDCs to pass on their lower costs to customers in the form of lower prices. In essence this means finding a way to develop competition within gas distribution, between LDCs and other potential new gas suppliers to end users.

A number of approaches are possible, including direct competition (competing networks), competition through TPA, indirect competition through franchising, and regulatory methods such as price caps. These options need to be weighted in terms of how effectively they will meet policy objectives (economic efficiency, security of supply, social and environmental objectives). They also need to be evaluated against the different backgrounds of individual countries (including their degree of energy self-sufficiency). It is clear that the countries studied differ in the relative weight they put on various policy objectives and their assessment of how best these objectives can be achieved. Each country will have to develop its own reform path.

RECOMMENDATIONS

☐ Economic incentives to improve efficiency in gas distribution should be introduced. The current lack of competitive pressure means that costs, profits and prices are higher than they need to be.

☐ The most effective approach should involve two steps. The first is to consider making LDCs fully eligible for access to the high pressure transmission grids, giving them a choice of supplier and hence opportunities to purchase gas at least cost. The second is to consider how to introduce competitive pressure within gas distribution, so that LDCs are encouraged to pass on the benefits of their lower costs through lower prices to end users.

☐ In assessing the best approach to stimulating competition, governments should be clear about their underlying policy goals and priorities. Maximising economic efficiency needs to be balanced by considerations of the effects of competition on supply security, environmental policy and social equity. There may be trade-offs, for example, between lower prices (in the short term) and long term security of supply. Social objectives may need redefinition: in particular, the concept of "public service" should be made compatible with competitive markets.

☐ Related regulatory or market issues, including taxation, ownership (and privatisation), regulatory structure and the role of general competition law need to be addressed, in order to maximise the effectiveness of the chosen approach.

☐ Finally, governments should be prepared to engage in reform, even though this requires a significant reshaping of the regulatory framework and a careful approach to the transition from monopoly to competition.

II. THE GAS DISTRIBUTION MARKET IN CONTEXT

INTRODUCTION

The IEA has published two major studies on the gas market – "Natural Gas Transportation – organisation and regulation" in 1994 and "The IEA Natural Gas Security Study" in 1995. It is now time to assess the issues raised by market liberalisation and to provide a detailed analysis of the distribution sector (as neither of the previous studies considered this part of the gas chain).

A very significant (up to 50%) share of the value added in the gas chain is in gas distribution. It is, thus, important to take a critical look at economic performance of the distribution sector. Of course, this should not lose sight of other key issues, notably security of supply, social and environmental objectives.

Optimisation of economic efficiency begins with the scope to reduce costs. In a situation characterised in most European countries by exclusive rights to distribute gas in given geographical areas (i.e. local distribution monopolies) there is no clear incentive for cost minimisation.

The second element is the possibility that the gas distribution industry – or at least parts of it – is earning excess profit (i.e. more than would be earned in a competitive market – see further discussion in chapter 7). If there are excess profits, economic efficiency is not being optimised and end consumers are foregoing lower prices.

Eligibility for access to the gas transportation systems – whether the local distribution companies which currently supply most gas retail customers in Europe, and/or retail customers themselves should be eligible for third party access (TPA) and have a choice of suppliers – is a key issue of the current debate on market opening, and will determine to a large extent how much effective competition will develop across the gas market as a whole.

This new study is therefore undertaken first, to provide a firm basis for these policy debates; second, to engage the debates, at least as far as identifying the issues which need attention; and third, to provide some tentative recommendations on the way forward.

The study concentrates on the six major gas countries in Europe, Belgium, France, Germany, Italy, the Netherlands and the UK. These six countries account for more than 90% of total gas consumption in OECD Europe and are mature markets.

Annexed to the study are detailed reports on each of the six countries as well as on Canada, Japan, Turkey and the US.

This chapter sets the scene for market opening and gas distribution in Europe by considering the policy objectives of market opening, by reviewing the EU Gas Directive and national implementation, and finally, by summarising ongoing market developments in response to or anticipation of regulatory changes.

THE CHANGING MARKET: ECONOMIC EFFICIENCY

Globalization through free movement of goods and capital has put performance and efficiency – terms previously only used in the business domain – on the political agenda in most countries. Encouraged by larger prosperity gains, the world wide trend during the last decade has been and still is towards free trade and investment, thus embracing global competition. Excluded from this trend until recently were the so-called utility sectors, including electricity and gas supply.

However, the pressures of competition on the open sectors of the economy are such that cost and efficiency optimisation has now also become an issue for the utility sectors on which the former depend. As a result, the utility sectors are all undergoing (or have undergone) a process of market opening and regulatory reform. The gas industry is no exception.

As stated, the main driver towards more competitive and open markets is the expectation of greater economic efficiency and hence competitiveness in the global market. The drive to optimise economic efficiency in natural gas supply is prompted by the question whether the current gas market organisation and regulatory framework is optimal for achieving this.

In the gas market, as for other utilities, the present organisation has evolved out of a regulatory framework which institutionalises perceived economies of scale and scope. It also gives priority to security and social (public service) objectives. Economic efficiency, so far, has not been a key priority – the consumer / end user was given an advantage in terms of energy diversification at an oil product price level. It can be considered that this was justified, and still is for nascent gas markets, under the goal of diversifying out of oil. In more mature markets, however, further gas penetration brings fewer benefits in terms of energy supply diversification. Here, allowing end-users a share of the gas rent brings new economic benefits.

Both, practical experience and economics have established the disadvantages of monopoly structures, the advantages of competition and the relative merits of the latter in maximising economic efficiency. This also applies to the gas industry, despite its specificities. Fundamentally, there is no automatic incentive for monopolies to minimise costs, maximise efficiency (or productivity) and reduce prices to consumers, in the absence of competition which would force them to do this in order to survive. The establishment of a new market framework with competing suppliers and consumers able to exercise choice will spur suppliers to look systematically for productivity gains and comparative advantages. This is a self reinforcing process – as energy markets become more competitive and more complex, new forms of competition emerge, and the structure of the industry will undergo significant changes. As new entrants come into the gas market, they will tend to disturb the rules of the game and reinforce competitive pressures. Examples are the competitive pressures already created by independent power producers using natural gas, and the entry of Wingas, a new pipeline operator and wholesale gas seller, into the Germany gas market.

Technology has also played an important role in shaping a new perspective, by driving down costs and providing incentives for competition. Arguably, the most significant example is CCGTs, which have transformed the economics of using gas as a fuel input to electricity generation. Power producers are at the forefront of wanting greater choice of gas supplies. Another example is measuring/metering technology which removes a key obstacle to gas-to-gas competition in end user markets.

There is a broad consensus therefore that the underlying economic rationale for a monopoly structure needs revisiting. Regulatory reform and the dismantling of institutional barriers to competition is now underway as a consequence. The box below summarises key features of US and UK markets where the reform process is underway.

Key Features of the UK and US Markets:

☐ Presence of several competing supply companies

☐ Self-supplying in natural gas (though significant US imports from Canada)

☐ Third party access (TPA) enforced by clear regulation

☐ Independent regulator (OFGAS and FERC)

☐ Private ownership throughout the gas chain

☐ Important price reductions for gas consumers

The economics of gas distribution are in particular need of reassessment. Gas distribution has traditionally been viewed as a monopoly. However, the current assessment is that only the physical distribution of gas remains a natural monopoly, in other words, it would not be economically efficient to have competing pipelines. Other activities making up what is normally associated with gas distribution are contestable. Metering and billing could potentially be undertaken by companies separate from the company responsible for the physical transportation of gas. In North America and Great Britain, the supply (or merchant) function of distribution companies is being or has been separated from that of the transportation function.

The consequence has been drastic structural changes in Great Britain and in North America and the same can be expected in Europe if competition is fully engaged. The unbundling of services and the cost savings that took place after deregulation of high pressure transmission in North America have been dramatic. Instead of one market for gas (bundling transportation and supply), separate markets for supply and transportation have arisen. The same developments at distribution level i.e., unbundled supply and transportation, can be observed in Canada which has taken liberalisation further than the US.

With gas becoming more of a normal commodity in these new markets, price volatility increased. This has raised a need for price management, resulting in the development of gas futures and derivative markets. The use of such markets is now widespread among LDCs in North America. Gas suppliers in the UK use the short term markets for gas that have arisen, and will probably also start using the hedging possibilities now being offered in the futures market. In a contract recently established between sellers in the UK and a buyer in Continental Europe there is a clause in the price chapter saying that the pricing over time could be changed to take into account the development in spot prices for gas.

THE CHANGING MARKET: OTHER POLICY OBJECTIVES

It is clear that economic efficiency has to be coupled with other policy objectives. In gas/energy supply, the most important are:

- **security of supply**

   (Open energy markets need to be able to guarantee security of supply. This deserves particular attention with import dependent gas markets such as those of the European continent. But it does not constitute an argument per se against regulatory reform. It is possible to envisage a situation in which at least the larger end-users will be empowered to a larger extent to take responsibility for their own security of supply – in other words, a bottom-up market approach to short-term security can replace the current top-down security arrangements. It is less clear how long-term security (investment, gas/energy supply diversification) will evolve, and governments should, therefore, monitor this.)

- **social objectives**

   (Energy policy has always also been an instrument to pursue social policy objectives – e.g. supply at affordable terms to poor people, but also industrial/structural policy, regional planning, employment. In some countries, the public service aspect of energy policy is very strong. Here, gas distribution has traditionally been considered a public service. In some countries this implies that all consumers have a right to be served and sometimes on terms that are equal for all.

   The choice of objectives to be maintained in a competition environment is a matter for debate at national or even local level. Obviously, trade-offs between social and economic objectives will have to be and can be found.)

- **environmental and climate protection**

   (The question is, how greater competition will affect the tools and methods to meet environmental objectives. What works under the current regulatory frameworks (such as demand-side management) will have to be redesigned for competition where consumers and producers rather than Governments make the decisions.)

This study, however, focuses on the economic efficiency aspects of gas distribution.

THE CHANGING MARKET: EU LEGISLATION

The central element of the regulatory reform process in Europe is the European Union Directive on the creation of an internal market in natural gas (the so-called "Gas Directive"), and its implementation by the European governments. The objective of the Directive is to open up the gas markets of the European Union (EU), both within and across national boundaries. The Directive sets minimum requirements for market opening, in particular concerning Third Party Access (TPA) to the gas transportation systems, enabling large gas consumers (but not necessarily distribution companies) to contract directly with a choice of suppliers for their gas.

It should be noted that the Directive aims at creating an internal market for natural gas in the European Union, which means opening the national gas markets to one another. As such, it obliges the EU Member States to open their gas markets at least to the extent required by its provisions and rules. However, Member States can go further than the Directive's provisions if they wish.

So far, the discussion at European level has concentrated more on gas transmission than on gas distribution. One consequence is that the Directive does not specifically designate LDCs as eligible for TPA - thus not explicitly recognizing for them the benefit of lower wholesale gas purchasing prices (by giving LDCs a choice of suppliers). This issue should get more prominent attention in the decisions which have to be taken at national level on implementation of the Directive and on how far to go in market opening.

In fact, several Member States have already decided or are considering going much further in terms of market opening (for example, setting lower or no thresholds for TPA eligibility, making sure that LDCs are eligible) than is required by the Directive.

This promises to be a fundamental change: at present, in most of the EU the supply of gas is based on exclusive relationships between customer and supplier. In a fully competitive market, customers would purchase gas from a choice of suppliers; and suppliers would have a larger potential customer market for which they would need to compete with other suppliers.

THE CHANGING MARKET: NATIONAL LEGISLATION

Several principles set out in the Gas Directive are not new to European countries, some of which already comply with parts of it. The most conspicuous example is Great Britain, which has legislation providing for competition all along the gas chain (described in more detail in the country annex). Germany already has legislation in place to allow for negotiated TPA and providing for freedom to supply and to build pipelines. The latter two elements did generate competition in the wholesale gas market through the construction of major transmission pipelines by Wingas. Germany has now taken a further step by abolishing the right of gas supply companies to demarcate their gas supply areas from each other, and by abolishing the exclusive character of concession agreements with local authorities. In Italy, there is legislation for limited

TPA to the transmission system which enables gas producers to use own gas in affiliates, and some power producers to buy gas directly from producers. Distribution companies in Austria and Switzerland can request access to the transmission system, and some of them import gas directly, but there is no general TPA.

Several European countries have already adopted or are considering reforms in anticipation of the Gas Directive.

■ Spain has recently adopted TPA legislation which is now under implementation. It was originally intended to be a form of negotiated access, but work is now underway to establish tariff regimes for third party shippers. The threshold limit for eligibility is high (around 0.5 bcm a year), which will make only very large industrial users and power producers eligible. The intention is to lower the threshold over time.

■ Ireland has adopted legislation providing for negotiated TPA which is under implementation.

■ The Netherlands is preparing legislation for reinforcing its provisions on negotiated TPA (in order to make it more effective), and which resembles the UK legislation in that it foresees the creation of licensed public gas suppliers.

■ In Germany, new legislation to reform both the electricity and gas sectors has just been passed.

All that said, for the present, LDCs still have few or no choices between gas suppliers (and most of them remain dependent on one transmission company only).

A key current obstacle in most countries is a (de facto) import monopoly held by one transmission company. Germany is the only practical exception of this in the five continental European countries studied.

A second obstacle is transportation. So far, there are few obligations to transport gas on behalf of another party. The vast majority of transport for another party covers transit and is based on voluntary commercial agreements. Recently, German transmission and regional distribution companies have negotiated contracts with sellers in the UK. They rely on voluntary transportation agreements with the Belgian transmission company. Easier access to transportation would be needed to see more of this.

A third potential obstacle could be access to storage facilities and other means to provide for seasonal flexibility (high winter demand and low summer demand) which characterizes the European markets.

The Gas Directive will change this, though not uniformly and not at the same time in all countries. As stated before, the degree of market opening and the pace of change is to a large extent to be decided by each EU Member State.

THE CHANGING MARKET: INDUSTRY DEVELOPMENTS

More competitive markets will challenge the specialisation of companies as well as their behaviour. A redefinition of the company's business area may be necessary or desirable. The drivers may be:

■ temptation to capture profits in activities located upstream or downstream in relation to present activities. The involvement of gas producers in power generation and in gas distribution is one example; power producers going into gas production is another.

■ competitive pressure to take up new activities in terms of selling new forms of energy to respond to customer needs which tend to be multi-energy based. The customer wants to minimise his energy bill and does not necessarily care whether the energy comes from oil, gas, electricity or coal. In North America and Great Britain there are developments towards "one stop energy shops" that is, companies whose aim it is to satisfy the customer's complete energy needs or even directly the energy related service, e.g. heating or cooling (arguably, Continental European countries also provide numerous examples of this mostly in the form of LDCs).

■ a perceived need to invest in areas outside the energy area (mostly in utilities related networks) in the expectation that the energy business alone may be less profitable. The typical diversification for companies in some European countries is telecommunications. Some companies define themselves as being in the "comfort business" providing not only energy but also entertainment and communication and possibly also other utility services like water supply, sewage, waste collection and treatment.

Alliances, joint ventures and partnerships of various kinds are likely to increase as a means of reducing risk and exposure in the changing market conditions, and because economies of scale, such as in securing gas supply, could appear.

In anticipation of market opening on the European continent, many gas companies are already moving: transmission companies are seeking upstream and downstream participation, and to expand in to gas supply areas abroad (e.g. GDF, Ruhrgas, Enagas/Gas Natural); distribution companies are testing the limits of legislation in order to diversify their supplies (e.g. EnTrade and Delta in the Netherlands), planning to raise private capital on the stock markets (e.g. SMA in Germany) or engaging abroad in similar distribution activities. Where competition has already hit (UK, Germany), customers have been lost to rival suppliers, and prices to customers have come down.

An interesting perspective on the current developments is provided by the European distribution companies themselves. The view is that the future will bring more competition and that margins will come under pressure in the future and costs have to be reduced. At the same time, questions are asked, like whether services that have been free so far should now be priced to the customer, and if so, how. On the other hand, at least where LDCs expect to become eligible for TPA, they also perceive opportunities in terms of supply diversification and purchase price reductions (through exercising market power).

KEY POINTS

☐ Fundamental changes have started to take place in the European gas market. Governments are looking to improve the gas industry's economic efficiency, as part of a general drive to improve macro-economic performance. Other policy objectives – security of supply, social policy, environment – will also have to be accommodated.

☐ The EU Gas Directive will set a minimum benchmark for market opening.

☐ In the Directive, distribution has received less attention than gas transmission. European Member States may need to redress the balance when implementing the Directive, letting LDCs participate in TPA, and by introducing competition into distribution.

ANNEX

SUMMARY OF EU GAS DIRECTIVE

1. The Directive intends to create a competitive market in natural gas by introducing minimum rules for all EU Member States governing the transmission, storage, distribution and supply of natural gas. It focuses on access by third parties ("eligible customers") to network facilities in order to enable them to buy gas from the supplier(s) of their choice.

2. The Directive will ensure that Member States, as a minimum, will make natural gas undertakings, power generators and large final customers eligible for network access, and give LDCs network access for the volumes of gas consumed by the customers in their distribution area that have been designated as eligible. The definition of eligibility of final customers is governed by threshold levels of gas consumption, but also by the total percentage share of market opening. On both, the thresholds and the market opening, the Directive sets precise rules: In a first step, final customers taking at least 25 mm^3/a should become eligible. After 5 years, this threshold should reduce to 15 mm^3/a, and again 5 years later to 5 mm^3/a. If these thresholds fail to achieve a total market opening of at least 20% at the start, 28% 5 years later, and 33% thereafter, Member States are obliged to lower them in order to reach at least these targets. On the other hand, Member States may apply higher thresholds if they wish to avoid exceeding market opening levels of respectively 30%, 38% and 43%. As an indication: averaged out across the European Union, the thresholds set by the Directive represent an effective market opening of respectively +/ 34%, 36%, 42% of total gas consumption.

3. Member States can choose between negotiated and regulated third party access (TPA). Regulated TPA implies a right of access to the system on the basis of published and fixed tariffs for use of that system. Under negotiated TPA, the parties are asked to engage into commercial negotiations for access. Gas companies are to publish their "*main commercial conditions*" for the use of their system. It seems likely that negotiated TPA will not provide the same access guarantee as a regulated TPA regime.

4. Natural gas undertakings may refuse access to their system on the basis of lack of capacity, or where the access to the system would prevent them from carrying out the public service obligations that are assigned to them by the Member State, or if this would cause serious economic and financial difficulties with take or pay contracts (derogation in respect of the latter being subject to Member State and subsequent Commission approval on a case by case basis).

5. As regards to access to upstream pipeline networks, Member States are merely required to observe "*the objectives of fair and open access, achieving a competitive market in natural gas and avoiding any abuse of a dominant position (..)*", and to put in place dispute settlement arrangements.

6. Member States are to designate competent authorities independent of the parties with access to the (internal) accounts of the natural gas undertakings to settle access disputes

expeditiously. Natural gas undertakings are required to keep separate accounts in their internal accounting at least for their gas transmission, distribution, storage and consolidated non gas activities "*as they would be required to do if the activities were carried out by separate undertakings*".

7. In addition to the access provisions, Member States have to guarantee a general freedom to build and operate natural gas facilities (e.g. via objective, non discriminatory and transparent authorisations).

8. The Directive allows Member States to derogate:
   • from point 1., in times of crisis;
   • from point 1. and 6., in case (and as long as) they are dependent on one main external supplier (market share >75%), and are not interconnected with the system of another Member State (Finland and Greece);
   • from point 1. and 6., if they would encounter substantial problems in developing a gas transmission infrastructure in an emergent gas region, or if (and as long as) they qualify as an emergent market (Greece and Portugal) and encounter substantial problems;
   • from point 6., if necessary in order to ensure the proper performance of public service obligations in the distribution sector, and/or until existing or proposed capacity of newly supplied distribution areas is not saturated.

III. MARKET AND INDUSTRY STRUCTURE

INTRODUCTION

This chapter summarises the information on natural gas distribution markets and industry structures given in the country annexes.

It briefly describes the structures of the natural gas market and natural gas industries in the countries concerned. It then focuses on distribution company ownership, market concentration in gas distribution, the degree of horizontal and vertical integration and on contractual arrangements for gas supplies.

MARKET STRUCTURE

Natural gas distribution is normally associated with the activity carried out by Local Distribution Companies (LDCs), that is distribution of gas to end users in the residential/commercial sector, including some industrial users. This study, however, considers all conveyance and sales of gas to end users downstream of the high pressure transmission system as gas distribution for the following reasons:

■ in many countries, the borderline between transmission and distribution of natural gas is not entirely clear in that both the transmission company (in most European countries there is only one) and the LDCs convey gas to industrial customers and to power generators;

■ both in physical and accounting terms, supply of natural gas to large users is often one integrated business activity. To understand the economics and the creation of value added in the whole gas chain it is, therefore, necessary to include the end user segments served by the transmission companies. This is also an important point when comparing the economics of gas transmission and gas distribution;

■ in physical terms, gas distribution as undertaken by LDCs usually takes place at a pressure of less than 15 bar. National transmission takes place in high pressure networks (60 to 80 bar). The pressure in regional transmission is typically between 40 and 15 bar.

The importance of including the large user segments is illustrated by the fact that the share of gas consumption supplied by the LDCs in the countries dealt with in this part of the study varies between 46% and 70% of total supplies, the weighted average being about 60% when Great Britain is excluded. The following table shows the percentages for the six countries:

Figure 1   Share of Total Gas Supplies Distributed through LDCs (1995)

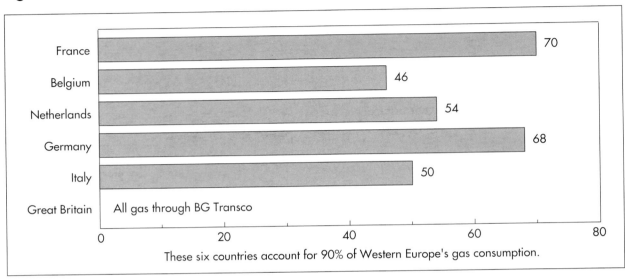

These six countries account for 90% of Western Europe's gas consumption.

Source: International Energy Agency
The figures for France include volumes sold through the LDCs held by GDF as well as the independent LDCs. In Great Britain, one cannot really speak of LDCs since all gas is physically distributed by BG TransCo.

The following table gives a concise description of the gas market structure in the six countries.

Table 1   Gas Share in Energy Balances – 1995

Gas Share in Total Primary Energy Supply (Mtoe)						
	Belgium	**France**	**Germany**	**Italy**	**Netherl.**	**UK**
Gas production	–	2.7	14.8	16.3	60.4	63.6
Gas imports	10.4	28.1	55.3	28.5	2.7	1.5
Gas TPES	10.6	29.6	66.4	44.6	34.0	64.9
Gas Consumption by Sector (Mtoe)						
Residential / commercial	4.4	15.3	26.7	18.5	8.6	25.2
Industry	4.1	12.4	20.8	15.7	8.7	12.7
Electricity and heat production	2.0	0.5	15.1	9.37	8.8	11.2
Gas Penetration in each Sector (%)						
Residential / commercial	34.8	27.1	30.4	50.1	64.6	54.7
Industry	28.5	26.8	26.5	37.0	45.3	29.9
Power generation	13.8	0.8	8.1	19.8	51.9	17.5
Gas share of total TPES	20.3	12.3	19.6	27.7	46.5	29.3

Source: International Energy Agency
TPES = Total Primary Energy Supply, Mtoe = Millions of tonnes of oil equivalent.

The gas share of TPES varied in 1995 from 12% in France to 46% in the Netherlands. In all the countries, the residential sector accounts for the largest share of gas consumption. The share of gas in industry varies from 26% to 45% across the countries. The share of gas in electricity production is still consistently lower, except in the Netherlands which have an exceptionally high share of gas in power generation.

The rate of gas penetration in households varies a lot in the six countries covered. The Netherlands has 97% of all households connected. In the UK, 89% of all households are located within 25 yards of the mains and 92% of these households use gas. In Belgium, 65% of all households can be reached by the grid, but only 55% of these are connected. In France, 28% of the households are connected. In Germany, 37% of all households in the old Bundesländer are connected, whereas the figure for the new Länder is 18% (here, most of the district heating is fuelled by natural gas). In Italy, about 70% of the population has access to natural gas.

The density of customers along the pipeline network and the size of their consumption are important parameters for the economics of the companies involved. Table 2 gives an overview of the number of customers and the consumption per kilometre of pipeline.

Table 2   Customers and Consumption per km of Distribution Pipeline in 1995

	Germany	**Belgium**	**France**	**Italy**	**Netherlands**	**UK**
Length of distribution pipelines (km)	237.000	36.770	129.358	150.000	105.100	277.000
Total sales (mill. GJ)	2637	446	1276	1858	1509	2551
of which to domestic sector	1014	196	732	784	696	1522
Domestic customers per km of pipeline	62	59	72	90	56	74
Non-domestic customers per km of pipeline	2.5	2.1	3.8	5.4	2.6	3.5
Total sales (mill. GJ)/km	11.1	12.1	9.9	12.4	14.4	10.3
of which domestic (mill. GJ)/km	4.3	5.3	5.7	5.2	6.6	5.5
Sales per domestic customer (GJ)	70.0	89.8	79.0	58.1	119.4	82.9
Sales per non-domestic customer (GJ)	2705	3218	1110.0	1326.0	3011	1202

Source: consultant

Surprisingly, the Netherlands is the country with the lowest number of domestic customers per kilometre of distribution pipeline. This is compensated by the fact that it has the highest consumption per domestic customer. Its domestic consumption per km of distribution pipeline is slightly larger than that of other countries (6.6 TJ/km instead of 5 to 6 in other countries). The number of non-domestic customers per km and the average consumption per such customer are influenced by the distribution of such customers between LDCs and the transmission companies, which is most of the time the result of a division of the market giving transmission companies an exclusive right to serve customers beyond a certain size.

INDUSTRY STRUCTURE

The downstream gas industry in most European countries typically consists of a transmission part and a distribution part. The typical pattern is that the transmission part consists of one dominating transmission company, and distribution of a larger number of companies (except Great Britain and France). In all six countries, the transmission companies also sell gas directly to large end consumers (mainly in industry and power generation). The share of total gas supplies delivered directly to such users varies from 30 to 54%.

The LDCs buy their gas almost exclusively from the transmission companies in their country. Today, there are very few examples of European LDCs importing their own gas. The number of companies involved in gas distribution varies a lot from one country to another.

Great Britain is a special case in that British Gas (and now BG plc) has historically taken care of transportation (transmission and distribution) to end consumers. LDCs in the form existing elsewhere in Europe have never existed in Great Britain. After the demerger of British Gas in February 1997, transportation has been separated from marketing of gas. All the companies marketing gas to end consumers in Great Britain have to use the transportation network belonging to BG plc. There are now 64 supply licence holders that are authorised to market gas.

The following table is an attempt at describing the structure of the gas industry in the five remaining countries.

Table 3   Share of Transmission and LDCs in Deliveries to End Consumers (1995)

	Number of transmission companies	Transmission company share of total gas deliveries to end consumers	Number of LDCs	LDC share of gas deliveries to end consumers
Belgium	1 (Distrigaz)	54%	23	46%
France	1 (Gaz de France) + GSO, CFM	30%*	Gaz de France + 15	70%
Germany	18	32%	673	68%
Italy	3 Snam, Edison, SGM	50%	813 **	50%
Netherlands	1 (Gasunie)	46%	35	54%

Source: Company Annual Reports and Figaz, Syndicat Professionnel des Entreprises Gazières Municipales et assimilées (SPEGNN), Bundesverband der deutschen Gas- und Wasserwirtschaft (BGW) and EnergieNed.

*   share of total deliveries includes deliveries by GSO and CFM (note that in France, the 1946 nationalisation law initially envisaged the creation of regional distribution companies - 'EPRD', i.e. 'Etablissements Publics Régionaux de Distribution'; this was suggested again by the Mandil Report in 1994, but never implemented).

** in 1996: 798.

The table needs comment about the situation in each country:

■ In France, there are two companies in addition to GDF that could be considered as transmission companies (GSO and CFM). GDF has majority ownership in CFM and large interests in GSO. Both companies act more or less as regional transmission companies in that they supply to end users as well as to LDCs own locally produced gas (GSO), and gas purchased from GDF or from abroad. In addition to GDF, 15 régies and non-nationalised distribution companies supply gas to the residential/commercial sector and to small industrial customers. In 1995, their share of total gas deliveries to the market as a whole was only about 2.8 %. They purchase their gas from the three transmission companies. In total, GDF had a market share of about 88% of total gas consumption in France in 1995.

■ In Germany, Ruhrgas is the leading transmission company. About 70% of total gas supplies to the German market pass through this company. There are, however, 17 other transmission companies. The German gas industry is much more complex than that of other countries in that there is no clear distinction between different types of gas supply companies in the gas chain. Many companies mainly active in distribution are also involved in transmission and vice versa. Categorisation of regional gas suppliers into transmission or distribution is particularly difficult as many of them have important parts of their business in both.

■ In Belgium, the Netherlands and Italy: one transportation company has a de facto monopoly on high pressure transmission, selling about half the total gas deliveries directly to large users and the other half to LDCs.

In terms of the number of LDCs, Germany and Italy are different from the other countries in that they have a very high number of companies. In the Netherlands, the number of distribution companies has come down from 158 in 1985 to 36 in 1995. In Belgium the number of companies is 23.

The difference in the number of companies per country is reflected in average sales per company:

Figure 2   Average Sales per Company – 1995 (million m³)

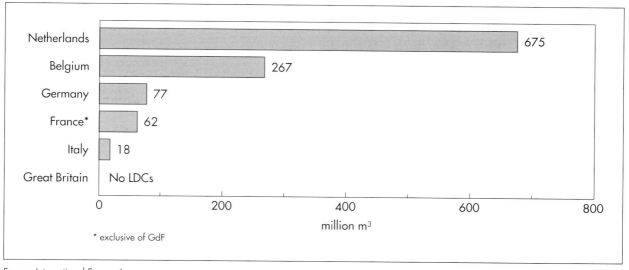

Source: International Energy Agency

In the calculation for France, only the 15 régies and non-nationalised companies existing in 1995 have been taken into account. No figure has been calculated for Great Britain since the marketing companies existing in 1995 only sold gas in the contract market, i.e. mainly serving industrial customers.

OWNERSHIP IN DISTRIBUTION

Ownership is an important policy issue because public and private ownership create different incentive structures. Private owners, to the extent that they are not limited by regulation, may be interested in maximizing profits by reducing costs and keeping up prices. Public owners may be more interested in lowering prices to end consumers, or - on the contrary - inclined to maximize profits in order to cross-subsidise other public services. Public owners (specially state owners as in France), may have only little incentive to reduce (labour) costs because of employment and trade union policies.

Ownership also has implications for the distribution of profits: with private companies, these are likely to remain in the private sector. With public ownership, dividends will accrue either to the state (e.g. in the case of Gaz de France) or to local governments (as, for instance, in the Netherlands). In the case of Belgium (mixed ownership), there are clearly separated dividend streams to local governments and to private partners. Chapter 7 seeks to shed some light on the distribution of profits in local distribution.

In most countries, gas distribution has historically been perceived as a public service and therefore associated with public ownership. Private ownership, however, is becoming more widespread. In the six countries examined, ownership varies from purely private to purely public:

In Belgium there are two categories of company: the "intercommunales mixtes", that is companies originally owned by the municipalities but where private interests have been allowed to take ownership shares; and the "intercommunales pures" - companies that are entirely publicly owned. The first category of companies dominate, holding about 90% of the distribution market. The private shares in 19 mixed companies varies from 42% to 99%. In 17 of them the private share is equal to or higher than 50%. It is important, however, to note a distinction between financial participation and voting rights. The public shareholders always keep the majority of votes on the "Conseil d'Administration" of these companies.

In France, GDF is entirely state owned, and most of the non-nationalised companies are publicly owned. The exceptions to public ownership are the private interests in GSO and CFM (Elf and Total) and minority private interests in three of the non-nationalised companies.

In Germany, it is difficult to get a precise figure for private ownership of the 673 LDCs. The main rule, however, is that the German LDCs are publicly owned, typically by municipal authorities. Less than 25% of the companies have some degree of private ownership. This seems, however, to be more widespread in the large companies than in the smaller ones. The LDCs recently set up in the new Bundesländer have all a mix of public and private ownership.

Italy has a higher private ownership share than Germany. Almost 40% of the companies are private, but these companies account for about 54% of LDC sales and serve 72% of the municipalities. However, Italgas, the largest distributor, is 30% owned by Snam, itself part of the partly privatised state group ENI.

In the Netherlands, all 36 LDCs are in public ownership. As the companies are almost all organised as public limited companies, they are run more like private companies in spite of the fact that regional and local authorities still hold the shares. This is, however, a situation that could change with market liberalisation over the next few years.

In Great Britain, British Gas plc, which physically takes care of all gas distribution, was privatised in 1986. There is very little public capital in the 64 companies that are supply licence holders.

Ownership of LDCs in Europe is thus predominantly public, albeit often with private participation. This is gradually changing, and private participation (though not necessary exercised control) is increasing. The behaviour of LDCs is also slowly changing from public service to market attitudes. This is particularly palpable in Germany and Italy where competition in gas is increasing. The latest example is that of Stadtwerke Mannheim AG (SMA), part of the MVV group and one of Germany's 10 largest mixed local distribution companies: it plans to float 25% of its shares on the German stock exchange in order to finance new expansion activities into distribution, energy trading and telecommunications in Germany and abroad. Another example is Aem Milano, which distributes 850 Mmc/year and was going to float 49% of its shares on the stock exchange on 22 July 1998.

CONCENTRATION AND MARKET SHARES

A high degree of concentration means that there are few but relatively large companies. With captive consumer markets, this is not especially relevant. But under a competitive regime the size of companies can become relevant: large, dominant companies may be more difficult to compete with than smaller ones. The structure of the gas distribution sector also has implications for how many companies may be candidates for direct supply, that is for the number of potential new actors on the scene.

Using the market share of the biggest distribution company as an indication of concentration for each country, the following picture emerges:

Figure 3   Concentration in Gas Distribution Measured by Share of Biggest Company, 1995

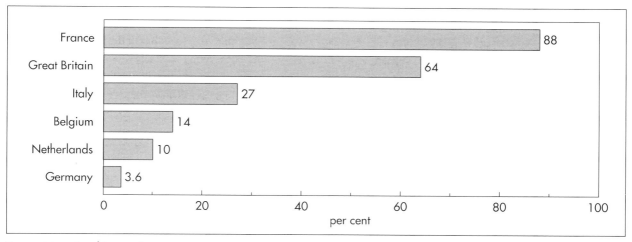

Source: International Energy Agency

France has a monopoly for gas distribution with some exceptions, and consequently the degree of concentration is high. If Centrica were considered a distribution company, it would have a market share of 64% in 1995, coming down from 100% at the beginning of the 1990's. The degree of concentration is much lower in the remaining countries. The biggest company in Italy, Italgas, holds a market share of about 24%, and the second largest about 5%. In 1994, the 30 biggest companies accounted for 58% of total LDC gas sales and served 64% of customers.

In Belgium, the biggest company has a market share of about 14%. The 7 biggest companies account for about 60% of total LDC sales. The three biggest companies in the Netherlands each have about 10% of the market. They are followed by four companies having about 6% of the market each.

Concentration is lowest in Germany: the biggest distribution company only has a market share of 3.6%. This is expected to change: with the energy reform bill now adopted, introduction of competition will rather lead to a concentration process in the German distribution sector.

HORIZONTAL INTEGRATION

Horizontal integration is defined as LDC involvement in activities other than pure gas distribution.

Horizontal integration is an indicator for the potential scope for cross-subsidies between different activities. In some cases, revenues and profits from one form of activity are used to subsidise other activities. Examples include the use of gas distribution revenues to cover deficits stemming from the operation of public baths and public transport. Also, horizontal integration in captive markets may be a limiting factor for inter-fuel competition.

France is a somewhat special case in that GDF is also a gas distribution company but has a close co-operation with EDF (the French electricity monopoly) on marketing. The majority of LDCs not belonging to GDF are also involved in other activities than gas distribution, the most usual combinations being gas, electricity and water distribution.

In Belgium, only 6 of the 23 distribution companies are pure gas companies. Some of these pure companies co-operate closely with electricity companies. The most widespread combinations are gas, electricity and cable TV signal distribution.

In Germany, only 16.2%, i.e. 89 of all the 549 LDCs in the old Bundesländer are pure gas distribution companies. The majority (53.8%) distribute gas, water and electricity, whereas 21.1% distribute gas and water. In the new Bundesländer, the share of pure gas distribution companies is higher: 52 out of 124 companies.

In Italy a high share of companies are also involved in other activities, water distribution being the most frequent. Some of the largest city companies are also involved in electricity distribution. The companies in Federgasacqua, an association of gas and water companies, account for 35% of all gas distribution in Italy. Among its 340 members (1996) 50 companies are pure gas companies and 120 combine gas and water distribution. A high share of the rest of the companies are also active in other public utility services.

In the Netherlands, only 11 of the 35 LDCs are pure gas companies. Another 11 companies distribute electricity, gas and heat. The biggest gas distributors are found among the multi-energy companies. In recent year, the LDCs to a large extent have diversified into activities beyond pure energy distribution.

Great Britain has had a tradition of pure gas distribution, although, as already noted, distribution companies in the Continental sense have never existed here. British Gas has never distributed anything but gas (and some LPG). However, after the introduction of competition, quite a few electricity companies have gone into gas marketing, and gas companies are now entering the parts of the electricity market already opened to competition, as well as other service sectors (e.g. insurance, telecommunications, banking).

The degree of horizontal integration in the LDCs of the countries examined is high. Companies distributing gas exclusively are the exception rather than the rule. What is more, most of these cases in Continental Europe represent supply combinations of e.g. gas, electricity and/or heat. The LDCs concerned enjoy monopoly positions in their supply areas.

VERTICAL INTEGRATION

Vertical integration is defined as ownership links between LDCs and other parts of the gas chain.

Vertical integration creates scope for transfer pricing between different parts of the gas chain and insider transactions. Effective competition will require unbundling and transparency of the different activities (for example, through accounting separation). Vertical integration can also be a factor of inefficiency in that it may prevent detailed cost allocation – the precondition for efficient cost management.

On the other hand, from a security of supply perspective, vertical integration may be seen as positive, especially between gas producers and gas distributors in the wide sense of the word. A producer who holds an investment in distribution could have stronger economic incentives to keep the gas flowing than in an arms-length contract. An example where the security argument has been used to justify vertical integration is the joint-venture between Neste in Finland and Gazprom in Russia, Finland's only gas supplier.

In the Netherlands, there are almost no ownership links between the LDCs and Gasunie, the transmission company, nor between the LDCs and the gas producers. The only exceptions are the minority shares (10%) held by Gasunie in two LDCs (Intergas and Obragas).

In Belgium, which does not have any gas production, the situation is the same: there are no ownership links between Distrigaz, the transmission company, and the LDCs. Indirectly, there are close links, however, in that Electrabel, the dominant electricity company in Belgium, which has a majority ownership share in most of the LDCs, has partly the same owners as Distrigaz.

In Germany, most of the transmission companies have ownership interests in LDCs. Over the last few years, there has been a growing tendency for transmission companies to go further downstream. Some of the gas producers have ownership interests both in transmission and distribution companies. The majority of these cases, however, cannot be qualified as physical vertical integration (they are rather of a financial nature).

In Italy, vertical integration is limited to the ENI group which covers both gas production and imports (Agip and Snam) as well as distribution (Italgas which is partly owned by Snam). The Italgas group holds about 27% of the market. Apart from these links, no LDCs in Italy are vertically integrated.

The degree of vertical integration in France is high in that GDF accounts for 88% of the market and is an integrated transmission and distribution company. The two other gas transmission companies in France are owned by Elf, TOTAL and GDF. Elf is a gas producer in France and imports gas (swapped with GDF) for its CFM operation. Some of the LDCs not belonging to GDF have ownership links to the three companies that have been mentioned.

The gas distribution system in Great Britain has historically probably been the most highly integrated in Europe. In addition to being a considerable gas producer, British Gas was completely integrated from the beach to the burner tip until the early 1990's when competition was introduced. However, transportation and marketing of gas has now been separated through the demerger of the company. Apart from this, some degree of vertical integration has been sustained by gas producers going downstream to market gas, however, in competition with the former monopoly and each other.

LDC CONTRACTS FOR GAS SUPPLIES

The general rule in continental Europe is for LDCs to have long term gas supply contracts with the transmission companies. These contracts typically have a duration of up to 15 or 20 years, and give the LDCs considerable flexibility in volume off take. Mostly there are no take-or-pay clauses in the contracts, which means that the transmission companies handle the volume risk through storage, interruptible contracts and the flexibilities built in to their long-term contracts. In the majority of countries studied, the LDCs have little or no choice of supplier – they have to buy their gas from a monopoly transmission company. This is the case in Belgium and France. In Germany and Italy, distributors have now more choice. In Germany, the abolition of the demarcation contracts has opened the markets to LDCs which can no supply themselves from third parties. In Italy, some LDCs are able to buy gas from Edison and SGM, in abolition to the supplies received from Snam.

Specific points should, however, be noted:

Belgium

In Belgium, the LDCs are obliged to take gas from Distrigaz, the only transmission company. The LDCs have a standard, but individualised ever-green contract which runs for three years and is then renewed. The volume flexibility is virtually unlimited. In principle, the LDCs can buy gas from another supplier if they can get gas at a lower price and are willing to share the price advantage with the other LDCs. Distrigaz must, however, have an opportunity to respond by improving its conditions, and a contract with another supplier would also need government approval after advice from the Comité de Contrôle.

France

In France, as GDF is vertically integrated, only the very few LDCs that are not part of GDF (15 in 1995) have gas supply contracts. These contracts have a minimum duration of three years, but most of them vary between 5 and 15 years. No independent LDCs has the option to choose between GDF and GSO, the other transmission company selling gas to LDCs. GSO's contracts with LDCs are renewed every three years. The contracts between transmission companies and LDCs do not contain take or pay clauses.

Germany

Gas supply contracts between transmission companies (national or regional) and LDCs in Germany typically have a duration of 20 years. Standard contracts are used but terms and conditions vary from one company to another. Because of the demarcation contracts (which disappeared with the adopted energy law reform) most LDCs had no choice of supplier. But since recently and after the entry of Wintershall on the market, there is now potentially more than one supplier in most areas. Up to a few years ago, the contracts between transmission or merchant companies used to be supply contracts with an obligation for the LDCs to take all present and future gas volume under the contract with the merchant company. In recent years, in expectation of the energy law reform (or perhaps the expansion of the Wintershall network) new contracts have typically been limited to specified and sometimes smaller volumes, leaving some freedom for LDCs to take gas from alternative suppliers when the possibility arises.

Italy

In Italy, framework agreements between Snam and the various gas associations are negotiated on a national level. This means that most LDCs have standardised long-term contracts with Snam with no take or pay clauses. Most companies have very little choice of supplier in this context. In the south, though, some LDCs enjoy a choice of supplier due to the presence of Eni and Edison pipelines. Up to now, the choice of supplier has not resulted in price differences but in other contractual aspects. Current contracts are valid for 6 years but economic conditions are renegotiated after three years.

The Netherlands

In the Netherlands, all LDCs buy their gas from Gasunie. In certain cases, however, LDCs can turn to alternative suppliers (this has, in fact, happened recently). Each LDC has its contract with Gasunie, but the contracts are standardised. Duration varies for the different gas streams: for gas to industrial customers it is 5 years; for gas to small consumers it is 10 years. Notice to terminate a supply contract with Gasunie differs as well: LDCs must give 3 years on 5 year contracts and 10 years on 10 year contracts. The off take flexibility is in principle unlimited.

Great Britain

Up to very recently, gas distribution in Great Britain was covered by one integrated company, and consequently one could not talk about specific gas supply contracts for distribution. This has not really changed with the demerger of British Gas and the entry of a number of new gas suppliers on the market. The companies supplying end consumers buy their gas at the beach or from intermediaries under a variety of contracts. A general feature is that these contracts tend to have a shorter duration than before.

Conclusions

Existing contractual arrangements may turn out to be crucial for the speed of change towards a more competitive regime. Given that many of the LDCs have long-term supply contracts, it may take time for these LDCs to change supplier if they are to respect existing contracts. In most countries, however, the LDCs would in principle be free to contract additional volumes from suppliers of their own choice if they were to become eligible for third party (TPA) access.

It will be a matter of legislation to go further and address problems of stranded contracts (or when contracts may not be honoured by LDCs), which may well arise given the long-term nature of present supply contracts between distribution and transmission companies.

KEY POINTS

☐ In 1995, the gas share of TPES varied from 12% (France) to 46% (Netherlands). The rate of gas penetration in households varied considerably from the Netherlands (97%) to Germany (18%). 96% of gas customers across the six countries are households.

☐ The gas industry typically consists of a transmission and distribution part, except for Great Britain and France. LDCs generally buy their gas exclusively from one transmission company (with the exception of Germany). The share of gas consumption supplied by LDCs in the countries studied is significant (in 1995 46-70% of total supplies). Transmission companies also sell gas to end users (30-50% of total consumption). The number of companies in gas distribution varies considerably between countries.

☐ LDC ownership is predominantly public, though it varies across the countries. There are signs that public ownership is gradually changing towards more private ownership, and/or that the 'public service mentality' is slowly turning into more 'market actor' behaviour (e.g. Germany and Netherlands).

☐ LDC concentration and market share also varies significantly. France (state owned, vertically integrated monopoly company) and Great Britain (formerly state owned monopoly with British Gas) have significant concentration, the other countries have much lower concentration levels.

☐ Horizontal integration in LDCs is widespread, and may hinder inter-fuel competition in cases where gas, heat and electricity supply are concentrated within one company. By contrast, the degree of vertical integration is low, with the notable exception of Gaz de France (which may, among other things, hinder introduction of effective gas-to-gas competition in France).

☐ LDCs in most countries have long-term supply contracts with their transmission company. These tend to contain only few and/or hard escape clauses (with the exception of Belgium), and thus bind the distribution company to the transmission company. Recently, however, in anticipation of market opening LDCs are imposing shorter and more flexible terms in contract (re-negotiations (e.g. Germany).

IV. REGULATION

INTRODUCTION

This chapter recalls the main objectives of regulation, and provides a summary of general gas industry regulation in the countries studied. More specifically, it reviews regulation at distribution level, including regulation via concessions/authorizations, tariff setting and taxation.

DEFINITION AND OBJECTIVES OF REGULATION

Regulation can be defined in the widest sense as intervention by governments (state/federal, regional or local) in the market place with a view to framing the behaviour of market participants, and thereby, the market outcome. It, therefore, encompasses general competition law among other interventions.

Ideally, governments only regulate in cases where the market, if left on its own, would (presumably) deliver unsatisfactory results, and where the benefits of governmental intervention are presumed to offset its costs. More specifically, this can be in cases of market failure, or when a market by itself cannot deliver specific results, or in the case of an emergent market.

A government's perception of whether certain conditions justify or make regulation necessary may vary over time in relation to political, economic and technical changes and not least of changes in the market itself.

As such, the nature, place and time of regulation is very much subject to the discretion of governments, which are nevertheless bound by legal and political frameworks (e.g. at regional, national or supranational level).

Regulation can take different forms. A key distinction is between direct and indirect regulation, the former designating any action directly aimed at a market actor (or a group of actors), whereas the latter can be anything affecting the general economic and market framework for all market participants and potential entrants alike (competition law).

Natural gas distribution is an economic activity where government regulation currently plays an important part in the market. The regulation usually takes the following forms:

■ structural regulation determining which firms can or must engage in particular activities, e.g. The evaluation in certain countries of the need for new infrastructures, the award of distribution concessions;

■ price regulation (directly through price/tariff setting, indirectly through taxation or subsidisation); the price can also be regulated directly or through competition law by ensuring that no monopoly or oligopoly power exists;

■ the use of standards in such areas as health, safety and pollution control (they are not the focus of this study) as well as the use of command and control mechanisms for addressing issues which are not always easily assimilated by the market, such as social or environmental policy.

Regulation is not an end in itself. It exists in order to promote public policy objectives. The main policy objectives underpinning the gas market in the countries studied can be grouped under the following headings:

• economic efficiency

• security of supply

• social objectives (e.g. social equity)

• environmental objectives

Up to now, with the exception of the UK, regulation of the gas market in Europe appeared to assume that overall efficiency is maximised by a regulatory framework which promotes monopolistic structures; with security of supply and import dependency best addressed by a bundled demand (through monopsony or oligopsony, e.g. by way of exclusive rights); and that this is also the best framework to deliver desired social objectives (e.g. tariff equalisation).

However, the introduction of competition in some countries has provided substantial benefits to consumers, and is challenging current regulatory systems. The question has to be asked how current systems can be changed so as to achieve a more competitive environment that delivers better economic benefits without ignoring other essentials, such as security of supply, environmental and social policy objectives. This is a fundamental question for Continental Europe, which unlike the countries that so far have introduced competition (Great Britain, US, Canada, Australia, New Zealand) relies significantly on imports from remote sources and few suppliers.

OVERVIEW OF GAS SUPPLY REGULATION

Gas industry regulation in the countries reviewed in this study reflects the importance traditionally attached by the state to the public service character of gas transportation and distribution. It also reflects the economic importance of the industry and the high share of gas in the energy balance of most of the countries.

The country that has the most competitive gas sector of the countries studied, Great Britain, also has the most comprehensive regulatory system. Germany, the country with the second most competitive gas industry, probably has the least comprehensive and interventionist regulatory system. These two examples show that the relationship between competition and regulation is not straight forward.

With one exception, the LDCs in the countries studied operate on the basis of an exclusive right to supply a particular geographical area which most of the time is organised through a concession system. Great Britain is the exception where gas distribution is based on a licencing system which awards non-exclusive rights to distribute gas. In Germany, the exclusivity of the concession agreements between municipalities and distribution companies has just been removed.

The six countries differ significantly in the extent to which they regulate gas prices. In continental Europe, tariffs are set to make gas competitive with other fuels, mostly oil products, and to a smaller extent, electricity, and are subject to regular reviews. Belgium has a stated social objective of "peréquation des tarifs", i.e. all consumers in the same category having the same characteristics should pay the same price wherever they are located. This is also the case in France at regional/local level, but the majority of domestic tariffs are not equalised throughout the country. Equalisation of tariffs is not a stated goal in the other countries. In Germany and the Netherlands, both countries with more liberal attitudes to markets, market value is generally seen as at the basis of gas pricing. However, none of the pricing approaches used can be said to be pure: in countries apparently using a cost plus approach, market conditions are taken into account in various ways, not least the fact that the important gas purchase component is likely to be market value based; and in countries using a market value approach one often finds elements of cost plus pricing. In the UK (and in Germany), where there is gas-to-gas competition, prices are influenced by cost-plus. In these cases, however, prices are more the result of the workings of a competitive gas market rather than of regulation. The UK also has a detailed regulatory tariff regime for the use of its national transport network (which is not subject to competition).

In addition to national gas price regulation, LDCs in some of the countries are also subject to more or less explicit local regulation, typically by municipalities. In many cases, municipal authorities have the final word in tariff matters, and it is clear that decisions are influenced by specific political circumstances, which for instance leads to reluctance to increase gas prices to end consumers just before elections.

Price regulation also has to be seen in a wider political and non-energy context. Taxation, subsidisation (particularly for new customers) and tolerance/support of cross-subsidisation are equally important forms of regulation, aimed, for example, to promote social or regional objectives. They can have an important bearing on the end-price of natural gas, on interfuel-competition as well as on the promotion/discouragement of market actors.

Cross-subsidisation, if required by law or encouraged by authorities, can constitute a form of price regulation. The European Commission defines cross-subsidisation as follows[1]: "Cross-

---

1. See cross-subsidisation in Community Law Volume 13 by Leigh Hancher. Series of publications by the Academy of European Law in Trier Bundesanzeiger Verlagsges. mBH Köln, 1995.

subsidies means that an undertaking allocates all or part of the costs of its activity in one product or geographic market to its activities in another product or geographic market."

The definition of cross-subsidies for the purpose of this section will be distinguished from other forms of subsidy by the fact that they are financed from internally generated funds as opposed to some external source (whether state funds or some other form of direct or indirect financial benefit).

It should be noted that in most European countries, public sector firms are mandated by law to cover their costs (variously defined) and to avoid undue discrimination. This is extremely general and it is of course difficult to determine what constitutes 'discrimination' (e.g offering the 'same' service at different prices) and 'undue' (i.e. when such discrimination is unjustifiable). The question is whether charging different tariffs for the "same" output where costs of distribution differ constitutes "undue" or "due" discrimination. This leads to the difficult issue of balancing potentially conflicting public policy goals. From the perspective of economic efficiency, tariffs should reflect the wholesale gas price plus the costs for its supply and security of supply. However, the route predicated by the policy goal of economic efficiency is not necessarily compatible with other public service or social goals. In particular, should differences in cost of provision be translated into higher tariffs? Should the utilities be able to discriminate between consumers on the basis of demand characteristics? The first of these questions raises the issue of tariff equalisation while the second raises the issue of preferential tariffs. Within the European Union these issues have been tackled in several Commission decisions and rulings of the Court of Justice. On the whole, it seems that tariff equalisation based on cost averaging across geographical areas is viewed as perfectly legitimate at regional or local level, and that preferential tariffs have been accepted only when they have been objectively justified.

In that respect, a key question raised by the new UK government is that of social equity. Broadly speaking, smaller consumers have the highest unit costs (because of the need to recover all fixed costs of distribution on smaller gas unit sales), but they cannot be entirely recovered through domestic tariffs (since the energy bill for low income consumers tends to be very high already in proportion to their income, even with equalised tariffs). There is no perfect answer to this social problem. It may warrant some kind of subsidy in order to recover the fixed costs of the investment in the distribution network.

Cross-subsidisation in the gas sector between different gas customers and gas customer groups seems to be widespread in Europe, particularly in gas distribution. Unfortunately, to give an estimate of the size of the total cross-subsidies in the gas distribution sector in Europe or per country is beyond this study's scope. This would at least require full access to (and understanding of) the internal accounts of most gas supply companies in Europe (assuming that they have the relevant information at all available).

A general observation is that in countries or distribution areas with declared gas price equalisation, publicly supported/encouraged cross-subsidies in favour of users in remote areas take place by definition. Cross-subsidies are also likely to be present where disadvantaged groups (disabled, elderly, poor) obtain a rebate on the gas price they pay for their own consumption.

Besides cross-subsidisation between gas customers, there is often also a form of sectoral cross-subsidisation relating to gas distribution. Where local governments have ownership shares or exert majority control in a local gas distribution company (which by the nature of its supply network is a monopoly), gas tariffs can be regulated in order to maximise the company's income so as to support local community needs. The income is used to subsidise/finance other activities (e.g. water distribution, public transport, etc.). This is often the case with German gas distribution companies which are majority-ruled by local government. The reverse can happen too. In some cases, economically unjustifiable distribution network extensions have received economic support from local authorities.

Energy taxation is another potentially very important source of price distortion. Energy taxation in the six countries varies significantly both in terms of types of taxes and of levels. For small gas customers, the share of total taxes in gas end user prices varies from about 50% (made up of many different taxes) in Italy, to only 5% (V.A.T.) in the UK. V.A.T. is the only tax on gas in the UK. In most of the countries in Europe, taxes on gas oil and heavy fuel oil are at least as high if not higher than taxes on gas. This has no doubt facilitated the penetration of natural gas. Today, despite the environmental advantages of natural gas over other fossil fuels, the tendency seems to be in favour of increasing taxation on natural gas both in absolute as well as in relative terms to other fuels. This would have a negative influence on gas penetration in the EU.

The following table give an overview of the level of taxation for gas and for its main competitors, gas oil and heavy fuel oil, in the six countries:

Table 4   Taxes in ECU/GJ as of 1 January 1998

ECU/GJ (net calorific value)	B	F	D	I	NL	UK
use of natural gas in residential/comm. sector	0.37	–	0.51	5.46	1.66	–
use of gas oil in residential/comm. sector	0.37	2.18	1.04	10.78	2.85	0.91
non-deductible VAT (%)	21.00	20.60	15.00	20.00	17.50	5.00
use of natural gas in industry	–	0.33	0.51	0.45	0.20	–
use of gas oil in industry	0.50	2.18	1.04	10.78	1.79	0.91
use of heavy fuel oil (–1% sulphur)	0.15	0.46	0.34	0.56	0.83	0.71
heavy fuel oil (+1% sulphur) in industry	0.46	0.61	0.34	1.15	–	0.71
use of natural gas in power generation	–	0.33	0.51	–	0.20	–
use of heavy fuel oil (–1% sulphur)	0.15	0.46	0.62	0.36	0.83	0.71
heavy fuel oil (+1% sulphur) in power generation	0.46	0.61	0.62	0.36	–	0.71
use of coal in power generation	–	–	–	–	0.36	–

Source: Eurogas 1998 (taxation figures given for natural gas can vary in the cases of Italy and the Netherlands, depending on region or volume of consumption)

The following observations can be made on this table:

■ Gas use in the residential/commercial market is generally more heavily taxed than gas in industry and power generation.

■ V.A.T., which is non-deductible only in the residential/commercial sector, varies a lot across countries: from 5% in the UK to 21% in Belgium.

■ In all the countries, taxes on gas oil for residential/commercial are equal (Belgium) or higher to those on natural gas.

■ Without exception, taxation of gas oil for industrial use is heavier than for gas. With one exception (Germany), taxation of heavy fuel oil for industrial uses is heavier than for gas.

In addition to having different levels of taxes on gas, the countries studied also pursue very different objectives with energy taxation, reflecting different policy objectives. Some of the taxes have a revenue raising purpose. In some cases taxation is used as an instrument for economic development. In some countries new taxes are introduced explicitly to pursue environmental goals.

Table 5   Gas Rent in Terms of Taxation, 1998
(ECU/GJ)

	Belgium	**France**	**Germany**	**Italy**	**Netherlands**	**UK**
Residential sector	0	2.18	0.53	5.32	1.20	0.91
Industrial sector	0.15 - 0.5	0.13 - 1.85	0.11 - 0.53	0.36 - 10.33	0.63 - 1.59	0.71 - 0.91

Source: Eurogas

In the EU, energy taxation still lies in the hands of Member States (decisions on taxation are taken by unanimity). This makes any significant progress in energy tax harmonisation in the EU very difficult. The European Commission put forward a modest draft Directive in 1997 proposing to increase the taxation of natural gas in relation to its competing fuels. The effect would be an automatic reduction in gas rent for the downstream gas supply companies. There are no direct benefits to consumers (most consumers without dual-firing capacities would be likely to face a higher gas bill).

COUNTRY-OVERVIEW

Belgium

General
Regulatory
Framework

There are a number of specific laws concerning the gas industry on which Government intervention in the gas sector is based. Legislation clearly defines who is allowed to play what role in the market. Government influence over the gas industry is no longer based on extensive ownership, although its golden share in Distrigaz is of importance. The Energy Department of the Ministry of Economic Affairs is responsible for the gas sector. A separate regulatory body, the Comité de Contrôle, composed of representatives from the gas industry, various consumer groups and the Government, administers the price regulation decided by the Ministry of Economic Affairs.

Regulation
of Distribution

There is no concession system. Generally, though not always, the municipality is the owner of the grid. The LDCs have exclusive agreements with the municipalities where they operate. On the other hand the LDCs have an obligation to supply subject to economic constraints. The agreements with the LDCs last for 20 to 30 years and can in principle be renewed. The LDCs pay fees to the municipalities.

Belgium (5)+

Price Regulation

Prices in all parts of the gas chain are fixed by the Comité de Contrôle and approved by the Government. The prices set are considered maximum prices and the companies are supposed to follow these (which is of course a necessity if customers are to pay the same price all over the country). The approach to pricing looks like a relatively pure cost-plus approach: the elements added to the import price to arrive at the price to the end user contain no direct reference to the prices of alternative fuels. That said, oil prices do have an influence on end user gas prices through import prices since these are linked to a large extent to crude and oil product prices. Prices to end consumers in the distribution sector consist of three elements: the gas import price, a transportation element and a distribution element. The latter two elements are supposed to cover costs and profits in the transportation and distribution part of the chain, respectively. Prices to large energy consumers and power producers served by Distrigaz, the monopoly transmission company, are composed of only the first two elements. The gas import element is based on the market value of gas less the transportation and distribution cost elements.

The tariff system features a social tariff in favour of certain disadvantaged groups. The price reduction obtained by these groups (which is not a large share of the normal price) can be characterised as a cross subsidy. Large industrial users seem to be favourably treated in terms of gas prices: some of their volumes are kept outside the general calculation of import costs, and interruptible volumes are exempted from contributing to transportation costs. This could indicate an implicit cross-subsidization of such users.

Taxation

The main tax on gas is an energy tax introduced in 1993. As the taxation policy is one of fiscal neutrality between fuels, this tax is the same for gas as for oil products. More indirect taxes on gas are a special withholding tax on income earned by gas distribution companies, and payments to the municipalities for the right to distribute gas. Some municipalities also levy specific local taxes.

France

Regulation is primarily the role of the Ministry of Industry, Posts and Telecommunications. The Ministry of Economics is also closely involved in price regulation. Regulation also partly takes place through the state ownership of GDF, which in turn is based on specific legislation. The tri-annual 'contrat d'objectifs' between the Government and GDF has in recent years become an important instrument of regulation. Legislation clearly defines who is allowed to play what role in the market.

Regulation of Distribution

France has a fully-fledged concession system where both GDF and the independent distribution companies hold concessions which give them an exclusive right to sell gas and an obligation to supply under certain conditions. A model concession agreement imposes upon the 'concessionaire' a number of rules in terms of obligation to supply, connection and delivery conditions, tariff setting and investment criteria for network extensions. The concession period is thirty years or more.

Price Regulation

Prices to all gas customers are regulated, although prices to large industrial customers beyond a certain size less so than prices to other consumer categories. The relevant legislation lays down an apparent cost-plus principle allowing the gas seller to cover gas purchase and operating costs. In practice, the gas seller is allowed to take the market situation into account by pricing the gas in relation to competing fuels. Also, price regulation is relatively light-handed in that gas suppliers deposit their tariffs with the Ministry of Finance, and they become valid if there are no objections from the Ministry. GDF is also subject to regulation through the 'Contrat d'Objectifs' with the Government: it stipulates that tariff adjustments shall take place according to a price cap formula under which half the productivity gains realised shall be reflected in tariffs.

Taxation

Natural gas is subject to a gas consumption tax and a special levy to fund the Institut Français du Pétrole, both applying to consumers using more than 5 million kWh/year. Gas for domestic use and for feedstock is exempt, gas for co-generation is exempt for the first five years of operation of the installation.

Germany

General Regulatory Framework

Gas industry regulation is to a large extent based on two broad framework laws, the Energy Act and the Act against Restraint of Competition. Both have recently been reformed in order to implement the "EU Electricity Directive", with significant consequences also for the gas sector. Nevertheless, further gas-specific reforms are expected in order to insure compliance with the EU Gas Directive. The Federal Ministry of Economic Affairs is responsible for the gas industry, but plays no dominant role for instance in terms of price regulation, issuing only fairly general guidelines. The Competition authorities, either at federal of Länder level, have a fundamental role in price monitoring and in preventing abusive and discriminatory pricing. A licence system contained in legislation defines the conditions for market entry.

Regulation of Distribution

Germany has a two-part system in that a licence from the Land and a concession is needed for a company to engage in gas distribution. The licence does not grant a monopoly or exclusive rights. Up to now, exclusivity of supply derived from the concession contract, a private law agreement between the municipality and the utility, but the reform has now outlawed this. As a result, all such exclusivity agreements will fall. German LDCs pay significant concession fees to the municipalities. They have an obligation to supply; there is even legislation providing for payment of damages to the customer by the LDC in the event of interruptions or irregularities in supply. The LDCs may for economic reasons be exempted from their obligation to supply new areas.

Price Regulation

There is no direct regulation of the level of prices as such, which are primarily based on the market value principle which means that prices are set to stay competitive with alternative fuels. In the prices paid by LDCs to transmission companies there is, however, also an element of cost plus. There is some legislation concerning the structure of tariffs to small consumers. Prices are monitored by the competition authorities, which are empowered to intervene if they consider prices too high. The principle of price equalisation is alien to the German concept of energy supply, and gas (as electricity) prices can vary significantly between companies and regions. Lately, however, the federal competition and cartel authority has put a number of distribution companies under pressure to align their tariffs on lower levels offered by comparable companies.

A minority of LDCs distribute only gas. Most of the distribution companies are also involved in other energy distribution activities and non-related activities like public transport and public baths. In some cases gas distribution revenues more or less directly subsidise other activities, which means cross-subsidization of one category of public service users by another. The concession fees paid by different types of gas users are differentiated (in a way that does not necessarily reflect economic criteria.)

Taxation

The main tax on gas is an excise tax introduced in 1989, currently 3.6 DM/MWh. German gas companies also pay concession fees to the municipalities which in the end are carried by the consumers. Gas producers also pay royalties to the Länder governments.

Italy

General Regulatory Framework

There is no single framework law applying to the gas sector. Traditionally, regulation has been the responsibility of the Ministry of Industry, Commerce and Crafts, partly exercised through state ownership in the ENI companies. In 1995, however, a new joint regulatory body for gas and electricity was set up to promote competition and regulate prices, based on Law no. 481 dated 14 November 1995. In parallel, the state has reduced its ownership in the ENI group.

Regulation of Distribution

Italy has an exclusive concession system. The attribution of concessions is not necessarily but frequently subject to public tender. The tendering procedure is based on criteria like rent to the municipality and pace of development (because it increases the rent generated), but rarely quality of service and prices to end consumers. Each LDC typically has a concession in several municipalities. The normal duration of a concession is between 20 and 30 years. In principle, a concession can be renewed. The concession holder is generally obliged to supply all customers within +/-10 metres of the mains (varies between concessions).

Price Regulation

Price regulation, especially at the LDC level, was formerly handled by the "Comitato Interministeriale Prezzi" (CIP - Interministerial Pricing Committee). The CIP was abolished in 1993 and its functions taken over by the Minitry of Industry, Commerce and Crafts. In principle, general guidelines for pricing could be laid down by the Ministry as part of energy policy. But the regulatory authority is independent of the Ministry and has full responsibility for tariffs setting in electricity and gas. Prices agreed between Snam, LDCs and other customers are supervised or monitored. The prices applied by LDCs are set through a detailed method which uses a kind of cost of service approach akin to the method used in North America cost of service regulation. It is not a pure cost plus approach in that there are also elements of price capping and rate of return regulation in the method. Since liquidation of the CIP, the function of monitoring LDC prices and LDC compliance, has been under the responsibility of the provincial offices of the Ministry, but this role will expire at the end of the year and will be fully taken over by the Authority.

The Italian gas tariff system features some examples of cross-subsidies between customer categories that are particularly evident between the domestic and the small industry sectors.

Taxation

Italy applies a complex tax system on natural gas. Gas taxation differs between sectors, and also within the residential/commercial sector. Here, gas is subject to different excise taxes depending on the use made of the gas (cooking & water heating / cooking, water & space heating / other, which in practice is based on the level of consumption of each household) and on the region where it is consumed, to additional regional taxes, and to VAT (until mid 1997 also differentiated according to region and type of consumption). Natural gas for industrial use is subject to an excise tax of 20 Lit/m^3 and to a regional tax of 10 Lit/m^3. Gas used in power generation, in refining processes, and as raw material in chemical plants is exempted from taxation.

The Netherlands

General Regulatory Framework

The Government bases its intervention in the gas industry on several agreements with the major actors in the industry and on ownership in some of them. These agreements do not define exclusive rights as such. The Ministry of Economic Affairs covers the Government's interests in the gas industry. The Agreement between Gasunie and the State defines certain competences of the minister of Economic Affairs towards Gasunie. Other agreements define rules for government, for example in the case of the Groningen field. Explicit regulation is not a salient feature of the Dutch system.

Regulation of Distribution

Some LDCs in the Netherlands had a local concession, but the concession system was abolished when the "Wet Energiedistributie" came into effect, ruling out the possibility for local governments to give concessions. In the new legislation under discussion in the Netherlands, a state licence system will probably be introduced giving existing gas suppliers a temporary monopoly on the supply of approximately 50% of gas consumers. As soon as the market is liberalised completely these licences will be abolished altogether.

Price Regulation

Price regulation is relatively light handed. Under an agreement between Gasunie and the Dutch government, the Minister of Economic Affairs has the power to approve the selling prices charged by Gasunie to its customers. Under the Natural Gas Price Law, the Minister is authorised to lay down maximum prices if prices agreed upon by the parties are considered not to reflect the market value. The Minister also has the power to make binding recommendations to individual gas companies on their tariffs. In principle, however, no legal restrictions apply to prices charged by the LDCs.

End user prices are set by applying the market value principle. The price for gas paid by the LDC is negotiated by Gasunie and EnergieNed, the association of LDCs. The price is set by deducting a margin to cover LDC costs and profits from the market value price. One might therefore claim that from the point of view of LDCs there is an element of cost plus thinking.

Consumers in four provinces in the Netherlands benefit from tariffs that are lower than elsewhere in the country. Apparently consumers in these areas are cross-subsidized, but it is claimed that their price reduction reflects lower transportation costs because of geographical closeness to the Groningen field.

Taxation

Natural gas is taxed with an environmental levy and a so-called ecotax, in addition to VAT. The Dutch energy tax system can be considered as relative neutral to natural gas (no discrimination, no promotion).

United Kingdom

General Regulatory Framework

Explicit new regulation is a salient feature of the British gas industry after the privatisation of British Gas and introduction of competition in the early 1990s in order to promote and establish competition (the reason for regulation is therefore rather different than that of many other countries). The major role of the Department of Industry and Trade has been to put in place the necessary framework laws to promote competition. In parallel, a separate regulatory body for the gas industry, OFGAS, has been set up. This body monitors and regulates the industry on a detailed level. However, the intention is to reduce the level of regulation once a fully competitive and self maintaining market will have been established.

Regulation of Distribution

LDCs as such do not traditionally form part of the British gas industry structure. The Gas Act 1995 introduced a comprehensive licencing system in the British gas sector. Companies that want to market gas need a gas supply licence and also a shipper's licence where gas is to be delivered elsewhere than at the beach. A transportation licence is needed by companies that want to offer transportation services. Licenced transporters cannot, however, hold either of the other two licences. The holder of a supply licence has to accept a number of fairly strict conditions intended to protect captive customers. The licence does not give exclusivity in a geographic area, but implies an obligation to supply all customers within 23 metres of the mains. The licencing system covers a broad range of conditions aimed at securing social objectives and security of supply.

Price Regulation

In spite of the fact that it has the most competitive gas market in Europe, Great Britain has a heavy handed price regulation. OFGAS is the institution responsible for gas price regulation. At the beginning of 1997, price regulation was as follows:

- prices for end users taking less than 2500 therms per year are regulated through a price cap formula on British Gas.
- prices for customers above 2500 therms per year (but less than 25.000) are not regulated but are set according to a tariff schedule.
- prices for end users in the contract market (above 25.000 therms per year) are not regulated any more.

The prices applied by Centrica (formerly British Gas Trading) to end consumers are to be considered as maximum prices. The result of introducing competition in the various market segments has been to push prices down.

Transportation and storage tariffs are regulated through restrictions on the rate of return of Transco, the only gas transporter, and through a price cap formula. The price formulae both for end user prices and transportation may be said to be of a cost plus type. Many of the changes in regulation in recent years have been undertaken to make tariffs more cost reflective.

Traditionally there has been a strong link to oil prices in both beach supply contracts and in end user contracts in the UK, although less strong than in other European countries. With the emergence of gas to gas competition, this link has progressively disappeared. Many of the new contracts do not contain any reference to oil prices or other fuels at all.

Finally, there is the question of how best to eliminate or at least reduce the scope for cross-subsidies. Unbundling of activities or functions is key. Transparency – at least for the regulator – is also very important. Unbundling has, to a large extent, taken place in Britain through the separation of the transportation and marketing functions. In addition price regulation by OFGAS has actively forced both the transportation and marketing parts of British Gas to justify their cost allocation. It is still claimed, however, that there exists some scope for cross-subsidies in that Centrica (one of the companies coming out of the British Gas demerger) might subsidise its activity in the areas where there is now also competition in the domestic market with revenues from areas where competition has not yet been introduced. Yet this will disappear with the opening of competition in the whole household market.

Taxation

The UK has a very simple system in that there are no taxes on natural gas apart from an 5% VAT in the domestic sector.

KEY POINTS

☐ Regulation is not an aim in itself but has to promote policy objectives.

☐ Gas industry regulation in the countries studied is comprehensive. Regulation takes place through relevant legislation implemented by governments, through state ownership in the industry, through regulatory bodies more or less independent from governments and through taxation (e.g. excise taxes, environmental levies).

☐ Reform of the gas supply industry is likely to require, at least at the beginning, a significant level of 're-regulation' in order to provide new transparent and explicit rules to replace the former 'ad hoc' system.

☐ An important aspect of LDC regulation in most of the countries is the concession system. The concession typically gives the LDC an exclusive right to distribute gas in a specified area. In return, the LDC has an obligation to supply, in some cases subject to certain conditions of an economic nature. Only in very few cases are concessions subject to public tender. Linked to this is the fact that most concessions are - fully or largely - publicly owned (state, region, municipality).

☐ Price regulation in the gas sectors of the six countries is always based implicitly or explicitly on 'market value' to which - in varying degree from country to country - elements of cost plus pricing are added. Equalisation of prices, that is the same price for the same type of customers in all parts of the country, is a stated goal only in Belgium. This used to be the case in France also, but domestic tariffs are no longer equalised throughout the country (equalisation is still applied in some areas).

V. ECONOMICS AND COSTS

INTRODUCTION

This chapter reminds the reader of the natural gas distribution sector's significance in the national economies of the six west European countries, and in comparison with the natural gas transmission sector.

It then analyses the typical cost structure in gas distribution, and tries to identify cost items that might offer potential for cost reduction.

Finally, the issue of potentially poor cost efficiency in gas distribution is examined, based on the analysis that gas distribution companies seem not to exploit enough the possibilities for economies of scale and scope.

Warning

The analysis in chapter V. and chapter VII. draws to a large extent on figures provided by individual companies in their annual reports. It should be taken with some reservation.

The use of such figures for economic analysis is not without some difficulty as they are defined in each case to satisfy the purpose of publishing the financial and business performance results of an individual company in accordance with the respective national legislation in place. As such, these figures do not always carry the full information relevant to the economic analysis, and comparisons between companies and countries must be done with caution as definitons and perimeters may be different in each case. As examples can be mentioned: specific company activities can be outsourced or carried through by own workforce without accounting figures revealing the actual situation in each case; the definition of distribution can vary between companies from e.g. *'all volumes sold to small consumers' to 'low pressure gas transport'* (both definitions could result in substantially different figures on distribution for a same company).

ECONOMIC SIGNIFICANCE OF NATURAL GAS DISTRIBUTION

Natural gas distribution is a sector of considerable significance in economic terms. The value of total gas sales in the six countries comprised by this study amounted to some 76 billion US $ in

1995, which corresponded to about 1% of their combined GDP. This figure includes the total revenues from domestic sales. Total LDC revenues in the countries where LDC sales are easy to distinguish from transmission activity (Belgium, Netherlands, Germany and Italy) amounted to more than 32 billion US $ in 1995, or to 0,7% of the same overall GDP if one extrapolates figures for the UK and France (as reference: the European Commission estimates the benefits of European Monetary Union at 0,5% of the EU's GDP).

The share of gas distribution revenues in the economic value chain from the wellhead or the import point down to the final consumer accounts for about half of the average end price (see below):

Table 6   Value Added in Gas Distribution as a Percentage of Average End User Prices, 1995

Belgium	46%
France	52%[1]
Germany	53.7%[3]
Italy	37%
Netherlands	19.7%
UK	48.2 [2]

Source: IEA calculations based on data from various sources.

(1) Gas distribution as undertaken by the "régies" independent from GDF.

(2) Value added here also includes high pressure transmission.

(3) Including gas tax of 3,5Pf/m^3.

Differences between the countries can be attributed to different market structures and LDC customer bases, different ways of setting prices in the chain, and differences in costs.

The weighted average of value added in gas distribution in these countries is around 50% of end user prices. This means that the gas distribution part of the total natural gas chain represents a high, if not the highest share of the total cost of bringing gas from the wellhead to the consumer. Policy makers in Europe need to keep this in mind when designing reforms for implementation of the Gas Directive (which leaves distribution very much to subsidiarity).

The following is an attempt to review investment, costs and cost efficiency in natural gas distribution.

INVESTMENT IN GAS DISTRIBUTION AND TRANSMISSION

The natural gas supply industry is a very capital-intensive sector, as shown in the following table.

Table 7 Investment in Gas Transmission and Distribution, 1996
(million ECU)

B	D	F	I	NL	UK
206	3400	862	2190	540	375

Source: Eurogas

However, it is difficult to allocate the investment clearly to either gas transmission in high pressure networks or to gas distribution. Gas distribution is generally considered much more capital intensive than transmission, though. Below are two examples thought to be fairly representative of European conditions, those of Germany (the old Bundesländer) and Belgium.

Table 8 Investment in the Gas Chain in Germany and Belgium, 1995 (%)

	Germany (old Bundesländer 1994-1995)			Belgium (1991-1995)		
	Transmission	Public	Total Investment by Source	Transmission	Public	Total Investment by Source
Production	0.5	0.2	0.7	–	–	–
Storage	4.7	1.6	6.3	4.2	–	4.2
Compression reduction stations	3.7	4.3	8.0	2.9	7.0	9.9
Network: pressure stations	18.8	52.3	71.1	9.9	70.4	80.3
Meters	0.2	3.6	3.8	–	–	–
Other	3.4	6.7	10.1	1.9	3.7	5.6
Total Investment by Sector	**31.3**	**68.7**	**100%**	**18.9**	**81.1**	**100%**

Source: Consultant

Generally speaking there is broad agreement on the following figures:

■ Investment in transmission accounts for approx. 20 to 30% of total investment cost in the downstream gas chain (production excepted). It covers:
  • the network (pipelines): 60 to 75%.
  • storage: 5-7%.
  • compression, blending stations, pressure reduction stations, metering: 15-20%.

■ Distribution makes up 70 to 80% of the total investment. The main items are:
  • the network: more than 75%.
  • meters: around 5%.
  • compression reduction stations: around 10%.

In general, one can assume that total investment in gas distribution will typically be at least twice as high as in transmission.

Investment in gas distribution is also long-term in nature. This is illustrated by the fact that the depreciation periods in economic terms in some countries reach sixty years.

COSTS IN GAS DISTRIBUTION

The following table gives an indication of the relative size of the costs (and profits) in the natural gas chain that are generated in gas distribution:

Table 9   Representative Distribution Company Accounts (%)

	Germany [1]	Belgium [2]	Italy [3]	Netherlands [4]
Operating revenue	100.0	100.0	100.0	100.0
(Gas purchase)	(65.7)	(48.1)	(56.6)	(66.3)
Gross margin	**34.3**	**51.9**	**43.4**	**33.7**
(Salaries)	*(10.1)*	*(11.1)*	*(10.9)*	*(9.4)*
(Other operating costs)	*(9.4)*	*(10.2)*	*(13.8)*	*(9.0)*
(Depreciations and provisions)	*(10.7)*	*(8.9)*	*(12.1)*	*(7.9)*
Operating result	**4.1**	**21.7**	**6.6**	**7.4**
Financial result	(1.4)	4.1	(0.2)	(3.9)
Extraordinary result	2.2	(0.9)	0.3	(0.3)
Profit before taxes	**4.9**	**24.9**	**6.7**	**3.2**
(Taxes)	(3.2)	(0.1)	(3.5)	-
Net profits after taxes	**1.7**	**24.8***	**3.2**	**3.2**

(1) 33 gas only LDCs - 1994 data

(2) One gas only LDC - thought to be representative

(3) 32 distribution companies - 1995 data

(4) All gas, electricity and water distribution companies in 1994.

*   Net profit figures for Belgium and the Netherlands are not comparable to those of Germany and Italy. Belgian distribution companies do not pay concession payments to their municipal/regional authorities as in Germany and Italy (in the Netherlands some did and some didn't; since 1997 existing concessions disappeared). Instead they pay the equivalent in the form of dividends, hence the important profits. In Belgium, these dividend payments to public authorities represent on average about 80% of the profits made in gas by distribution companies.

The costs described here are the ones normally found in annual reports of gas distribution companies. The structure corresponds to the general format of the profit and loss accounts of LDCs.

The gross margin of the companies could be considered as the value added of the company. It is supposed to cover the operating costs and remuneration of the capital invested. The three major classes of costs are as follows:

■ salaries, including personnel salaries, social charges and pension provisions;

■ other operating costs, which include a wide variety of items like purchase of goods apart from gas and investment goods, purchase of services from sub-contractors; insurances; public relation costs; post, telephone and other communication costs, etc.;

■ depreciation and provisions which are costs incurred as a result of distributing investment costs over the economic lifetime of installations needed for the gas distribution and to cater for certain risks that the company is exposed to. Depreciation is a function of investments undertaken and the depreciation rules that are applied. These rules vary considerably from one country to another, and are one of the factors that make international comparisons of cost and profits difficult.

The following shows how these costs relate to gross margins and operating results in the same countries.

Table 10 Distribution Costs: Share of Operating Profits (%)

	Germany	Belgium	Italy	Netherlands
Salaries	29.5	21.4	25.1	28.2
Operating costs	27.4	19.7	31.8	26.7
Depreciation	31.2	17.1	27.9	23.4
Subtotal	88.1	58.2	84.8	78.3
Operating profits	**11.9**	**41.8**	**15.2**	**21.7**
Gross margin	100.0	100.0	100.0	100.0

Source: Consultant based on annual reports

Table 11 Distribution Costs: Relative Weights (%)

	Germany	Belgium	Italy	Netherlands
Salaries	33.5	36.8	29.6	36.0
Operating costs	31.1	33.8	37.5	34.1
Depreciation	35.4	29.4	32.9	29.9
Subtotal	100.0	100.0	100.0	100.0

Source: Consultant based on annual reports

From these tables, the following observations can be made:

■ on average, for a European gas distribution company, its non-gas costs correspond to about 50% of its gas purchase costs - in other words, 2/3 of the costs are related to gas purchase and 1/3 are fixed costs;

■ on average, in the distribution companies observed in the four countries, these costs seem to be structured in a similar way, i.e. salaries, operating costs and depreciation costs roughly account for one third each of the pure gas distribution cost;

■ salaries or total personnel cost seem to constitute about 10% of total operating income (this does not, however, prevent salaries from varying a lot in absolute terms).

The comparison also reveals that gross margins as a share of total revenues vary a lot between the countries, as well as operating profits as a share of gross margin (from 11.9% in Germany to 41.8% in Belgium). This will be further analysed in chapter VII.

When looking at these tables, the question arises whether the costs generated by gas distribution companies are excessive or not. In other words, is there a significant scope for cost reduction? The following comparison of costs in transmission and distribution attempts to give an indication.

The table below illustrates the cost structure (transmission and distribution) in German gas industry. It should be borne in mind that distribution companies account for 68% of total gas sales to end consumers (the remainder being directly sold by transmission companies), and that this share varies from one country to another. Therefore, the figures in the table are not representative for the other five countries studied. Profits are not included in the figures except in the average LDC sales price.

Table 12   Cost Structure in the German Natural Gas Industry, 1995

	Cost in Pfennig/cubic meter (Pf/m³)*	In % of average end user price from LDCs
Transmission Company		
Gas purchase cost	11.5	29.3
Gas tax	3.5	8.9
Non-gas cost[1]	7.9	20.1
Total cost in transportation	19.4	49.4
Distribution Company		
Gas purchase cost	21.1	53.7
Non-gas cost [1]	15.0	38.3
Total cost in distribution	36.1	91.9
Average end user price from distribution companies	39.3	100.0

Source: Rheinisch-Westfälisches Institut für Wirtschaftsforschung

*Average input price: 12.6 Pf/m³ (according to Bundesministerium für Wirtschaft, the average import price was 13.2 Pf/m³).

(1) Non-gas costs include: capital cost, labour cost and other costs.

Nevertheless: these figures show that total non-gas costs in the high pressure part (transmission, storage and interruptibility costs) of the chain (including the gas tax, which in Germany is paid by the producers and importers) constitute about 20% of the price to end consumers. After deduction of the gas tax, the percentage is reduced to slightly more than 10%. Compared to this, the total non-gas cost of the distribution part constitutes about 40% of the average sales price.

This means that the costs that could potentially be influenced by the LDCs are more than three times higher than the costs on which the transmission companies can act directly. It suggests that there is a larger potential for cost reductions in the distribution sector than in the transmission sector.

Assessing this potential is of central importance. Any reform aimed at eliminating 'fat' and increasing efficiency in the gas chain should start by concentrating on costs. It should, therefore, focus at least as much on gas distribution as on transmission. Bringing competition to distribution is, therefore, essential.

COST EFFICIENCY IN GAS DISTRIBUTION

It is normally assumed that the nature of gas distribution should allow some economies of scale (larger companies should be able to enjoy lower distribution unit costs). It can also be argued that horizontal integration in the distribution of gas and of electricity, water, heat, etc. should allow decreasing unit costs (economies of scope).

Inductive analysis confirms this: about 25% of costs should be influenced by scale economies, and some 30% by scope (horizontal) economies (see annex to this chapter).

Statistical analysis, however, shows that this potential for both economies of scale and of scope is only poorly realised in gas distribution (again, see annex):

Statistical analysis has been carried out for three countries: the Netherlands, Germany and Italy. For each of these countries, the performance of LDCs of different sizes and in particular those of pure gas distributors (as opposed to horizontally integrated distribution companies) have been compared in order to establish whether the potential for economies of scale and scope is generally realised.

The data used for the statistical analysis consisted of

- the financial reports (1995 figures) of large samples of German LDCs (ranging from 32 to 368 companies) and Italian LDCs (60 companies);

- data from "Centraal Bureau voor de Statistiek" for the Netherlands.

Regression analysis on samples of gas distribution companies in Germany and Italy does not reveal identifiable causality links between parameters such as:

- average distribution costs and company size;

- operating profits (expressed as a percentage of operating revenues) and operating revenues;

- operating profits (as a percentage of fixed assets) and the size of the fixed assets;

- profits after tax (as a percentage of equity capital) and total equity.

For the Netherlands, the relationship between distribution cost and company size was studied. The results contradict the hypothesis that large companies should have lower average distribution costs per m³ than smaller ones. In fact, it revealed the opposite, namely that large Dutch gas distribution companies tend to have higher unit costs than smaller ones.

In order to study the effects of horizontal integration, a profitability analysis of the accounts of 191 German companies was made (profitability was measured in various ways: operating profits as a percentage of operating revenues; operating profits as a percentage of fixed assets; net profits after tax as a percentage of equity capital). In none of the cases was it possible to demonstrate any positive effects of horizontal integration on profitability.

The foregoing analysis indicates that economies of scale seem not to be fully realised in gas distribution, and that horizontally integrated distribution companies do not perform better than pure gas companies, although in theory they should have an advantage of scope.

These findings are an indication that overall cost efficiency is rather poor in gas distribution.

The most probable reason for this is that LDCs and/or the public authorities involved rarely see an advantage in cutting down expenses, in particular when this could be detrimental to local employment or economic activity. This attitude is likely to be reinforced by the absence of pressures or incentives to cut costs, that would exist in a competitive situation.

The introduction of competition in gas distribution would create the necessary incentives to realise more fully economies of scale and scope, and also to cut costs generally. Efficiency would, therefore, be improved, and this would set the scene for all consumers (including small consumers) to benefit through lower prices.

KEY FINDINGS

☐ Gas distribution has considerable economic significance. The value of total gas sales in the distribution sector corresponds to over 0.7% of GDP in the six countries.

☐ Gas distribution is capital intensive, even more so than gas transmission, hence fixed costs play a very large role in small end-user tariffs.

☐ Non-gas costs of LDCs appear to be significantly higher than those of gas transmission companies. National policy should go beyond the EU Gas Directive by introducing competition in distribution in order to draw out cost savings potential.

☐ Whereas the nature of gas distribution should offer potential for economies of scale and of scope, statistical evidence shows that these are poorly realised in continental gas distribution companies.

☐ The most likely explanation is that both distribution company shareholders (often municipalities) and management are not (really) committed to cost saving and efficiency. This attitude is only possible to maintain in an environment of little or no competition.

☐ The introduction of effective competition in gas distribution should create the necessary incentive to realise more fully economies of scale and scope, thus to reduce costs, and hence customer prices.

ANNEX

ANALYSIS OF ECONOMIES OF SCALE AND SCOPE IN GAS DISTRIBUTION

Does gas distribution become more economic or do unit costs of distribution decrease with increasing company size or with the introduction of simultaneous distribution activities (such as electricity, heat, water) ? And, if yes, do distribution companies fully exploit these economies?

We tried to answer these questions by two approaches: one inductive, one statistical.

The inductive method gives reason to believe that the increase in the size of the companies or the addition of other activities should potentially lead to economies of scale and scope.

But the statistical approach, based on the financial or technical data of large samples of gas distribution companies in the Netherlands, Germany, and Italy shows that these potentials are not fully exploited.

Moreover, evolution of the distribution sector in the west German Länder (table A.2.) - increase in the number of distribution companies, increase in market share of pure gas distribution companies - reinforces the observation that economies of scale or scope are not being fully realised.

Is there potential for economies of scale and scope in gas distribution?

Table A.1. seeks to highlight the possibility of economies of scale. It seeks in the same way to highlight the favourable influence on costs of horizontal integration of various public utility activities (distribution of gas and electricity, district heating, water supply as well as cable television).

The table should be read as follows: economies of scale affect slightly (first column), fairly (second column) and in a significant way (third column) indicated costs.

The same approach was used with regard to the effects of horizontal integration.

Table A.1. gives average distribution costs (except financial costs). It also indicates the importance of the influence on these costs of company size (scale effect) and of horizontal integration.

Unit costs related to monitoring and control are partly fixed and should be able to be reduced when increasing company size or with integration horizontal.

Above the approximate size of 5.000 customers (gas distribution company), maintenance costs of equipment and pipelines become quasi proportional to total pipeline length of to the number of meters. This suggests that concerning maintenance costs there are no benefits of size or of horizontal integration.

The costs related to the displacement of installations are not influenced either by company size or horizontal integration.

Lastly, royalties are influenced neither by the scale effect nor by horizontal integration.

On the other hand, the unit costs of technical services, in particular the costs of management and administration, can present economies of scale and can also be reduced with horizontal integration.

With regard to the costs of customer services, it is necessary to distinguish between their nature:

- meter reading: very weak potential for economies of scale except for small sized companies; but there should be potential for economies of scope because one employee can read several meters at the same time (gas, electricity, water);

- invoicing: potential for economies of scope because only one invoice per customer necessary;

- recovery of the invoices: no economies of scale and only small economies of scope possible;

- contract resignment formalities: no economy of scale possibility, but potential for economies of scope.

The costs of information and publicity should be sensitive to both scale effect and horizontal integration.

Overheads are also sensitive to the scale effect and horizontal integration, at least when the characteristics of the different parts of the gas distribution networks do not differ too much from each other or from those of the other distribution systems.

On the other hand, depreciation is sensitive neither to company size nor to horizontal integration.

In conclusion, the scale effect could intervene for 25% of total costs, while horizontal integration may influence 30% of costs.

The following will show, however, that the statistical analyses made for the Netherlands, Germany and Italy reveal that economies of scale and scope are not, or at least not fully, realised.

This points at inefficient cost management. Its most likely explanation is the fact that gas distribution is not carried out in a normal market environment, but from a monopoly position.

Is this potential for economies of scale and scope fully exploited?

In analogy to a public transport grid system where output can be measure in traveller × km (in ton × km in the case of carriage of goods), the output of gas distribution should ideally be measured in m^3 × km between point of supply and customer (meter). However, this is not possible to measure. The off-take in m^3 can be recorded by meter, but cannot be weighted by the distance between the point of supply and the meter, as would be the case when a railroad ticket is sold to a traveller, each ticket being representative of a distance between two stations.

One is thus led to quantify the output either in units of gas distributed, or by the number of customers of a gas distributor, or by the length of the distribution network or by the company's turnover.

This is exactly the kind of analysis that was made of available data on the Netherlands, Germany and Italy.

The Netherlands

In the case of the Netherlands, we used data of "Centraal Bureau voor of Statistiek" and represented in graph A.4. The regression between the average cost expressed in NLG/m³ and the number of customers of the distribution companies divided into five classes. As the average sales (m³/client) and the average length of the conduits (km/client) are not identical in each class, it is necessary to see whether differences are likely to modify the classification in terms of a number of customers.

Example: a company with 1.000 customers consuming 100 m³ each should be larger in terms of m³ than a company of 1000 clients consuming only 1 m³ each. In our calculations, this is not the case because sales by customer, sales by km of conduits as well as pipeline lengths per customer vary less than the average cost of distribution (table A.3.).

The relation between the average cost in NLG/m³ and the size measured by the number of customers consequently provides us with a reliable indication on the existence or not of economies of scale in gas distribution. As the linear regression (see graphic A.4.) between the average cost in NLG/m³ (y) and company size << ... >> presents a positive slope, we can conclude that economies of scale in gas distribution are in fact underachieved.

The regression is written:

y = 0,05642 + 0,09 X (values t of 23,05 and 5,77 significantly higher than 2,23)

In other words, statistically the cost of distribution increases with company size.

This assertion is consolidated by the regressions of graphs A.5., A.6. and A.7.

In graph A.5., we note that the number of employees by 1.000 customers (it is an indication of the level of productivity) does not decrease with the size of the distributive firm measured by the number of customers. Indeed, the regression is written:

y = 1,74 + 0,02 X

The coefficient of (X) does not differ however to a significant degree of zero (value from the statistics t: 0,41 and thus lower than the threshold of significance of 2,23).

In graph A.6., we note that the number of employees by 1.000 km of pipeline conduits (approximate measurement of the productivity) grows with distribution company size, measured by the number of customers. This indicates that there are no economies of scale.

Table A.3. The regression is written:

$$y = 86,97 + 0,17 \ X \ \text{(values t: 8,12 and 2,58 the higher than 2,23)}$$

We note a positive slope, i.e. a growth of the number of employees by 1.000 km of conduits with the size of the company.

Graph A.7. representing the number of employees per distributed million m^3 (used like indication of productivity) in relation to company size, measured by the number of customers, goes in the same direction. Indeed, the regression is written:

$$y = 0,39 + 0,08 \ X \ \text{(values t: 9,77 and 3,17 the higher than 2,23)}$$

We note a positive slope, i.e. a growth of the number of employees per million m^3 with increasing company size.

The conclusion of this analysis are:

■ cost of the distribution increases with company size (graphic A.4.),

■ employment by 1.000 customers is increasing (graphic A.5.), though not much,

■ employment by 1.000 km of pipeline increases with company size (graphic A.6.),

■ employment by 10^6 m^3 sold also increases with company size (graphic A.7.),

■ add to this that the CBS has published a graph which shows that the distribution company's profits (in % of turnover d' affaires) decrease with company size, and this for each surveyed year - 1988, 1989 and 1990 (graph A.8.).

Germany

On the basis of the financial reports of 191 companies, the statistical analysis comes to the following results:

■ operating result expressed as a percentage of turnover does not relate to the importance of turnover,

■ operating result expressed as a percentage of fixed assets does not correlate with company size,

■ net benefits (expressed as a percentage of stockholders' equity) does not correlate with the importance of stockholders' equity.

Additional statistical analysis of the financial data of 32 pure gas distribution companies shows that cost of distribution does not correlate with company size.

On the basis of physical data of 368 companies published by BGW, we can also say that:

■ sales by employee do not correlate with absolute sales volumes,

■ employment per km of pipeline does not correlate with network length,

■ the number of meters per employee does not correlate with the total number of meters.

About horizontal integration, the statistical analysis of the operating statements of a sample of 191 German companies shows that horizontal integration does not have visible effects on their respective financial results.

Italy

We also studied the effect of size in Italy by a statistical analysis of the 1995 financial results of 60 gas distribution companies.

Just like in Germany, the analysis shows that there is no effect of size on the level of the financial results:

■ net benefits per equity capital does not correlate with the size of equity capital,

■ operating result expressed in percentage of turnover does not correlate with turnover size,

■ operating result expressed in percentage of fixed assets does not correlate with the size of fixed assets,

■ operation cost for a sample of 56 companies does not correlate with company size.

Table A.1 Potential of Economies of Scale and of Scope in Reducing Gas Distribution Costs (Sample of 7 Belgian Gas Distribution Companies, 1995)

	average share of costs (%)	economies of scale			economies of scope		
		high	medium	low	high	medium	low
Technical costs	32	28	4	0	29	3	–
monitoring, control	5	3	2			3	2
maintenance	12	12			12		
moving/installation	1	1			1		
technical service	12	10	2		11	1	
royalties	2	2			2		
Customer service	18	18			10	7	1
Communication	4		4			4	
General costs	16		11	5		11	5
Depreciations	30	30			30		
Total	**100**	**76**	**19**	**5**	**69**	**25**	**6**

Table A.2   Evolution of Gas Distribution in Germany (West) (10^6 kWh)

	1988					1995				
	number	%	average sales	total sales	% market share	number	%	average sales	total sales	% market share
gas	79	14.9	1279.0	10141.0	22.2	89	16.2	1831	162959	25.9
gas + water	134	25.3	422.0	56548.0	12.4	116	21.1	819	95004	15.1
gas + electricity	43	8.1	1225.0	52675.0	11.5	49	8.9	958	46942	7.5
gas + water + electr.	273	51.6	900.7	245891.1	53.9	295	53.7	1097	323615	51.5
Total	**529**			**456155.0**		**549**			**628520**	
						104			**138**	
Regiebetrieb[1]	4.0	0.8	91.5	366.0	0.1	0.0	0.0	0	0	0.0
Zweckverband[2]	5.0	0.9	568.0	2840.0	0.6	4.0	0.7	1001	4004	0.6
Eigenbetrieb[3]	232.0	43.9	396.0	91872.0	20.1	189.0	34.4	501	94689	15.1
AG Eigenges.[4]	151.0	28.5	927.0	139977.0	30.7	186.0	33.9	1077	200322	31.9
AG öffentl. Ges.[5]	33.0	6.2	2353.0	77649.0	17.0	35.0	6.4	1387	48545	7.7
gemischt[6]	76.0	14.4	1564.0	118864.0	26.1	105.0	19.1	2301	241605	38.4
privatwirtschaftl.[7]	26.0	4.9	861.0	22386.0	4.9	27.0	4.9	1345	36315	5.8
sonst.privatrechtl.[8]	2.0	0.4	1154.0	2308.0	0.5	3.0	0.5	1024	3072	0.5
Total	**529.0**			**456262**		**549.0**			**628552**	

Source: BGW

Public companies:
(1) public utility
(2) association of communes
(3) private company owned by commune
(4) public company owned by commune
(5) public company owned by several communes

Private companies:
(6) mixed company
(7) private company (AG or GmbH)
(8) other forms

Table A.3   Gas Distribution in the Netherlands
(all Distribution Companies, 1990)

	Number of clients per company				
	< 50.000	50 - 100.000	100 - 150.000	150 - 200.000	> 200.000
total sales (10^6 m³)	4249.0	5748.0	2404.0	2953.0	5725.0
clients (10^3)	863.3	1544.3	712.5	807.6	1865.1
network (km)	16013.0	24679.0	14364.0	13723.0	24497.0
sales per customer (10^3 m³/customer)	4.9	3.7	3.4	3.7	3.1
length per customer (10^3 km/customer)	18.5	16.0	20.2	17.0	13.1
sales per km (10^3 m³/km)	265.0	233.0	167.0	215.0	234.0

Source: CBS

Figure A.4   Cost of Distribution in Relation to Customer Number per Company

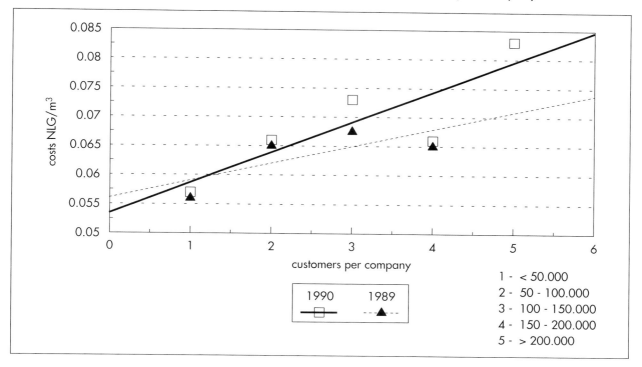

1 - < 50.000
2 - 50 - 100.000
3 - 100 - 150.000
4 - 150 - 200.000
5 - > 200.000

Figure A.5   Number of Employees per 1000 Customers, NL 1990

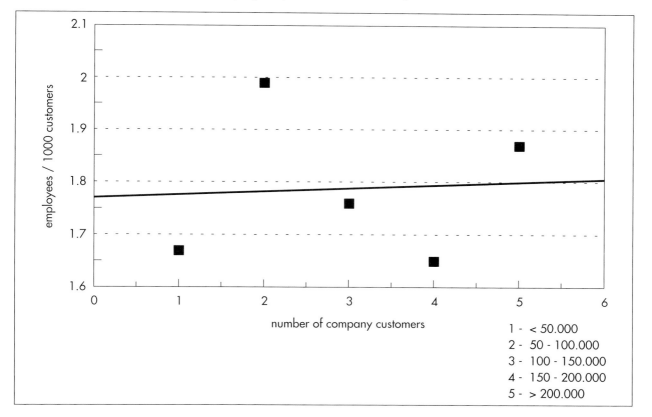

1 - < 50.000
2 - 50 - 100.000
3 - 100 - 150.000
4 - 150 - 200.000
5 - > 200.000

Figure A.6   Number of Employees per 1000 km of Pipeline, NL

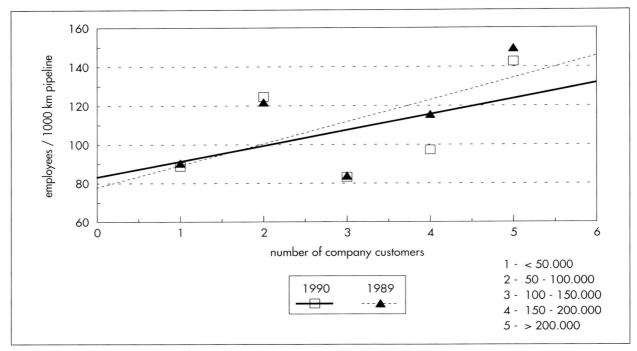

Source: CBS, statistiek gasvoorziening Nederland 1990.

Figure A.7   Number of Employees per Million cm Distributed, NL

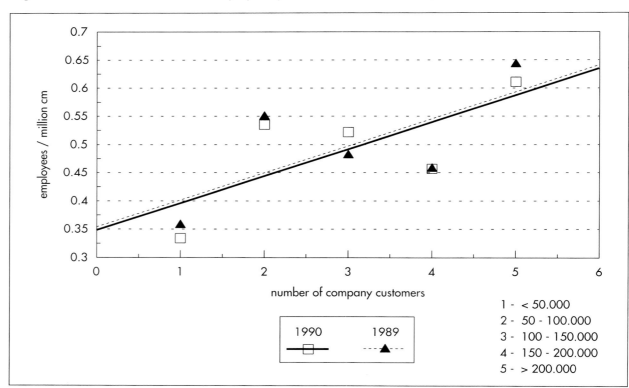

Source: CBS, statistiek gasvoorziening Nederland 1990.

Figure A.8   Profits in Relation to Number of Clients, NL

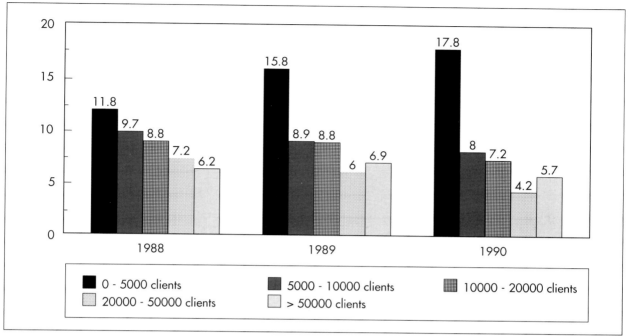

Source: consultant (CBS, statistiek gasvoorziening Nederland 1990).

VI.   COSTS AND PRICING

INTRODUCTION

Optimal pricing to maximise economic efficiency should reflect costs as accurately as possible. Costs, therefore, need to be identified and allocated directly.

Basically, gas distribution is about bringing the required gas volumes in its required form from point A to point B at the required time. Cost allocation raises the issue of what costs are incurred to serve a particular gas customer according to his needs?

This chapter identifies the basic cost drivers in gas distribution. It then provides a description of how these are reflected in prices and tariffs in gas distribution in Western Europe, including an analysis of the gas pricing systems in the different countries. The chapter gives a comparison of gas prices in Western Europe, and proposes general conclusions on pricing and economic efficiency.

WHAT ARE THE COSTS IN GAS DISTRIBUTION ?

Gas Purchase Costs

As seen in the previous chapter, the costs of gas purchasing represent the largest cost for a distribution company, and therefore deserve a detailed investigation. Broadly speaking, gas purchasing costs include the costs of the commodity, transport and flexibility (availability of the gas at required volumes at all times), plus some regulatory costs (e.g. taxation).

A survey of the six countries shows that, in general, a gas distribution company takes its contracted gas from its supplier at the entry point of its own system at a price which is supposed to cover the cost of that gas, plus the cost of transporting the gas from the wellhead or the import point and of providing supply flexibility, plus a profit element.

Where a single gas pipeline company is granted a monopoly over the transportation and sale of gas to LDCs and end-users within a specific geographic area, the company may in principle set gas prices on the basis of cost-plus (i.e the acquisition cost of the gas plus a mark-up for non-gas costs and a return on capital) or on the basis of the market value of the gas in competition with other fuels such as oil products. The latter approach, by definition, involves price discrimination according to the different demand characteristics of end-users which determine the practical alternatives to and cost of using fuels other than gas. Such price discrimination inevitably results in cross-subsidies. Often, the government or regulatory authorities limit the extent to which the gas company may apply market value pricing if this results in excessive profits, such that monopoly gas companies in many countries apply an amalgam of cost-plus and market value.

PTO

Table 13 Tariffs for Local Gas Distribution Companies

Tariff calculation	
Cost plus	**Market value**
Gas purchase cost (from national transmission companies) + transportation cost + profit element	Maximum price at which gas supply remains competitive with alternative supplies of energy, taking into account the costs of supply and a profit element
Tariff components	
These may include: Connection Charge Fixed/Standing/Subscription Charge Commodity Charge Capacity Charge	
Tariffs may be regulated and/or negotiated	

Source: IEA

In fact, there is an implicit market value element to the cost plus approach, in that the gas purchase cost element of the price paid by LDCs will reflect the price by the transmission company for gas from the wellhead/border, which is based on the market value approach. Similarly, there is some element of cost plus in the market value approach, in that the price paid by LDCs is normally adjusted to allow a margin to cover transportation costs (and a profit element).

This has its origins in the netback approach, which until the present day traditionally has been and remains the basis for gas pricing in Europe[2].

The Netback Market Value Concept

The netback market value of gas to a specific customer at the beach or border is defined as follows:

Netback = Delivered price of cheapest alternative fuel to gas to the customer (including any taxes) adjusted for any efficiency differences;

minus Cost of transporting gas from the beach/border to the customer;

minus Cost of providing flexibility (e.g. storage) to meeting the customer's seasonal or daily demand fluctuations;

minus Any gas taxes.

The weighted average netback value of all customer categories is used as the basis for negotiation of bulk prices at the beach/border.

---

2. Such a system developed because of the imperative for pipeline companies to recover the large captial costs involved in building the pipeline infrastructure. By pricing gas through the chain in relation to competing fuels, the pipeline company can ensure that throughput and per unit revenues are maximised, and that the gross margin is protected. Risk is effectively transferred to the producer, which may be compensated by a share in any economic rent available.

Thus, the price paid by the gas company to the foreign or domestic gas producer at the border or beach is negotiated on the basis of the weighted average value of the gas in competition with other fuels adjusted to allow for transportation and storage costs from the beach/border and any taxes on gas. There are in principle three different average netback market values corresponding to existing gas users, to new energy users (e.g. greenfield industrial plants) and to existing oil users with no dual-firing capability (the latter being the lowest because of the capital cost of fuel switching). The beach/border base price that is ultimately negotiated will correspond to a level between the highest and the lowest of the three values, weighted across the different end-user customer categories. The base price is usually escalated on oil product prices (usually heating oil and/or heavy fuel oil), or simply crude oil (on the implicit assumption that the ratio of crude to product prices will remain broadly constant), to ensure that effective prices over the life of the contract remain broadly in line with market values.

The cost of production and supply to the beach/border can be considerably lower than the lowest weighted average netback market value. In Europe, Dutch and Norwegian supply costs are thought to be lowest, followed by Algeria and Russia. In general, near-to-market producers are at an inherent advantage because pipeline costs are much lower. As a result, there may be considerable relative (geographical) rent (i.e., the cost difference between two suppliers – one closer to the market than the other) to be earned between the average netback market value and the supply cost. Negotiations at present typically result in a sharing of the absolute rent between the producer, any transit countries and the importer/transmission company. Once the border price is set, the importer may (and usually does) pass on a proportion of its rent by setting city gate and/or end-user prices that are below the average netback value for existing gas users to encourage new users or even existing oil users to choose or switch to gas, i.e., there is a trade-off between short term profitability and long term market growth[3] (see figure).

In practice, the notion of two competing fuels, light fuel oil and heavy fuel oil, is an oversimplification. Coal can play a role for new investments (e.g. in the Netherlands, one gas contract for a CCGT plant was set in order to insure that the full cost of electricity would match that of a coal-fired plant), and electricity can be an important competitor in office and water heating (hence the end-user price of gas in these cases could be higher in order to reflect the important netback value). Also, in reality there is a stronger differentiation between oil products themselves: HFO can range from heavy sulphur quality (3.5% sulphur content) to low sulphur quality (0.5% or even 0.2% sulphur), the latter coming very close to a light fuel oil.

A further element is that the pricing principles are not applied in their pure form – there is always an element of negotiation involved which will partly determine the price and, hence, the distribution of profits between the parties. Although the typical pattern is one of standard contracts, each LDC will have its individual contract, and in most countries prices to distribution companies are allowed to differ. Finally, the degree of regulation varies – in some countries prices are heavily regulated, in others not.

---

3. This arbitrage evolves over time. At the initial stage of a young market, the netback approach is important for market development, and can be justified on the grounds that it does not bring additional costs to the users while increasing energy security due to supply diversification. At maturity of the gas market, however, each increase in natural gas consumption delivers only marginal benefits in terms of energy diversification. Hence, it can be argued that at this stage the gas rent should be shared with the end-user via gas-to-gas competition.

Figure 4   Netback Market Value and Pricing
(example of an existing monopoly integrated transportation and distribution company)

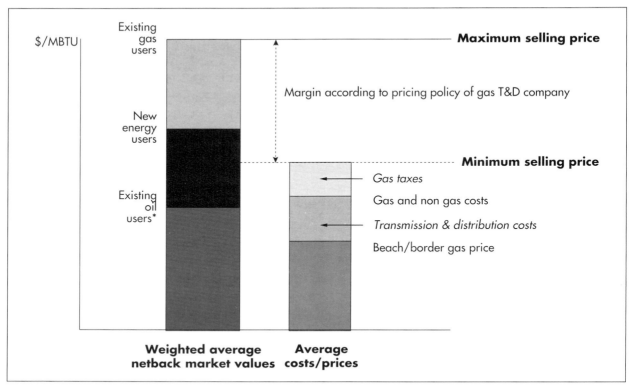

* Typically gas oil for domestic/commercial and small industrial consumers, and CCGT power generators; low or high sulphur heavy fuel oil for large industrial consumers and power generators (steam boilers).

Note: The gas company (for a given contracted level of beach price and non gas costs and taxes) has the freedom to set prices between the level set by its total costs plus gas taxes, and the level set by the highest netback market value (for existing gas users). If the company sets prices lower than the minimum, it would not be able to cover its costs. If it sets prices above the maximum, it would quickly lose market as its customers switch away from gas to less expensive fuels. In practice, the company would have to set prices no higher than the market value of new energy users to maintain its market and encourage market growth in the long term.

It is important to note, though, that regardless of the approach, the LDC will see a single price (no desegregation of costs or attempts to price elements separately).

In most cases, there is no take-or-pay clause in the contract between the transmission company and the LDC. However, one could say that in the cases where the price consists of both a fixed element and a proportional element, the fixed part could have the same effect (but the fixed element is normally such a low share of the total price that this is not the case).

In Great Britain, where the pure cost-plus approach is used, there is no reference to alternative fuels in the price formula. In Continental European countries, where the market value approach is used, the gas price is implicitly (or even explicitly) linked to the price of oil products (often gas oil). Most of the time, the price is digressive with increasing volume and higher load factor. Seasonal pricing is only found in France.

Country
Specifics

Belgium uses an apparent cost-plus approach. The price paid by LDCs is a commodity price including a calculated share of transportation costs to be carried by LDCs. However, tariffs are the same across the country; transportation cost is not distance related. The profit element is contained within the transportation cost.

France also uses an apparent cost-plus approach. However, the few LDCs apart from GDF's own distribution business which buy gas from GDF or from GSO only see one common price. The implicit price paid for transportation on the high pressure national transmission system is not distance related, but the prices on the regional transportation systems are. GDF's tariff system is sophisticated and features seasonal variations and volume rebates. The price consists of a subscription fee, a capacity charge and a commodity charge of which the commodity charge is by far the most important.

In Germany, contracts between transmission companies and LDCs are not regulated. This and the large number of distribution companies make it difficult to obtain a complete overview of the structure of these contracts although the use of standard contracts seems widespread. In standard contracts the market value approach is used but with reference only to light heating oil price. The price consists of a capacity charge and a commodity charge. The capacity charge is partly (normally 30% weight) linked to salaries and constitutes a cost plus element. The remaining 70% is linked to the light heating oil price. The commodity charge is exclusively linked to the light heating oil price. The escalation takes place with a lag. There is normally a mechanism built into the contract that makes the link to light heating oil weaker in times of decreasing oil prices. This means that the gas price becomes apparently more cost oriented in times of low oil prices (it can also be explained by competition with electricity in the household sector – e.g. cooking). There is, however, no corresponding mechanism to create a ceiling for the gas price when oil prices are high.

In Italy, the supply tariff to LDCs has a two-part (binomial) structure made up of a fixed charge and a commodity charge. The fixed charge is a flat rate updated annually according to inflation indices. The commodity charge (which accounts for about 80% of the total price most of the time) has the peculiarity of being differentiated according to the average annual consumption of the final consumers supplied by the LDCs and is updated every two months on the basis of heating gas oil price variations. The supply price faced by the LDCs increases with per capita consumption and is supposed to support LDCs that need low supply prices to develop their market. This mechanism is a way of cross-subsidising companies with high average cost (for instance caused by low population density) by letting companies with lower costs pay a higher price. However, it apparently ignores the element of electricity competition in the household sector (cooking and water heating).

The present price structure for sales from Gasunie to LDCs in the Netherlands is quite simple in that there is only a commodity price. This commodity price is differentiated according to the type of customer for which the gas is destined (market value concept netbacking the different gas users). The gas purchased by LDCs for small customers is linked to light heating gas oil whereas the price to large customers is linked to heavy fuel oil. The average price paid by an LDC is thus a function of the composition of its markets. To the extent that two LDCs have a different market structure, they will also pay different average prices for their total gas volume.

The prices along the gas chain from Gasunie to the final consumer are set based on market value and net-back. They generally allow LDCs to earn a margin sufficient to cover costs and profits. There is also a direct contribution from Gasunie to LDCs of 80 NLG a year per connection which is supposed to constitute a margin independent of volume.

Great Britain is very different. First, it has introduced widespread gas to gas competition. Second, the structure of the industry is very different. One can hardly speak about a gas price to LDCs since there are no LDCs with a structure similar to the ones found elsewhere in Europe. All gas suppliers or gas marketers buy their gas at the beach or from intermediaries. Following the demerger in February 1997, Centrica has taken over responsibility for marketing gas to end consumers from British Gas plc. This means that it has inherited the supply contracts entered into by BG. Wholesale or beach contracts have generally become shorter term and less linked to oil prices than they were before the introduction of competition.

Table 14 Pricing Approaches Under Different Gas Market Structures

	Interfuel competition only	Interfuel and pipeline-to-pipeline competition	Competitive wholesale market (TPA to high pressure system)	Full retail competition (TPA to entire system)
Pricing approach	Price discrimination between customers. Netback market value (pure or not)	Restricted form of discriminatory netback market value (depending on extent of competition)	Gas-to-gas competition (marginal fuel competition)	Same as for competitive wholesale market
Example	France, Belgium	Germany	United States, Canada	United Kingdom

Source: IEA

Gas Supply Costs

Matching supply and demand and guaranteeing security of supply is a task that requires a mix of instruments (infrastructural and contractual) that can not always be dissociated from one another, e.g., it is difficult to draw a distinction between what are security of supply measures and what are normal daily and seasonal balancing efforts. In some countries, this has been solved by characterising a specific share of either the storage and balancing costs or interruptible sales contracts as security costs. Some make a distinction between storage or interruptibility for load balancing purposes and storage for strategic reasons. This could give some guidance for allocation of costs, although the distinction would be quite arbitrary.

Gas demand fluctuates both on a daily and seasonal basis. The daily send-out on a cold winter day may be six or seven times as high as that on a warm summer day. The biggest fluctuations are found in the residential sector. In France, for instance, the ratio between the month with the highest demand to the month with the lowest demand is 6.6 in the residential sector whereas it is about 1.7 in the industrial sector. These differences in load needs give rise to differences in cost, but are not always clearly identified; nor are the corresponding costs calculated and allocated.

Gas storage costs are particularly difficult to allocate. Storage is needed for both load balancing and for strategic security. Either way, it is costly. A tradeoff with other means of flexibility, for instance interruptibility, or to some extent dimensioning of pipelines, could be made. Another tradeoff may be against volume flexibility in supply contracts. The greater the flexibility in the off-take provisions of a gas supply contract, the higher the price (flexibility at production fields is generally more expensive than the use of storage). The cost of storage can also be compared to the cost of applying interruptible contracts, on condition that these costs are identified and specified. It has to be realised, however, that these instruments cannot substitute completely for each other in all situations. In a supply crisis, interruptible contracts can only be used to redirect supplies; storage facilities can provide a net addition to available gas volumes.

Most of the countries dealt with in this study have security of supply standards. It is quite clear that the costs of fulfilling those standards are considerable. It is also worth noting that the calculation of those costs and their allocation are rarely explicit.

HOW DISTRIBUTION COSTS ARE PRICED TO THE END-CONSUMER

In natural gas distribution, allocating the different costs or even weighting them is a difficult task which cannot be done without some arbitrage. Consequently it is done differently in every country (if not by individual LDCs). The relative importance of the cost drivers also vary from one country to another. This inevitably leads to price differences.

In particular, the translation of pipeline costs into tariffs is far from straightforward. The complexity of the problem can be illustrated by the fact that at one stage there were 29 different method for the apportionment of demand or capacity costs in the US [UNECE page 18].

Basically, there are four regulatory approaches to pipeline pricing, combining the two purposes of cost recovery, and the need to set appropriate tariffs in order to send the right price signals.

- AAC (accounting average cost): takes only first element into account;

- AIC (accounting incremental cost): introduces marginality element into AAC;

- LRMC (long run marginal cost): enhancement of AIC in that it aims to recover the marginal long run cost);

- SRMC (short run marginal cost): is the purest price signal but does not guarantee full cost recovery of initial infrastructure.

In practice, actual price formulas always seek to combine both elements. In the US, where pipelines are regulated on an individual basis, the straight variable cost formula is a combination of AAC and SRMC with emphasis on the latter. In Great Britain, AAC is dominant with some added incentive to reduce the variable cost element, but TRANSCO has its own interest in optimising the overall system and hence in using a SRMC approach for its own desegregated accounting and tariff setting.

In Europe, there are only few attempts at directly cost reflective tariff structures, the main reason being that most countries use a market value approach which inherently focuses on revenue rather than cost. Belgium and France are the only countries where traces of cost reflective pricing can be found. In Belgium, such traces can be found only at the transmission level, which still has some implications for distribution since most of the load balancing take place in transmission.

The introduction of competition and unbundling of transportation/distribution and marketing of gas in Great Britain has resulted in a heated discussion on this issue. Transportation and distribution are subject to price regulation and one of the regulator's main objectives has been to make tariffs more cost reflective. Identification of the cost drivers has been the first step in this process; distinction between capacity charge and commodity charge the second. In an initial phase, the British discussion focussed on the transmission part of the chain, but the same discussion then also took place concerning distribution. It should be kept in mind that in Great Britain, the two parts cannot be seen in isolation from each other given the integrated network and that much of the load balancing takes place in transmission.

Tariffs for Domestic Customers

In Europe, prices to LDC customers are generally regulated, but in some cases the LDCs still have influence on their prices and thus on their profits. In countries with cost-plus pricing, LDCs tend to have less influence on their prices than LDCs in countries with market value based pricing. (Great Britain in this context belongs to countries using a cost-plus approach). The countries using the cost-plus approach also practice price equalisation so that all customers in the same category and with the same characteristics pay the same price all over one region or the country. This is generally not the case in countries using the market value approach. In these countries prices are more heavily influenced by considerations on a regional or local level (though price equalisation within one distribution area is a generally applied principle). Only in the UK is there an element of direct profit regulation. Price capping in various forms is found in three of the countries.

Prices to end consumers served by the LDCs typically consist of a standing charge (normally a fixed amount per year) and a commodity charge, sometimes supplemented by a connection charge. In most of the countries the fixed charge accounts for a low share of the total price so that a large of part of the distribution cost, which to a large extent consists of fixed costs, has to be recovered through the commodity charge. This is in stark contrast to practice in North America where fixed costs are mainly recovered through a fixed element. The tariff schedules normally distinguish between different uses of the gas. Seasonal pricing is non-existent except in Italy and for some customers in France.

LDCs in some countries make a distinction between tariff customers and contract customers. The major differences are that tariffs are always published whereas this is not always the case for contract prices, and that tariffs normally have no explicit reference to competing fuels which is more often the case for contract prices.

Prices for Industry and Power Generation

Gas for industrial use is sold both by LDCs and transmission companies. Industrial customers beyond a certain size are typically served directly by the transmission company, but in some countries the border line between transmission company customers and LDC customers is not entirely clear and consistent. In some cases the customer has a choice between the two. This section deals exclusively with large industrial customers served by transmission companies.

Pricing in relation to other fuels is more widespread in the industrial sector than in the segments served by the LDCs. Even in countries using a cost plus approach, industrial prices in some cases have an explicit reference to other fuels. The major competing fuel is heavy fuel oil, but some contracts also have a reference to gas oil. In all the countries studied interruptible contracts in some form are offered. In some countries industrial contracts have take-or-pay clauses, but this is not very widespread. A low load factor is more often "punished" through a higher price resulting from the structure of the price formula. All over, the gas price is digressive with volume.

As a general observation, it could be said that industrial prices are less cost oriented than prices to customers served by the LDCs, the main reason being that a greater part of the industrial market is exposed to competition from other fuels. Hence, the market value approach to pricing is widespread.

Generally, prices for power generation in the countries studied are often similar, if not identical, to industrial prices. In some countries, however, there are special prices to cogeneration and to independent power producers. The majority of gas contracts in the countries studied have a price linked to the heavy fuel oil price. In recent years, new pricing concepts using other escalators have emerged in the Netherlands, Belgium and the UK.

Taxation

Of course, in addition, taxation of the use of natural gas can have an important bearing on end-consumer pricing (see Chapter IV., Regulation). In principle, gas suppliers tend to react to taxes by readjusting their net-prices in relation to the inter-fuel competition.

COMPARISON OF GAS PRICES (DOMESTIC AND INDUSTRIAL CUSTOMERS)

Not surprisingly, the different supply situations, energy taxation systems and pricing practices lead to differences in gas prices that can be substantial, in particular prices for smaller customers.

Domestic Prices

The following figure compares prices to various categories of domestic customers in the six countries. Consumption varies from 2326 kWh (7.93 MBtu) annually for cooking and hot water to 290750 kWh (993 MBtu) annually for central heating in residential complexes. In the figure prices have been calculated in US$/MBtu.

Figure 5   Gas Prices ex-Taxes for Domestic Users (1 Jan. 1996)

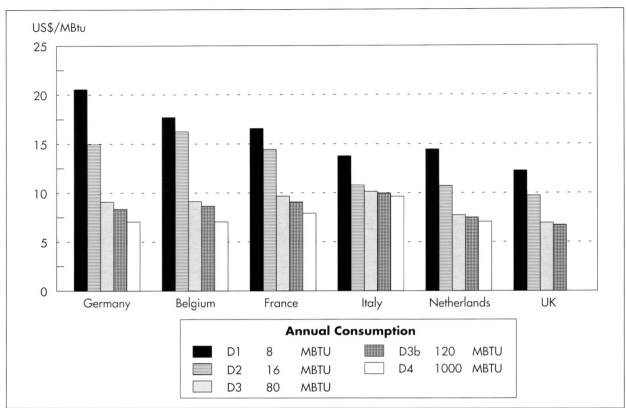

Source: Eurostat
Category:
D1   = domestic use: cooking and water heating; annual consumption up to 8 MBtu
D2   = domestic use: cooking and water heating; annual consumption up to 16 MBtu
D3   = domestic use: cooking, water and space heating; annual consumption up to 80 MBtu
D3b = domestic use: cooking, water and space heating; annual consumption up to 120 MBtu
D4   = domestic use: collective heating (at least 10 apartments); annual consumption up to 1000 MBtu

Industrial Prices

The next figure shows a comparison of industrial prices in the six countries. The following table shows the volume categories and the associated load factoring foreseen:

Table 15   Volume Categories and Associated Load Factor

Category (firm supply)	Volume in GWh	Volume in MBtu
I1	0.12	400
I2	1.2	4000
I3-1	12	40000
I4-1	120	400000

Source: Eurostat
I1   = small and medium sized customers; annual consumption up to 400 MBtu
I2   = small and medium sized customers; annual consumption up to 4000 MBtu
I3-1 = large industrial customers/power producers; annual off-take up to 40000 MBtu (firm)
I4-1 = very large industrial customers/power producers; annual off-take up to 400000 MBtu (firm)

Figure 6 Gas Prices ex-Taxes – Industrial Users (1 Jan. 1996)

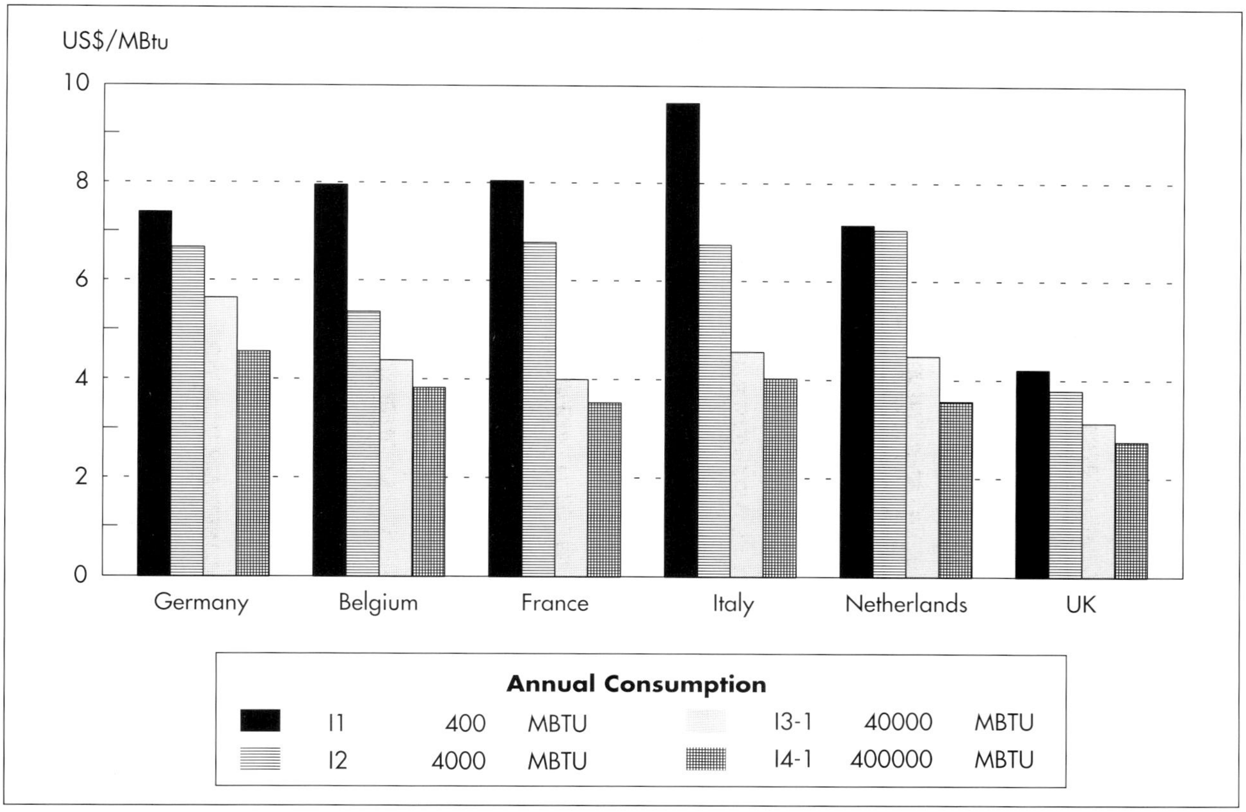
US$/MBtu
10
8
6
4
2
0
Germany
Belgium
France
Italy
Netherlands
UK
Annual Consumption
I1 400 MBTU
I2 4000 MBTU
I3-1 40000 MBTU
I4-1 400000 MBTU

Source: Eurostat

Industrial prices vary even more than domestic prices.

PRICING AND ECONOMIC EFFICIENCY

The pricing systems applied in most European countries do not allow for cost-reflective prices. An important reason for this is the inherent difficulty in gas distribution of cost allocation, and by implication the lack of cost transparency. Another issue is regulation in its various forms (as seen in Chapter IV.). But the differences in the price setting systems applied in the different countries, and indeed the large price differences, also suggest significant freedom for gas distribution companies in price setting.

This is most obvious where a market-value approach is taken. This approach presupposes cross-subsidisation between customer groups, and hence economic inefficiencies. By definition, the cost-plus approach would, in principle, be more likely to produce cost-reflective prices, were it not always distorted by regulation (e.g. price equalisation) and other factors (such as the possibility to introduce market value elements).

The cost-plus approach, when distorted by the netback approach, produces clear economic inefficiencies. However, concluding that the market-value approaches as known in Europe provide better results in terms of economic efficiency and low prices than the existing cost-plus approaches would be a mistake. Evidence from the price comparisons refutes this. Despite the clear gas price differences between the countries observed, none of the continental gas countries can claim to provide its gas consumers with the overall lowest and most cost-reflective gas prices.

It is Great Britain, of all countries in Western Europe, that consistently enjoys the lowest gas and most cost-reflective prices in all consumer categories. This has not always been so. Before the introduction of effective gas-to-gas competition, when the UK applied a pricing system similar to the continental ones, UK gas prices used to be situated in the upper category of European gas price comparisons, despite full independence from imports.

This only confirms that effective competition (meaning introduction of gas-to-gas competition in addition to inter-fuel competition) is the most important factor leading to price reductions, cost cuttings and increased economic efficiency.

KEY FINDINGS

☐ Cost identification and allocation is inherently difficult in gas distribution, but also generally not well developed (due to the regulatory environments).

☐ In Continental Europe, prices for LDCs and for end users do not reflect the true cost of serving each customer category. There are important cross-subsidies, both regulated and non-regulated. Indeed, the underlying approach to tariff setting is not cost-reflective. This can be sustained as pricing systems are not transparent.

☐ This does not provide incentives for cost-cutting, and, in fact, gives rise to economic inefficiencies, leading to higher costs and prices than necessary.

☐ This is confirmed by empirical evidence:

• Prices for end users vary considerably in continental Europe, with only the largest users and interruptible customers benefitting from tariffs close to what competition would bring.

• The consistently lowest gas prices are found in the UK, the only country with price transparency and widespread effective gas-to-gas-competition. This confirms that increased competition and price transparency leads to price reductions and cost cutting.

VII. PROFITABILITY IN GAS DISTRIBUTION

INTRODUCTION

In the (non-competitive) gas supply sectors of Continental Europe, pricing systems are based on cross-subsidisation between supply activities to different customer groups, and remain hidden to the customer's eye. This lack of transparency provides no incentive for cost efficiency or cost reductions. The results are natural gas supply systems with higher costs (and hence prices) than necessary.

The present chapter aims to give an analysis of the profitability of LDCs.

For this purpose, the IEA has analysed and compared financial data from large, representative samples of gas supply companies in Germany, Belgium, the Netherlands and Italy (see annex to this chapter).

Section II of this chapter explains the financial indicators used by the IEA for measuring a company's profitability.

Section III then summarises the results of the analyses per country, and where possible, complements this with a historical perspective. Distribution companies are distinguished according to whether they are gas-only distributors or mixed distribution companies. They are then set in relation to the financial performances of gas transmission companies.

Section IV provides inter-country comparisons of profitability in gas distribution.

Section V considers the issue of appropriate profits *and cost efficiency*. The inclusion of cost efficiency is key because analysing profitability alone would miss an essential point: independent of their profitability, gas companies in continental Europe, in particular gas distribution companies, appear to be cost inefficient (see chapter V). If gas distribution companies are making high profits an are also cost inefficient, both factors need to be addressed by reform in order to ensure that gas prices come down to acceptable levels.

FINANCIAL PERFORMANCE INDICATORS

Comparison of profitability and efficiency can be undertaken by using a series of indicators. Among the many possible indicators available, those which fall within the equity analysis method have been chosen by the IEA.

Equity analysis is often used by financial analysts in order to determine the financial position of a company, usually to invest. For purposes of this study, equity analysis is employed to determine the relative size of returns and the rate at which returns grow for natural gas distribution and transmission companies in Europe.

Equity Analysis of Company Performance	quantifies the returns which accrue to owners of investment
Profit	the capacity of invested capital to produce income to owners.

Ratios most commonly used to demonstrate financial performance include operating margin, net profit margin, return on assets and return on investment or equity. The following box displays the formulas of these ratios used in the next section.

Profitability Ratios

Operating Margin = Operating Profit / Total Sales Revenues
Operating Profit =
Total Operating Income - Total Operating Expenses
Total Operating Expenses =
total gas purchases + salaries + depreciation
+ provision for liabilities

Net Profit Margin on Sales = Net Income / Total Sales Revenue
Net Income = Net Profits after Taxes

Return on Total Assets = Net Income / Average Total Assets [(1994+1995)/2]
Return on Equity = Net Income / Average Equity [(1994+1995)/2]

Ratio Interpretations

- *Operating Margin* : this ratio is used to indicate the level of operating income earned from total sales revenues. A high result for this ratio implies that revenues are relatively sizeable compared to operating expenses (gas purchases, salaries and depreciation, provisions for liabilities).

- *Net Profit Margin On Sales* : reflects the amount of profit in total sales revenues after all expenses and taxes have been deducted.

- *Rate of Return on Total Assets* : indicates the net profit generated per investment in assets.

- *Rate of Return on Equity* : is widely employed to give a measure of return on investment. It is used by equity analysts to estimate the return to owners on equity interest in a company.

PROFITABILITY IN GAS DISTRIBUTION (AS OPPOSED TO GAS TRANSMISSION)

This section focuses on comparisons of categories of companies within individual countries. The analysis is based on the figures published in the annual reports of the reviewed companies, as well as on external work in the case of Germany and Belgium.

Three categories of companies were considered: pure gas distribution companies, companies with mixed distribution activities, i.e. having gas distribution as one of several activities; and gas transmission companies.

Germany (West)

For the sake of comparability, the review was limited to west German companies only (in 1995, most east German companies were still in or just came out of a heavy restructuring and investment phase).

Since in Germany, many regional gas companies are engaged in gas transmission as well as in gas distribution, a distinction between gas transmission and gas distribution companies is not always straight forward. Therefore, an effort was made to take into account only companies that could be clearly identified as either having their main activities in distribution or in transmission.

For these reasons, the samples chosen consist of:

■ 15 pure gas distribution companies (including 8 of the 20 biggest),

■ 10 mixed distribution companies (all among the 20 biggest in their category), and

■ 11 out of the 18 companies registered as transmission companies (Ferngasgesellschaften) in 1995.

The sample group of pure gas companies and the group of mixed companies each accounted for about 12% of total gas deliveries in Germany in 1995. Tables A1 and A2 in this chapter's annex give the results of the calculations undertaken company by company.

The following table gives a summary of the profitability figures for the various categories of companies:

Table 16 Financial Performance Figures for (West) German Transmission and Distribution Companies, 1995

Company category	Average operating margin	Average net profit margin on sales	Rate of return on assets	Rate of return on equity
Pure Gas Distribution	9%	4%	4%	18%
Mixed Gas Distribution	7%	3%	2%	8%
Pure Gas Transmission	4%	3%	5%	19%

Source: International Energy Agency calculations based on company annual reports

In addition to the information in the table it may be noted that:

■ of the 15 pure gas distribution companies, only 2 companies had a net profit margin of 6% or more; 5 of them had a rate of return on equity of 15% or more;

■ only 2 of the 10 mixed companies had a net margin of 4% or more; and equally 2 of the 11 transmission companies attained this margin level.

The following observations can be made, bearing in mind the fact that the company samples are not necessarily fully representative of the whole country (in particular since companies from the eastern part of the country have been left out):

■ in 1995, pure gas distribution companies were generally more profitable than mixed distribution companies (this corroborates the conclusion in chapter V that LDCs make no economic gains from horizontal integration);

■ on average, in 1995, LDCs have made more revenues in relation to operating expenses than transmission companies (in other words, profitability in distribution was higher than in transmission);

■ on average, in 1995, return on investment (measured in RoR on assets and equity) in pure gas distribution is almost equivalent to that in transmission, whereas it is significantly lower in mixed distribution companies.

Studies made by the Rheinisch Westfälisches Institut (RWI) in Essen, Germany, provide a historical overview of profitability in the German downstream gas industry:

Table 17   Gross Margins in Gas Transmission and Distribution in Germany (billion DEM)

	1985	1990	1991	1992	1993	1994	1995
Transmission							
Total revenues	23.8	14.6	20.3	18.4	18.7	18.6	19.3
Total costs	24.9	13.7	17.3	15.8	16.8	17.0	17.8
Gross margins	− 0.9	0.9	3.0	2.6	1.9	1.6	1.5
Distribution							
Total revenues	19.9	15.3	20.1	20.0	21.4	21.7	23.3
Total costs	20.1	16.0	19.3	19.1	20.4	20.7	21.4
Gross margins	− 0.2	-0.7	0.8	0.9	1.0	1.0	1.9

Source: RWI

From the table above, the following observations can be made:

■ over the last few years, the transmission part of the gas chain has generated significantly more profits in terms of gross margins (total revenue - total costs) than the distribution part. Over the period 1990 to 1995, the transmission part generated gross margins of some 11.5 billion DEM, whereas the LDCs generated gross margins of 4.9 billion DEM.

■ The gross margins generated in transmission are on a downward trend while in distribution they are on an upward trend. This is shown very clearly in the following graph which shows profits per cubic metre gas sold by the two sectors.

Figure 7   Germany: Gross Margins in Gas Transmission and Distribution

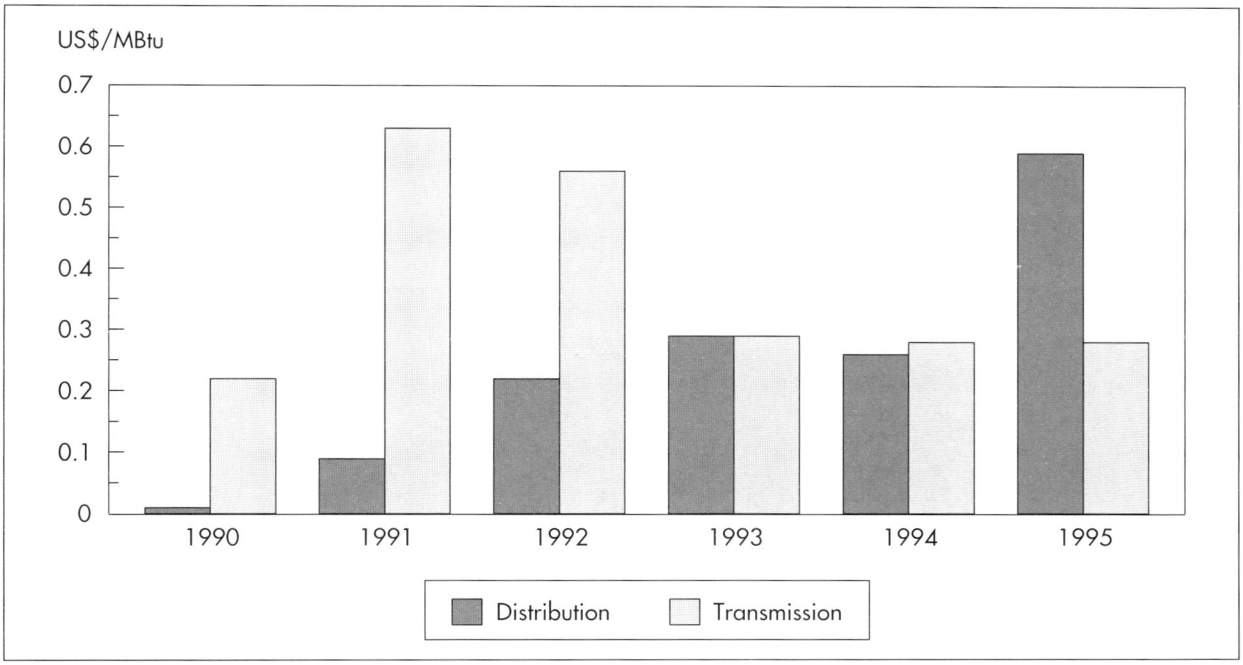

Source: RWI

This trend coincides with the arrival of gas transmission/trading company Wingas on the German gas market in the early 1990's. Since then, gas-to-gas competition has emerged in the areas to which Wingas extended its transmission grid. However, with Wingas targeting its competitive gas sales mainly on large customers (industrial consumers, distributors and other transmission companies), this largely spared the distribution level from the new competition. Hence, the distribution companies, for the largest part, could continue to exploit their monopoly situation to maintain high end-user prices, while enjoying the benefits of gas-to-gas competition on the wholesale level.

The trend could stop in 1999. During 1997/8, the Federal Cartel Office has put several regional distribution companies under pressure to align their tariffs and prices at the lower levels offered by comparable suppliers in Germany. In several cases, this resulted into significant price cuts. Should the Federal Cartel Office maintain the pressure and extend it to the remaining bulk of German gas distribution companies via cooperation with Regional Cartel Offices gas price cuts could become generalised throughout gas distribution in Germany. And, obviously, profitability would go down again.

Furthermore, the planned reforms of Germany's legislative framework for the energy sector will abolish the existing demarcation contracts and the exclusivity of concession agreements. This alone will liberate a huge potential for competition between gas companies on all levels (both transmission and distribution) with inevitable cuts in prices, and hence, profitability.

Belgium

Of the 23 LDCs that existed in Belgium in 1995, 17 companies - of which 7 pure and 10 mixed - have been analysed along with Distrigaz, the transmission company. The companies in the sample accounted for 85% of total gas sales in Belgium in 1995. Results for the whole sample are shown in table A3 of this chapter's annex.

Table 18

Financial Performance Figures for Belgian Transmission and Distribution Companies – 1995

Company category	Average operating margin	Average net profit margin on sales	Rate of return on assets	Rate of return on equity
Pure Gas Distribution	29%	24%	6%	16%
Mixed Gas Distribution	23%	21%	13%	29%
Distrigaz (Transmission)	6%	2%	2%	27%

Source: International Energy Agency calculations based on company annual reports

In addition to note:

■ 11 of the 17 companies have a profit margin of more than 20%;

■ 6 of the companies have a rate of return on total assets of 15% or more;

■ 9 of the companies have a rate of return on equity surpassing 15% and often far beyond.

This table shows that both pure and mixed distribution companies have higher profitability than the transmission part of the chain (except when measured as return on equity).

Also, sales profitability is higher in pure gas distribution companies than in mixed ones (as in Germany (west)); however, return on investment (both in terms of RoR on assets and equity) is higher in mixed distribution.

The profitability figures for Belgium may seem extremely high at first sight. It should be noted that, in Belgium, distribution companies do not calculate concession payments to municipalities/regional authorities as costs (unlike e.g. in Germany, the Netherlands and Italy). Instead they pay the equivalent in the form of dividends, hence the important profits. On average, these dividend payments to public authorities represent about 80% of the profits made in gas by Belgian distribution companies.

Historic overview: The annual calculations of the Comité de Contrôle enable a good overview of profits and profitability in the Belgian gas distribution sector since 1990:

Table 19   Profits in Belgian Gas Distribution

	1990	**1991**	**1992**	**1993**	**1994**	**1995**
Total revenue (BEF/GJ)	248.5	258.9	244.3	238.3	236.9	233.0
Total cost (BEF/GJ)	208.1	217.6	203.9	193.5	196.2	191.5
Operating profit (BEF/GJ)	40.7	41.4	40.6	45.0	40.8	41.7
Net profit (BEF/GJ)	8.1	5.6	5.9	9.4	8.5	7.1
Net profit ($/MBtu)	*0.23*	*0.16*	*0.17*	*0.26*	*0.24*	*0.23*
Net profits (Million BEF)	1399.0	1127.0	1165.0	1922.0	1705.0	1519.0
Net profits (Million $)	*41.9*	*33.0*	*36.2*	*55.6*	*51.0*	*51.5*

Source: Comité de Contrôle

The Netherlands The Netherlands is the only country where it has been possible to analyse all the LDCs (33 in 1995). These companies sold 23.4 bcm of gas in 1995. Table A6 in this chapter's annex gives the results of the analysis company by company.

11 of the 34 Dutch LDCs are pure gas distribution companies. This does not mean that 100% of their revenues come from gas sales, but gas is the only form of energy they distribute. The following table summarizes the results for pure gas companies, mixed companies and for Gasunie, the transmission company:

Table 20   Financial Performance Figures for Dutch Distribution Companies – 1995

Company category	Average operating margin	Average net profit margin on sales	Rate of return on assets	Rate of return on equity
Pure Gas Distribution	11%	7%	7%	39%
Mixed Gas Distribution	10%	5%	4%	74%
Gasunie (Transmission)	2%	0.5%	1%	20%

Source: International Energy Agency calculations based on company annual reports

On all four indicators, the distribution companies on average perform very well, and much better than Gasunie. It can be added to the table that:

■ 10 distribution companies had a net profit margin of 7% or more;

■ 14 distribution companies earned a rate of return on total assets of 5% or more;

■ 9 distribution companies earned a rate of return on equity of 40% or more.

As for Germany (west) and Belgium, pure gas distribution companies have higher sales profitability and operating margins than mixed companies.

And as in Germany (west) and Belgium, the profitability indicators are generally higher in distribution than in transmission. This goes for both pure gas companies and for mixed companies compared to Gasunie.

Traditionally, gas distribution in the Netherlands is more profitable than gas transmission – both in terms of absolute profits and of rates of return. Part of the explanation is found in the way the two sectors are regulated. The transmission company Gasunie (50% state owned, 50% Esso/Shell) is committed to cost-efficiency and is regulated so as not to allow large profits[4]. In comparison, total profits in the LDCs (all distribution companies including also those selling other energies than gas) that year were 686 million NLG.

Italy

Sufficient information was available from only 10 distribution companies. Given that the sample includes Italgas and Camuzzi, the two biggest distribution companies, this still covers more than one third of total gas sales in 1995. The results, company by company, are given in this chapter's annex.

There are only one or two companies in the sample that could be characterised as pure gas distribution companies. However, their share is too small, and their indicators cannot be considered as reliable. Hence the choice of an aggregated approach. The following table gives the average figures for Snam, the transmission company, and the total sample of 10 distribution companies:

Table 21   Financial Performance Figures for Italian Transmission and Distribution Companies – 1995

Company category	Average operating margin	Average net profit margin on sales	Rate of return on assets	Rate of return on equity
Gas Distribution (both mixed and pure)	9%	6%	3%	17%
of which Italgas	5%	2%	1%	9%
Snam (Transmission)	20%	10%	7%	112%

Source: International Energy Agency calculations based on company annual reports

Similar analyses have been done on a sample of 60 other LDCs in Italy for 1995. These gave the following results:

■ Operating margin was found to be 5.3% for the sample as a whole.

■ The average rate of return on equity was found to be 22.5%.

■ The rate of return on fixed assets was 14.8%.

These results demonstrate that profitability in Italy is higher in transmission than in distribution, but the difference is not as significant as for instance a comparison between SNAM and Italgas (43% owned by SNAM) would suggest.

---

4. This is somewhat misleading. Gasunie is allowed under its current statutes to make profits of only 80 million NLG (40 million $) a year. However, the commercial structure between the up- and middlestream is such that the substantial real profits are recorded at the "Maatschap" (the so-called partnership between the Dutch state and the producers Exxon and Shell, which together also constitute the shareholders of Gasunie).

France

Given that GDF does not distinguish between transmission and distribution activities in its external accounts it was impossible to make a reliable analysis of profitability in distribution. Further, the 15 LDCs that are independent of GDF account for less than 2% of the gas invoiced to end consumers in 1995, so that in-depth analysis of these companies would not give a representative result. In addition, some of these companies have accounts that make it difficult to separate their gas sales from other activities. Finally, it has been difficult to access to the necessary information despite the fact that these companies are public. Nonetheless, some calculations are contained in this chapter's annex.

Only so much can be said: Anecdotal evidence indicates that distribution is considered more profitable than transmission.

Great Britain

The distinction between transmission and distribution in Great Britain is less relevant than in continental Europe. British Gas has historically taken care of both. The distinction between transportation and marketing of gas is more important. Over the last few years British Gas has been transformed from a fully integrated company into two main companies, one essentially covering transportation (BG plc), and the other, marketing of gas (Centrica plc).

It is possible to calculate some profitability indicators for 1995 for the companies now making up Centrica, but these are not very meaningful as an indicator of the "normal" level of profits in gas marketing since they also take into account activities that have little to do with gas marketing. Further, profitability in gas marketing will probably undergo further change.

Analysis of accounts up to the demerger in 1997, however, showed that gas transportation, as carried out by BG TransCo, was the main contributor to BG plc's total profits. In 1995, BG TransCo accounted for about 80% of total operating profits of the BG group.

In gas marketing, the recent years have shown very little profits, *if at all*. In Great Britain, there are many companies trying to establish themselves in the gas trading business. This has led to and will continue to lead to harsh competition in which marketers are making losses or operate at low profitability. Out of this, there are signs of a concentration process. The regulatory system will need to insure that concentration will not lead to an unbalance of market power.

This study has not looked in detail at the profitability of UK gas companies for the following reasons:

■ since the physical distribution of gas still is taken care of by BG TransCo (after the demerger called TransCo UK, part of BG plc), profitability figures for the marketing companies would not be comparable with those of LDCs in Continental Europe;

■ the difficulty of separating gas marketing from other activities in UK gas companies.

Conclusions

From the foregoing it can be concluded that in 1995 on continental Europe, profitability is higher in gas distribution than in transmission, with the exception of Italy. The historical perspective, looking back to 1990 or even 1985, confirms this, except for Germany where transmission has traditionally been more profitable but is now exposed to growing gas-to-gas competition.

In Belgium, Germany (old Bundesländer) and the Netherlands, pure gas distribution companies are more profitable than mixed distribution companies. Anecdotal evidence allows a similar conclusion for France (not enough is known about the Italian distribution sector to allow for conclusions; Great Britain is in a totally different situation, and cannot be compared).

The measure of return on equity, among all countries is considerably higher than other measures of profit. This indicates that potential returns to owners of equity capital are sizeable.

INTER-COUNTRY COMPARISON OF PROFITABILITY

This section makes an attempt at comparing profitability indicators across countries.

This type of comparison is fraught with statistical, accounting and methodological difficulties, not least that samples on which the figures are based are not necessarily representative of the whole population. Four different performance indicators have been chosen in order to give a broader basis for comparison and to reduce the impact of different accounting standards. It should still be underlined, however, that the results must be interpreted with caution.

Problems with Inter-country Profitability Comparisons

The ratios chosen for the analysis use some of the least ambiguous and most informative financial figures, like revenues, equity, assets, net income, etc. However, methodological problems generally arise in cross-country financial comparisons because of differences in accounting standards. To illustrate this, three examples will be given:

☐ The notion of equity varies from one country to another. It is generally composed of two parts, equity as such and reserves. The rules for setting aside reserves vary a lot; one consequence is that the composition of total equity varies considerably. In Continental Europe, companies tend to have a relatively low level of equity capital and high level of reserves. This phenomenon is most pronounced in Germany and the Netherlands, one of the reasons being that distributed earnings are taxed higher than retained earnings. These companies have an incentive to maintain earnings in the company rather than to seek external capital sources. To the extent this influences the financial structure of the company in the direction of higher equity share and lower debt ratio, it will have implications for the rate of return on equity.

☐ The rules for asset depreciation vary from one country to another. The depreciation period for gas distribution pipelines varies from 12 years to 60 years in the countries studied. Analysis of the depreciation of tangible assets in the transmission companies shows that accumulated depreciations in 1995 varied from 43% to 72%. Annual depreciation, which is a function of the age structure of the assets, the investment profile of the company and the depreciation rules, is a substantial part of total cost and therefore has a significant effect on the bottom line result. Differences in depreciation could therefore explain some of the differences in profits across countries.

☐ As already stated, payments by gas companies to public authorities in return for their right of supply and/or right of way differ: in Germany and Italy, these are paid in the form of concession fees and thus included in the costs; in Belgium, dividends are paid instead, thus costs do not appear and profits seem larger; in the Netherlands prior to 1997 some LDCs made concession payments, other not - today the concession system has disappeared.

Earlier analysis shows that there is usually a significant difference between pure gas companies and mixed companies in terms of financial performance. This distinction will be maintained, another reason being that the number and the share of pure gas companies vary over the samples and would make comparisons biased if all the companies were lumped together.

The following table shows the most interesting financial performance figures for the pure gas distribution companies in three of the countries.

Table 22   Financial Performance Figures for Pure Gas Distribution Companies – 1995

Country	Average operating margin	Average net profit margin on sales	Rate of return on assets	Rate of return on equity
Belgium	29%	24%	10%	15%
Netherlands	11%	7%	7%	39%
Germany	9%	4%	4%	18%

Source: International Energy Agency calculations based on company annual reports.
Italy not included because almost the entire sample consists of mixed distribution companies.

The ranking of the countries is not consistent on all the performance indicators. Belgium shows the highest score on all but one indicator, and the Dutch companies score slightly higher than the German ones on all indicators. The main reason for this is that German gas distribution companies pay concession fees to local authorities, whereas Belgian companies do not (instead an equivalent is supposed to be contained in their dividend payments to their municipal shareholders). In Belgium, these "concession dividends" represent on average 80% of profits. Real net profits in Belgian pure gas distribution companies can be estimated at somewhere just above German levels. Estimates for Dutch companies are difficult (some companies paid concessions in 1995, some didn't). Their profits net of any kind of "concession dividends" are thought to be in the range of those in Germany.

The corresponding figures for mixed distribution companies are shown in the table below:

Table 23   Financial Performance Figures for Mixed Distribution Companies – 1995

Country	Average operating margin	Average net profit margin on sales	Rate of return on assets	Rate of return on equity
Belgium	23%	21%	13%	29%
Netherlands	10%	5%	4%	74%
Italy	9%	6%	3%	17%
Germany	7%	3%	2%	8%

Source: International Energy Agency calculations based on company annual reports.

In this table, a sample of some of the largest LDCs in Italy has been included (the sample represents more than one third of the total gas sales, but only a small share of all Italian LDCs).

It can be concluded that, generally, the performance figures for mixed distribution companies are lower than for pure gas distribution companies.

The pattern of the preceding table is more or less repeated: Belgian and Dutch mixed distribution companies outperform German companies on all indicators (Italy: all but one), though figures are difficult to compare due to above mentioned reason. Estimates for Belgium of corrected net margins put it at level with Italy, where the larger LDCs seem to perform better than German mixed distribution companies.

An attempt was also made to look into the distribution of profits made by gas distribution companies in 1995:

Belgium (17 companies out of 23): Of total net profits, 36% went into reserves, 28% were distributed to the municipalities and Electrabel, a private electricity company having ownership interests in the majority of the Belgian LDCs, received 28%. Electrabel receives dividends that are less than proportional to its ownership share.

Netherlands (26 of 35 companies): 58% of net profits were added to reserves. 37% were distributed directly to municipal shareholders and the remaining 5% to other distribution companies who are shareholders. There are no private shareholders in Dutch LDCs.

Germany (west) (gas-only distribution companies - sample of 15 companies): The largest portion of net profits was paid into reserves (73%). Other distribution companies received 11%, other private companies 11%, transmission companies 1% and municipalities 1%.

Thus, in Germany (west), the largest part of the profits made have been assigned for reinvestment in the business, whereas in the Netherlands about half and in Belgium well over half of the profits have gone to the shareholders (of which by far the main share went to municipalities).

Taking into account the conclusion in the previous section that returns on equity are sizeable, this confirms that public gas distribution companies constitute an important source of income to local/regional public authorities.

APPROPRIATE PROFITS AND EFFICIENCY

A natural monopoly can be regulated to earn a "normal" profit by charging a tariff equal to long-run marginal cost. According to theory, such a tariff would maintain the solvency and stimulate the efficiency of the pipeline or the LDC. It would be fair and reasonable to users, and, since one component of long-run marginal cost is a "normal" profit, would provide for such a profit. The problem is to determine what a "normal" profit is. If the company in question is publicly owned, one might argue the investment criteria should be the same as for other public investments as long as no particular externalities or significant differences in risk are involved. It is possible that such considerations play a role in countries where gas distribution is undertaken by government-owned companies, but the tendency appears to be to treat these companies as private in this respect.

The question of what constitutes an appropriate rate of return has been raised in the UK in the context of the MMC inquiries on the future organisation and regulation of British Gas. It is beyond the scope of this study to go into details on this discussion (see the 1994 IEA study "Natural Gas Transportation – organisation and regulation", where the issue has been dealt with at some length). Suffice it here to say that there is no objective and neutral answer to this question. In its discussions with British Gas, OFGAS came up with four possible methods of estimating a reasonable rate of return for the BG transportation business:

- Estimating BG's weighted average cost of capital, using the Capital Asset Pricing Model to arrive at the company's cost of equity. Since shareholders generally demand a satisfactory return or they will not invest, it is appropriate that the market-based return for BG should be the starting point for the assessment.

- Comparing market returns for similar foreign utilities, which provides valuable information on the returns required by investors in comparable utilities. The validity of many of these comparisons may, however, be questioned because utilities operate in very different regulatory environments.

- Comparing historic cost accounting returns for similar foreign utilities to seek an indication of the actual financial performance of these utilities. These returns are based on operating results and consequently form a large element in the market's forecast of profits. However, care must be exercised when comparing accounting returns because of differences in accounting standards and inflation rates. The regulatory framework will also have implications for the rate of return achieved.

- Comparing current cost accounting for UK utilities.

At the time, the MMC concluded that a rate of return of 4.5% on net assets (using current cost accounting) was a reasonable upper limit for BG's transportation activity. The rate of return and especially the capital valuation behind it remains, however, a bone of contention between OFGAS and BG.

Until recently, discussion about returns has focused on profitability in transportation and storage (which includes distribution in Great Britain). The proposals from OFGAS in the context of the 1997 Price Control Review of British Gas Trading to also regulate profits in gas marketing to domestic customers implied that profits should be limited to 1.5% of turnover in this activity.

Though the profitability indicators used in the earlier parts of this chapter are not directly comparable to those values discussed in Britain, they still indicate that Continental distribution companies make larger profits than could be considered appropriate. Nevertheless, the differences in the levels between real profits on the Continent and British maximum standards for profitability, nor the observed differences in real profitability indicators between the countries can fully explain the important differences in European gas prices (for identical consumer categories).

This is where costs and cost-efficiency in gas supply come into play. The conclusion from the above can only be that significant differences in cost-efficiency exist among gas companies in Europe. Together with the finding of chapter V that gas distribution companies appear not to be cost-efficient (e.g. mixed gas distribution companies do not take cost-advantages from horizontal integration) this points at the existence of excess costs in continental European gas supply.

The fact that comfortable to high profits are made in gas distribution while simultaneously cost-inefficiencies exist indicates that European gas LDCs enjoy considerable market power.

From all this, the following conclusions can be drawn:

■ Overall cost reduction should be as much a target of regulatory reform as avoiding excess profits. In this regard, particular attention should be given to distribution, where non-gas costs are double or more than in transmission (see chapter V).

■ Limiting the focus of regulatory reform to gas transmission (in accordance with the EU Gas Directive's minimum provisions) would not be enough and potentially counter-productive, as gas distribution companies could gain too much market power.

Introducing effective competition in gas distribution *as well as* (and preferably in the same time scale) in transmission, and ensuring a level playing field between competitors in both parts of the gas chain is the best means to meet both objectives – overall more appropriate rates of return and cost efficiency –, and hence better prices for Europe's gas consumers.

KEY FINDINGS

☐ Gas distribution is the more profitable activity in Belgium, Netherlands and probably in France. In Germany, gas distribution has only recently become more profitable than transmission. Only in Italy does gas transmission remain more profitable than distribution (which may be due to the market power of SNAM).

☐ Pure gas distribution companies clearly show higher returns than mixed distribution companies.

☐ On average, Italian and Belgian companies seem to perform best. The distribution companies in Germany (west) showed the lowest average performance figures. (Dutch figures difficult to compare – could be at German levels).

☐ The measure of return on equity among all countries is considerably higher than other measures of profit. This indicates that potential returns to owners of equity capital are sizeable.

☐ Profits are distributed among a variety of owners including municipalities, other gas distribution or transmission companies or other private companies. In most countries some proportion of profit is paid directly into reserves (most sizeably in Germany).

☐ Profits in gas distribution on the west European continent are to some extent higher than British maximum *standards*, but this in itself does not fully explain the high level of end-user prices on the continent as compared to Britain, an important additional factor being the excess costs in parts of continental gas distribution.

☐ This leads to the conclusion that regulatory reform on the European continent should not only focus on transmission but also on gas distribution. Introducing effective competition in gas distribution

• will bring incentives to reduce excess profits and costs in gas distribution, and

• should be a condition for full TPA eligibility of LDCs to the transmission grids in order to avoid LDCs gaining too much market power, and to ensure that the benefits of wholesale competition are passed on to end-consumers.

ANNEX

FINANCIAL PERFORMANCE INDICATORS, INDIVIDUAL GAS COMPANIES

Table A.1 Performance Figures for a Representative Sample of Distribution Companies in Belgium (1995)

MIXED	Op. Margin	Net Margin	ROA	ROE
IVEG	21%	21%	12%	60%
SIMOGEL	18%	17%	2%	6%
IEGA	27%	26%	10%	24%
INTERLUX	24%	25%	11%	16%
SEDILEC	15%	15%	10%	13%
WVEM	15%	15%	7%	14%
IDEG	24%	25%	15%	19%
IGEHO	17%	15%	1%	1%
IVEKA	38%	25%	45%	122%
IMEWO	30%	30%	13%	17%
TOTAL MIXED	**23%**	**21%**	**13%**	**29%**
GAS-ONLY	Op. Margin	Net Margin	ROA	ROE
PLIGAS	20%	20%	7%	9%
ALG	23%	27%	14%	19%
INTERGA	19%	16%	8%	10%
I.G.H.	24%	23%	6%	8%
SIBELGAZ	73%	37%	20%	33%
IGAO	27%	27%	8%	10%
GASELWEST	19%	18%	6%	16%
TOTAL GAS-ONLY	**29%**	**24%**	**10%**	**15%**

Source: IEA calculations based on company annual reports.

Table A.2 Performance Figures for a Sample of Distribution Companies in France (1995)

MIXED	Op. Margin	Net Margin	ROA	ROE
Colmar (Gaz)	6%	7%	5%	8%
Sté Monégasque	10%	16%	7%	22%
Gaz de Strasbourg	10%	4%	3%	14%
Grenoble (Gaz)	6%	3%	2%	8%
TOTAL MIXED	**8%**	**7%**	**4%**	**13%**

Source: IEA calculations based on company annual reports.

Table A.3   Performance Figures for a Representative Sample of Distribution Companies in West Germany (1995)

MIXED	Op. Margin	Net Margin	ROA	ROE
Kraftwerke Mainz-Wiesbaden	6%	3%	5%	16%
Südhessische Gas- und Wasser AG	6%	2%	1%	5%
rhenag	16%	7%	3%	8%
EWAG	4%	4%	4%	18%
Hannover Stadtwerke AG	6%	1%	1%	3%
SCHLESWAG AG	7%	2%	2%	7%
GEW AG	14%	3%	3%	5%
Düsseldorf Stadtwerke AG	5%	2%	2%	6%
Bremen Stadtwerke AG	3%	2%	2%	7%
TWS AG	3%	2%	1%	5%
TOTAL	**7%**	**3%**	**2%**	**8%**

GAS-ONLY	Op. Margin	Net Margin	ROA	ROE
Stadtwerke Diez GmbH	13%	6%	4%	15%
Gasvesorgung Oberschwaben	2%	2%	8%	47%
Pfalz-Gas GmbH	9%	2%	1%	3%
Gasversorgung Gelnhausen GmbH	12%	4%	2%	8%
Gasanstalt Kaiserslautern AG	9%	5%	4%	12%
Gasgesellschaft Aggertal	16%	3%	3%	7%
Badische Gas AG	11%	5%	7%	13%
Hürth Gasversorgung GmbH (Rhein-Erft)	10%	2%	1%	6%
Fränkische Gas-Lieferungs-GmbH	8%	2%	2%	9%
Gasvers. Südhannover Nordhessen	8%	4%	4%	23%
Energieversorgung Mittelrhein GmbH	12%	4%	4%	7%
Erdgas Südbayern GmbH	5%	2%	2%	11%
Ortsversorgung Westfälische Ferngas AG	4%	13%	15%	81%
Maingas AG	9%	4%	4%	12%
Hamburger Gaswerke GmbH	6%	4%	3%	15%
TOTAL	**9%**	**4%**	**4%**	**18%**

Source: IEA calculations based on company annual reports.

Table A.4   Performance Figures of the Transmission Companies in West Germany (1995)

Ferngasgesellschaften (West Germany)	Op. Margin	Net Margin	ROA	ROE
Ferngas Salzgitter GmbH	5%	3%	6%	26%
Gas Union GmbH	2%	1%	4%	13%
Ferngas Nordbayern GmbH	3%	1%	3%	11%
Saar Ferngas AG	3%	1%	3%	7%
Bayerngas GmbH	1%	2%	4%	7%
Gasversorgung Süd Deutschland GmbH	2%	3%	7%	25%
Westfälische Ferngas GmbH	4%	13%	15%	81%
BEB Erdgas und Erdöl GmbH	3%	2%	3%	4%
Thyssengas GmbH	4%	2%	4%	17%
Wintershall Gas GmbH	3%	0%	0%	1%
Ruhrgas AG	10%	5%	7%	20%
TOTAL	**4%**	**3%**	**5%**	**19%**

Source: IEA calculations based on company annual reports.

Table A.5   Performance Figures for a Sample of Distribution Companies in Italy (1995)

MIXED	Op. Margin	Net Margin	ROA	ROE
SICILIANA	3%	0%	0%	1%
TRENTINA	4%	3%	2%	5%
ITALGAS	5%	2%	1%	5%
RAVENNATE	5%	5%	3%	5%
CIS-CONS	6%	7%	2%	3%
RIMINI	7%	2%	3%	5%
MODENA	9%	15%	9%	61%
BRESCIA	12%	13%	7%	26%
CAMUZZI	13%	3%	2%	48%
PARMA	26%	6%	2%	11%
TOTAL MIXED	**9%**	**6%**	**3%**	**17%**

Source: IEA calculations based on company annual reports.

Table A.6   Performance Figures of the Distribution Companies in the Netherlands (1995)

MIXED	Op. Margin	Net Margin	ROA	ROE
NV Energiebedrijf Zuid-Kennemerland	9%	6%	8%	3%
Nutsbedrijven Weert	16%	8%	6%	834%
NV Energiebedrijf Rijswijk-Leidschendam (ERL)	6%	3%	4%	22%
NV Nutsbedrijf Heerlen	8%	5%	5%	16%
NV Openbaar Nutsbedrijf Schiedam	16%	6%	3%	39%
NV REGEV	11%	8%	10%	102%
NV Nutsbedrijven Maastricht	9%	5%	4%	15%
NV FRIGEM	10%	5%	6%	36%
NV RENDO	17%	11%	8%	323%
NV Nutsbedrijf Regio Eindhoven	6%	1%	1%	20%
Centraal Overijsselse Nutsbedrijven NV (COGAS)	13%	7%	4%	24%
Energiebedrijf Midden-Holland NV	13%	4%	3%	70%
Nutsbedrijf Westland NV	6%	2%	3%	14%
Energie Delfland NV	11%	4%	3%	59%
NV EWR	9%	1%	1%	5%
NV Delta Nutsbedrijven	6%	7%	3%	6%
NV MEGA Limburg	6%	3%	2%	6%
NV REMU	12%	3%	1%	4%
NV ENECO, Capelle a.d. Ijssel	9%	3%	3%	21%
NV PNEM	6%	3%	2%	6%
NV EDON	6%	2%	1%	6%
NV NUON	9%	3%	2%	6%
TOTAL MIXED	**10%**	**5%**	**4%**	**74%**

GAS-ONLY	Op. Margin	Net Margin	ROA	ROE
NV Gasbedrijf Noord-Oost Friesland	11%	10%	14%	35%
Gasbedrijf Midden Kennemerland NV	12%	7%	6%	75%
Nutsbedrijf Amstelland NV	20%	7%	11%	49%
Gasdistributie Zeist en Omstreken	8%	4%	4%	18%
NV Intercommunaal Gasbedrijf Westerlo	14%	9%	7%	49%
NV GGR-Gas	6%	5%	7%	23%
NV Nutsbedrijf Haarlemmermeer	8%	5%	3%	12%
Intergas NV	8%	6%	7%	26%
OBRAGAS NV	10%	9%	9%	85%
Gasbedrijf Centraal Nederland NV	9%	5%	4%	15%
NV GAMOG Gasmaatschappij Gelderland	12%	6%	6%	39%
TOTAL GAS-ONLY	**11%**	**7%**	**7%**	**39%**

Source: IEA calculations based on company annual reports.

VIII. DEVELOPING COMPETITION IN GAS DISTRIBUTION – POLICY ANALYSIS

GENERAL POLICY IMPLICATIONS OF THE STUDY'S FINDINGS

The introduction to this report set regulatory reform and market liberalisation in the context of the key policy objectives of:

■ Maximising economic efficiency;

■ Sustaining security of supply;

■ Meeting environmental goals;

■ Meeting social objectives.

Whilst the desire to optimise economic efficiency is a key driver in the current efforts to introduce competition into the gas supply chain, other policy goals remain important. Energy security remains a fundamental policy objective. In the case of the environment, the policy pressures are growing to accommodate objectives ranging from the need to curb greenhouse gas emissions internationally, to a range of national and local objectives. Social and quality of service goals remain part of the agenda although some redefinition of these goals may be appropriate. Competitive markets will therefore have to deliver across the range of policy goals, not just economic efficiency.

The analysis in this report of the economics of the European gas distribution industry suggests that economic efficiency is not optimised under the current institutional structures. The report draws this out in a number of ways. First, the chapter on gas economics and costs noted that whilst potential for economies of scale (from the vertical integration in the industry) should exist in theory, they are not realised in practice, i.e., average unit distribution costs do not decrease with output. In one of the examples analysed, costs go up rather than down with company size. Similarly, gas distribution companies appear not to have exploited advantages resulting from horizontal integration. It does appear, therefore, that costs are higher than they need to be.

The chapter on pricing shows that current prices do not reflect costs. A key explanation is in the fact that the current tariff mechanisms are not intended to minimise prices to end users. "Market value" pricing, in particular, is intended to sustain the value of gas in competition with oil in order to promote investment in gas (see chapter 6, part II) and as a means to diversify energy supply portfolios away from too high a dependence on oil. In other words, it was designed to support another key energy policy objective - that of energy security. Hence, cross subsidies are prevalent to the meet the objective of increased gas penetration as well as a variety of non-economic goals such as tariff equalisation. Furthermore, costs of security of supply are not made explicit in non-competitive markets. Only in Britain, which has taken significant steps to open its markets, are security costs (e.g. for storage) made explicit. Cost reflective pricing is fundamental to a properly functioning market in which economic efficiency is optimised.

Finally, the chapter on profitability notes that there is a wide variation in the gross margins and the operating profits of the gas distribution industry. Using the UK benchmark of the rate of return which has been imposed on British Gas (and accepting that there are issues about the validity of the comparison) it is reasonable to suppose parts of the continental distribution industry are currently earning excess profits. This, again, implies economic inefficiency, as the excess profits could be redistributed into lower prices for end consumers of gas.

However, the high level of end-user gas prices on the continent as compared to those in Britain cannot be fully explained by differences in profits alone. To a large part these price differences stem from cost inefficiencies, in other words "excess costs" in continental gas distribution.

This combination of excess costs and profits in gas distribution has arisen from the prevalent monopolised distribution structures (giving distribution companies considerable market power) and the lack of transparent prices.

The reasons for the current situation may be largely attributed to the regulatory framework which was designed to favour penetration of natural gas. This has proved effective, however, at the expense of economic efficiency (which has never been an objective - the price guarantee to the consumer was that he would pay gas prices in line with those of competing fuels). In a monopoly situation there is no incentive - as there would be in a competitive market - to minimise prices as costs can be passed on to consumers.

Today, most west European countries have mature gas markets, thus, their initial objective of gas penetration is fulfilled. It is now important to engage a debate about the best means of introducing incentives into the gas distribution industry that will encourage economic efficiency, whilst at the same time sustaining the other policy objectives.

THE NEED FOR WHOLESALE COMPETITION

An important stage of the debate is to consider the general framework for introducing competition across the whole gas supply chain.

The issue is the extent to which market players (not just numbers but also variety) - producers, intermediate suppliers and purchasers - can be brought into the market so as to establish the conditions for effective competition where supernormal profits are driven out, and prices are driven towards, and more closely aligned with, real costs. A first key to establishing competition is eligibility for TPA for a variety of gas purchasers, including distribution companies, to the high pressure transmission system. Variety is key to ensuring that there will be adequate incentives for real gas-to-gas competition. TPA to the high pressure transmission grid (without which it is very difficult to take advantage of a choice of suppliers – since the chosen supplier's gas must be transported to the customer) is central to introducing competition. If the eligibility criteria for TPA are drawn too tightly, there will be inadequate competition to force any real changes towards a more competitive approach to their markets by upstream gas producers and suppliers.

The EU Gas Directive obliges EU Member States to provide TPA eligibility only for large gas end-users (power producers and industrials), and largely leaves with them the decision about eligibility of LDCs and smaller consumers. At this stage, the issue is still under debate in most Member States. The majority of LDCs would prefer market access because of the possibility of buying cheaper gas. But there is concern at the possible implications for social and security policy objectives of introducing too much competition too fast – in terms of consumer protection, and/or in terms of allowing long-term take-or-pay contracts to become stranded, or in terms of the risk of jeopardising the further development of gas networks.

For countries deciding to go beyond the minimum requirements of the Directive, full eligibility of LDCs for access to the gas transmission system will make a very significant difference to competition in the gas supply chain. In this case, it will become important to address the issue of current long term contracts engaged by LDCs and to consider if and how they might be released from these contracts in order to take advantage of TPA. Simultaneously, long-term take-or-pay contracts of transmission would then become stranded[5].

Greater competition in the wholesale gas market – in other words, distribution companies operating in a liberalised transmission environment – has potentially beneficial effects on prices to end consumers (in that LDCs can start to shop around for gas and be in a position to get a better deal for their gas purchases which could be passed on to *their* customers). However, this will not necessarily happen if there is no incentive for LDCs to pass on these benefits. It would primarily put pressure on gas producers and transporters, but could considerably increase market power of LDCs. Therefore, LDC eligibility for a choice of supplier may be a strong condition for promoting end user benefits but is not in itself sufficient. A pre-condition needed to ensure that incentives exist for LDCs to pass on the benefits of wholesale competition to their customers is competition within the gas distribution sector. The next section explores more fully this issue.

---

5. The issue of stranded long term contracts can be tackled through market mechanisms. For example, if a gas company would like to be relieved of one of its long term contracts, the corresponding obligation could be taken by the state which could then auction the contract to all market participants. Any difference in value arising from the sale, as compared to the initial contractual price clauses, will then become a stranded cost to be recovered from the totality of the gas users (or the tax payers).

POSSIBLE MODELS OF COMPETITION FOR THE GAS DISTRIBUTION MARKET

It should be noted, as background to the discussion on competition models, that gas distribution is only partly a natural monopoly. Only the physical transportation has natural monopoly features. In principle, the other activities of the gas distribution business – those not concerned with the physical transport of gas, i.e., the supply function – are contestable either in aggregated or desaggregated form, i.e., they could be undertaken by others. Some caution is needed, however: the market barrier of the cost of information (sophisticated metering) needs to be addressed; economies of scope need to be considered; more important, the incentives to continue making long term investment in the distribution networks must remain in place – which goes directly to a major policy goal underlying the gas market – security of supply; lastly the potential issues raised by competition for social objectives (for example, a tendency to "cherry pick" customers) need to be addressed.

All that said, if the benefits of competition are to flow through to all end customers of gas, it is necessary to consider how competition might be introduced within the distribution market. There are several options, which could be combined with each other.

One option is direct competition, i.e., competition through physically separate competing networks. There are, however, economies of scale in physical gas distribution which in most cases will make competing networks economically inefficient. But it is worth noting that this statement might be challenged if the ongoing weight and difficulty of regulating the market to sustain other forms of competition is such that the long term sustainable benefit lies in direct competition. That said, the capital intensity of distribution networks (greater than for the high pressure transmission grid) makes this conclusion highly unlikely. Where several pipelines already exist, the situation might, however, be exploited by having different operators for them.

Another approach which will be reviewed only briefly has the starting point that there are other ways of promoting economic efficiency than competition. These primarily revolve around some form of price control, in other words, a managed, regulated attempt is made to define prices to end consumers such as they should be if there were an effectively functioning competitive market. It could be said that the present regulatory arrangements in most of the countries reviewed are a managed attempt to define prices, with the important difference that the objectives of current regulatory reform are not necessarily focussed on the promotion of economic efficiency.

One such "non-competitive" approach would be price caps. Price caps typically put a ceiling on end-user prices (or they can set a limit on the amount of price increases). The formulae for defining price caps can vary. They are usually set as a function of the evolution in inflation and productivity, thus forcing a company to reduce its costs and increase its productivity. A classic example is the formula used in UK, the RPI-X formula. Similar formulae have been imposed on GDF in France and the operation cost element of LDCs in Italy. Price caps may be established either as substitute for direct competition or to support the transition to a more competitive market (a period which might continue to require controls until competitive pressure forces prices down).

Yardstick cost comparisons may also be used as a surrogate for competition. Under this approach a regulated firm's prices would be based not on its own costs but on those of comparable regulated firms operating under similar technical and economic conditions. The approach is designed to encourage cost-effectiveness. The firm can only increase its earnings by cutting costs or boosting productivity. An important aspect is that the regulated firm would be allowed to keep any extra earnings that result. The potential for extra earnings provides the incentive to become more efficient.

There must be a question as to whether the latter two approaches will achieve much on their own. Price regulation on its own makes no use of market mechanisms to achieve the desired result. Yardstick competition is very difficult to apply effectively, as it relies on the establishment on an appropriate comparison group and identical accounting procedures.

This leaves two main (non-exclusive) options (both of which encompass a number of sub options, or which could accommodate a phased approach to the introduction of competition). Both are predicated on the assumption that generally speaking there is a single distribution network:

- The introduction of ongoing competition for supply.

- The introduction of competition at regular intervals through a franchise or concession system for supply *and* transport.

In both cases a key sub option revolves around whether (or when) to introduce full structural separation of transportation from supply. Another set of variations revolves around desegregation of supply itself (for example, metering and ancillary services).

Ongoing Competition

This is the approach that has been developed in Great Britain and North America and may be described as competition through TPA to the network. In its first phase it is characterised by the introduction of more than one supplier to distribute gas for customers, but with the incumbent network owner also retaining the right to supply.

To work effectively this requires a number of regulatory steps to be taken, equivalent to those which are addressed in promoting TPA to the high pressure transmission grid:

- TPA to the network needs well defined terms and conditions. Regulated access – which implies that access should be granted provided certain conditions are met, and that a regulator oversee the fair operation of TPA – is likely to be more effective than negotiated access, where the new entrant must negotiate a deal with the incumbent. (Given the market power of the incumbent, this could be very difficult, unless there is a strong and effective competition authority.)

- The incumbent LDC's physical transportation and supply functions need to be unbundled. This is essential to ensure that there is fair competition (e.g., no cross subsidies) between the LDC's supply function and supply competitors. It may be achieved through accounting separation, if this is backed by clear rules and a strong regulator. A stronger approach is to separate functions into subsidiaries (with "chinese walls"), and the strongest approach of all is the creation of separate companies (full structural separation).

- The incumbent LDC keeps its physical transportation monopoly but its monopoly over supply must be removed.

- In order to ensure a level playing field for new entrants to supply, the incumbent LDC must apply comparable transport tariffs to itself (in its activity as a supplier) as to others (which underlines the need for accounting separation, in order to make this effective).

- Retail price regulation is very likely necessary, at least in the transition to sustained and effective competition. The regulation may be necessary to safeguard consumer and social policy interests and the emerging competition before competition is fully effective. For example, prices could drop too much because the incumbent could engage in predatory pricing which forces out the new entrants (against the long term interest of consumers, which is best served by competition) More likely, prices could remain too high, because competition is not yet fully effective in reducing costs, removing supernormal profits, and forcing greater efficiency to be reflected in lower end prices.

As accounting separation (good enough to guarantee that the transport business cannot subsidise the supply business) is very difficult to achieve in practice, structural separation may be ultimately the most effective deterrent to cross subsidisation or undue discrimination in favour of itself by the incumbent (as it is simply no longer an option). This consideration was fundamental to the 1993 Monopolies and Mergers Commission (MMC) report which recommended the demerger of British Gas. This subsequently took place in February 1997, when transportation (both high and low pressure) was structurally separated from supply and marketing, the former becoming Transco and the latter becoming Centrica.

It should be noted, however, that the British approach differs in one fundamental respect from the conceptual framework elaborated here, namely that no distinction is made between the high pressure and low pressure transportation functions (Transco covers both)[6]. By contrast, the conceptual approach taken here is that it is helpful to develop the analysis on the basis of separating out the low pressure distribution market from the rest of the gas supply chain. This is likely to simplify and clarify the regulatory framework and help to put responsibilities (e.g., for sustaining supply) in the right place.

The second phase of this model might therefore be to move to structural separation of transportation and supply. This would still feature TPA - or more accurately Access - to the network by competing suppliers but the critical difference is that the network owner would no longer have the right to supply.

---

6. Great Britain is in fact unique in that it has separated the physical gas flow from commercial transaction.

Competition Through Franchising

This approach would involve setting up a franchise or concession system in which companies need to compete, at regular intervals, for the monopoly right to the whole distribution function (transport and supply). This, however, might be a preliminary phase. A second phase might consist of structurally separating the physical (transportation) function from the supply function, and to put both – separately – out to franchise competition (for the same period of time). A third phase might even combine the franchise and "ongoing" competition models by keeping a franchise competition for the transport, and moving to the introduction of ongoing competition for supply.

A number of issues will need addressing here too:

■ Public ownership could deter the introduction of competing franchises, hence privatisation is likely to be necessary. This would help to facilitate franchises, and create a level playing field with elimination of cross-subsidies.

■ An effective bidding process needs to be established, perhaps encouraging international competition through straightforward and standardised bidding procedures.

■ Consideration needs to be given to the potential transfer of the management of the infrastructure at the end of the concession (using the experience of other industries where franchising bidding is carried out).

■ Linked to the above asset transfer (if the incumbent has to hand over to a competitor) needs careful management in order to ensure that incumbents have an incentive to maintain the assets in good condition, and to expand them if opportunities and need exist.

A franchise bidding system has recently been introduced in Mexico, as set out in the following frame:

Mexican Franchise System

Permits (or concessions) for gas distribution are granted through a two step competitive bidding process and the winner is granted a 12-year exclusivity period. Bidding procedures are engaged by the regulator when it finds that there are sufficient factors to justify development of a distribution project, or as a result of a statement of interest from an interested party. Bidders must meet certain conditions defining minimum coverage, technical specifications, safety standards and quality of service. The bidding procedure involves two evaluations, one technical, one economic. The technical evaluation is done on a "pass" or "fail" basis; no ranking is made between the different proposals. The main criterion in the economic evaluation is the tariff for the service offered, defined as the maximum average revenue yield per unit that the company can obtain in its first five years of operation. (This revenue yield will be used as the initial value for the proposed price cap regulation.) All proposals within a 10% difference from the lowest tariff will be declared as tied, in which case the winning proposals is the one with the highest coverage. Coverage is defined as the expected number of users which the bidder expects to be connected to the system in the first five years.

Distribution concessions are also the subject of public tendering in Italy – to 40% of LDCs and 55% of sales (though municipalities may, and frequently do, decide to set up their own joint stock company with both public and private capital, thereby ungoing public tendering). The tendering process, however, is based on criteria like rent to the municipality and pace of development, rather than quality of service and prices to end consumers.

CRITICAL ANALYSIS OF THE OPTIONS

The chosen approach must above all else be capable of delivering enhanced economic efficiency. But it must also be measured against the other policy objectives - notably energy security and social objectives.

Economic Efficiency

The first option in principle carries a strong incentive to maximise this - provided adequate competition to the incumbent develops. There is, however, an issue of the long term sustainable performance of this model. Ongoing regulation - possibly quite heavy - may be necessary to ensure that competitors are not pushed out of the market by the incumbent, who may continue to benefit (despite regulatory efforts) from being in the transport and supply business at the same time. Structural separation of transport and supply will address this, but even for this, a fairly heavy regulatory overhead is likely to continue - access regulations, accounting separation regulations and (quite possibly) price regulations.

The second option, beyond the bidding process itself, in its first phase (franchise bidding for transport and supply too) contains some incentive to reduce costs, but none directly to pass on the benefits to consumers. As with the first option regulation (price controls and/or for the franchise bids to be evaluated on the basis of their commitments as regards end user prices) is likely to be necessary.

The approach may require less ongoing regulatory overhead than Option (2). The overhead will be concentrated in the bidding process.

In a second phase, where supply is structurally separated from transport, economic efficiency is likely to be higher. However, the issues of sustaining interest in investment in the transport remain.

Social Policy

How well do the options meet social objectives? These encompass objectives such as universal service (obligation to connect; obligation to supply), tariff uniformity (at regional or national level), and support to disadvantaged customers. Decisions on tariffs, geographical coverage, special arrangements for disadvantaged customers will be taken, under the first approach, in the context of whether they are economically justifiable and make competitive sense (unless regulations put back in what may be taken out by competition). The likely result is that current social arrangements will be eroded. For example, price variations are likely to emerge across geographical locations. Broad tariff uniformity (as currently promoted in Belgium) will not be sustainable.

The first question to ask, however, is what social policy goals are relevant to this stage of the evolution of energy markets? This is an issue for governments to determine, preferably before they embark on reform, in order that they have a clear view of what needs to be achieved in the new market conditions. Many social objectives were established ten or twenty years ago and there is scope for considering why they differ widely, even at present. Each element of social policy needs to be reviewed on its merits and in its national context. Thus, whilst Belgium practice tariff equalisation, Germany does not, with the result that the small consumer in some areas of Germany pays twice as much as consumers in other areas. The question is does this matter? And, if so, in what way? Tariff equalisation implies a cross subsidy from other customers located close to gas sources, to customers located further away. This was probably important in order to facilitate expansion of the network, but it is worth asking whether tariff equalisation benefits a mature market.

Universal service is another principle which is likely to be at risk in a competitive market although again, the question needs to be asked whether this matters. From a social policy standpoint, it is important to distinguish between the obligation to connect, and the obligation to supply. It might be considered – particularly in a more competitive context – that the latter needs to be enforced but not the former. The case for some form of universal service obligation in the gas market is also closely linked to the situation for electricity, given that in the domestic context electricity is substitutable for gas. An obligation in the electricity market might remove the need for a similar obligation in the gas market. Universal service across a national territory in any event has always been less of an issue for gas, than for electricity.

It can be safely assumed that social objectives of one kind or another will continue to be important. Given that the free play of competition in the first option is likely to have a "negative" impact on these issues, how can the issues be tackled? The UK's approach is to address the issues through the licensing system (which regulates both who is allowed into the gas supply market, and the terms and conditions which they must fulfill in the market). Thus the UK regulator may withhold a licence where he/she considers that the area which the new supplier wants to cover has been framed in such a way as to exclude a large number of pensioners and disabled people. The supply licence contains a number of conditions specifying the supplier's social obligations. There are a number of performance standards which must be met, e.g., customer contact, new connections, meter reading, and the obligation to keep a register of elderly and disabled customers and to offer certain services. It should be noted that significant monitoring effort is needed to check whether licence holders are complying with these conditions.

The second option does not automatically deliver social objectives either and some regulation may therefore also be necessary under this option.

Security of Supply

Security of supply remains a vital consideration. Gas security of supply was addressed in some detail in the IEA 1995 Gas Security Study. However, the study did not consider gas distribution in detail. As with the social objectives it is important to be clear what is meant by security. Security of gas supply can be divided into three risks: the short term risk of disruptions to supplies through the failure of markets to balance supply and demand adequately, the long term risk that inadequate investment will be made to secure future supplies; and inadequate diversity of supply sources (which will be left aside in this context as more relevant to the upstream supply chain decisions).

Short Term Security

The evidence that open wholesale markets on the TPA model cope well with short term security is encouraging based on the track record of those countries - primarily the US, Canada, and UK - which have opened their markets significantly. The markets in these countries are balancing supply and demand adequately in normal times as well as in exceptional times. There have been virtually no disruptions to firm customers, despite some exceptional weather conditions to test the markets. Retail competition is too recent to say whether the same result will hold when competition is extended. As pointed out in other IEA studies (see both the transportation study and the gas security study) these countries are basically self-sufficient in gas. Market liberalisation in countries that are dependent on imports would have to take additional security aspects into account.

A key factor reflecting the good short term security performance is that more competitive markets force companies to think about their costs and to identify these clearly. This is partly because of the competitive pressure to seek cost reductions. It is also because there are financial penalties for supply default. Security costs in the UK are beginning to be explicitly and separately identified. For example, there are now explicit tariffs for storage services, and transport and distribution services are priced to include overdimensioning of these facilities as a security feature, all this contributing to pricing that truly reflects the seasonality pattern of demand and the means to address this.

At the same time - as with social objectives - UK licences for public gas transporters, shippers and suppliers contain a number of conditions, based on British Gas' historic standards. These promote security of supply standards for domestic consumers, (larger commercial and industrial consumers can effectively choose their level of gas security by agreeing to have their gas supply interrupted). The general short-term security of supply criteria used in Britain are the same as existed prior to market liberalisation, i.e. That supplies are maintained on the coldest day which could be expected in 1 in 20 years and also over the coldest winter that can be expected in 1 in 50 years.

These rules do imply that, as with social objectives, the market - left on its own - might not be trusted to meet the standards. However, unlike social objectives, it may be that this is a transitional issue. As the competitive market grows in sophistication, security costs are more explicitly identified and consumers become more aware of the need to balance security against prices. Again, it is too early to come to definitive conclusions. However, the main point in this context is that it seems possible to promote a model of ongoing competition which can be trusted to meet key policy objectives - in this case short term supply security - through regulatory requirements (in the UK model - through licences) to be fulfilled by all market players alike.

Long Term Security

Under the franchise model it is, of course, just as possible to have similar rules both for assessing competing bidders and to regulate the winner.

Long term security of supply raises the further issue of investment in order to ensure that reserves are developed, production engaged and gas transported to the demand centres. The capital cost of high pressure transmission grids and distribution grids is very high and the incentive to continue investing in maintenance, replacement and grid extension must be sustained under competitive conditions.

The underlying issue for investment is prices , which in Continental Europe up to now translate into the related contractual commitments to buy gas. Traditionally, Europe has operated on the basis of take-or-pay contracts where the gas price is linked to the price of oil through "netback pricing". The introduction of wholesale competition – mostly through LDC eligibility for TPA, and then, potentially, through competition at the retail level – inevitably raises a question of what happens to the traditional arrangements. Will LDCs opt for cheaper short term gas? Will retail customers (if offered the choice) prefer cheaper gas even more than LDCs? Will this put traditional take-or-pay contracts under pressure, and what will be the implications for investment? Will price signals in the new environment be strong enough, and come soon enough, to encourage investment? The concern is perhaps not so much that such investment happens at all, but that it happens only with some delay, creating investment troughs at some point in the future.

The evidence of recent investments in European (and Russian) gas pipelines (as well as in other capital-intensive sectors) suggests that large infrastructure projects are possible without an outlet or price guarantee. In mature gas markets, the real question is that of price signals adequately reflecting the need to build, at benefit, new pipelines or other infrastructures.

In the context of retail competition the issue of charges for access to the distribution network is an important factor. The network owner/operator will want to ensure that they are covering their investment/depreciation costs. Under the first option this means that the cost of using the network – for both incumbent and supply competitors – should allow for new necessary and cost-effective investment.

There is also a strong connection with the tendency in a competitive market context for security costs to be made explicit. If suppliers are drawing out security costs in the prices of their products, cost reflective pricing and effectively functioning markets look as if they can and do result in the effective internalisation of security costs. It should, however, be noted that in itself, this does not prevent the possibility that some consumers could have difficulty signing for suppliers in a situation where suppliers' gas is already fully committed, i.e., that price signals coming from the market are to weak to mobilize new supply if the latter is too costly.

The big advantage of the second approach is that if there were any concern about the importance of long term contracts (at a certain, stable, price) it could be easier to regulate for this. In particular, it would be possible to mandate a franchisee to cover part of his supply by long term TOP contracts, and this could be linked to the duration of the franchise. There is a further potential advantage to this arrangement which links back to economic efficiency. Supply franchisees are likely to find it easier and cheaper to cover long term security costs (because of the stability provided through longer term contracts)[7].

---

7. One of the risks with a competitive market is the appearance of long-term cycles as known from e.g. the refining and petrochemical industries. High price levels set by interruptible customers against gas-oil might, with a delay corresponding to the lead time for new upstream production and midstream transport, create an investment boom. This could then well be followed by a bust as the additional gas quantities arriving on the market would push prices downwards. Investment would slow until the excess gas is absorbed, by which time the cycle could start again. In such a cyclical context, the key issue would be to insure that the bust period does not lead to gas shortages. One possibility to avoid such pronounced cyclicality would be to promote one way or another long-term supply commitments.

CONCLUSION

Competition and hence economic efficiency is probably well achieved under ongoing competition. The regulatory overhead, however, is high (especially in the early stages) and the "phase two" of the franchise model (where supply is structurally separated from transport and also opened to competition) is likely to be as effective. Both options can sustain social objectives through regulation. As regards security of supply, the key question is whether investment will be an issue. The franchise approach probably facilitates regulation to ensure investment is adequate, as efforts can be concentrated on building up adequate incentives and rules at a single level of the supply chain; it may also mean security is cheaper to provide for.

There is certainly a need for competitive pressure in retail distribution if lower prices to end consumers are to be achieved. It is important for governments to be clear about their underlying policy objectives, against which different approaches to competition can be measured. Policy objectives (e.g., tariff equalisation) are likely to be given a different emphasis in different countries.

Subsidiarity is likely to be important. The scope for competition and cost reductions varies between countries. Countries with access to a higher number of suppliers and a high market maturity probably have more to gain from ongoing competition than countries with few potential suppliers and an industry still under development (specially at early stages of development). Security of supply considerations may also differ. Countries have varying relevant physical and commercial characteristics. Countries with high self-sufficiency and a lot of storage may have a wider choice of models than others less well off.

Experience from countries that have introduced competition shows that a staged approach is often appropriate. The UK is an example of the gradual introduction of competition down the supply chain.

Finally, and very importantly, not all the options for competition are tried and tested yet. There is relatively little experience to go by. The franchise route does offer a relatively risk free approach to introducing competition in that it enables LDCs (or where these do not exist, their equivalents) to settle down into a new role where they can exploit the benefit of TPA to the high pressure system. If franchising proves inadequate, the model can evolve to a more competitive phase (as described earlier) where supply is opened to ongoing competition.

OTHER ISSUES

A number of other critical issues need to be addressed to promote competition, whichever option is chosen.

■ **Regulatory structure.** A strong, independent, regulator is essential to oversee and apply the rules.

There are three key components to effective natural gas regulation: system regulation (ensuring system integrity and reliability at all times); market regulation (guaranteeing a transparent, clear and unbiased clearing process); and competition regulation (maintaining a level playing field for all market participants).

■ **Ownership.** Governments need to make sure that ownership arrangements provide an incentive to reduce costs, improve efficiency, and lower prices to consumers.

Ownership has implications for the distribution of profits. Private companies will benefit their shareholders. With public ownership, dividends will accrue either to the State or to local governments. With mixed ownership, there will be separate dividend streams to local governments and private partners.

Ownership by the State may well raise difficult and contradictory pressures: to maximise efficiency in the gas market or to promote local employment (for example). Cross subsidies from one public activity to another should be avoided. Public owners may be interested in lowering prices to end consumers, or to maximise profits to cross subsidise other public services. Private owners are likely to have an incentive (if not limited by regulation) to maximise profits by reducing costs but keeping up prices.

Studies in the power sector indicate that privately owned companies are, in the long run, more efficient than publicly owned companies (there is no available evidence to this effect in the gas sector at this stage).

More competition in gas distribution could mean more risk. The trend towards privatisation could be reinforced if it is perceived that private companies are better placed to manage risk.

■ **Taxation.** This can be very distorting of cost reflective prices. Gas taxation – both in absolute terms and in relative terms vis-à-vis competing fuels (oil and electricity mainly) – is very heterogeneous across countries and there is an important issue of whether this is compatible with the economic efficiency objective. The rule should ideally be to have energy taxation consistently reflecting the full externalities of all competing fuels.

■ **General competition law.** The increasing horizontal integration at LDC level creates new possibilities for cross subsidies between unrelated activities and may blur the competition between fuels. Vertical integration between LDCs and gas suppliers is also growing in some countries. Whilst it may have positive aspects for security, some of this activity may be anti-competitive as, again, it raises the scope for cross subsidy. Finally, there is the fear that too much concentration at regional level could create excessive market power: the increasing pace of joint ventures and alliances deserves attention in the context of the efforts to establish more market players in competition with each other, not less. Hence the need to consider how general competition law should be applied to prevent anti-competitive behaviour. Indeed general competition law might be seen in terms of an option (though preferably associated with other measures) for approaching retail competition.

KEY POINTS

☐ There is some evidence that economic efficiency is not optimised under the current institutional structures of the European gas market.

☐ Thus, incentives for economic efficiency should be improved, whilst taking into consideration other important goals, notably security of supply.

☐ The introduction of competition in the wholesale part of the gas supply chain is an important first step. It must be effective, and serious consideration should be given to eligibility of LDCs for TPA.

☐ At the same time, consideration should be given to the introduction of competition within the gas distribution market. Various approaches are possible, including competitive concessions.

COUNTRY ANNEXES

BELGIUM

Statistical information

Natural Gas Supply/Demand Balance (Mtoe, 1996):

Indigenous production	0.0
Imports	11.9
Exports	0.0
Stock change	-0.1
Total natural gas supply (primary energy)	11.8
Electricity and heat production	2.2
Other transformation and energy use	0.2
Total industry	4.3
Residential	3.6
Commercial	1.6
Other	0.0
Statistical difference	- 0.1

Natural Gas in the Energy Balance (1996):

Share of TPES	20.9 %
Share of electricity output	14.6 %
Share in industry	30.6 %
Share in residential/commercial sector	35.9 %

MARKET AND INDUSTRY STRUCTURE

Belgium accounts for 3.8 % of total natural gas consumption in OECD Europe and is a mature gas market. It does not produce natural gas but has a well diversified portfolio of import contracts. It reexports gas to Luxemburg and acts as transit country for gas from the Netherlands and Norway to France and Spain. Transit volumes are higher than the country's own consumption. In 1995 it received pipeline gas from Norway, Netherlands and Germany as well as LNG from Algeria and Abu Dhabi. Zeebrugge is the location for the most important pieces of infrastructure; it already has an LNG import terminal and the terminal of the Zeepipe pipeline from Norway, and will be the landing point for the Interconnector from Bacton in the UK.

The natural gas market in Belgium can be considered as consisting of two distinct parts: the public distribution sector and the transmission sector. The public distribution sector comprises all gas distribution undertaken by the 23 regional gas distribution companies. The transmission sector distributes gas to large industrial users and to power generators. The two markets for gas are about equal in size. Total gas deliveries in 1995 were 493 801 GJ, of which public distribution accounted for 48.6% and the transport sector 51.4%. Gas deliveries to power generation, which is part of the latter sector, accounted for 17.2% of total deliveries, which in a European context is relatively high. Exports to Luxembourg are also considered part of the transmission sector.

Table 1 gives a breakdown of gas natural consumption in Belgium in recent years.

Table 1   Gas Consumption in Belgium in Recent Years (10^6 GJ)

	1991	1992	1993	1994	1995
Total public distribution	217.5	215.9	230.1	224.8	239.8
– residential, commercial and small industry	192.7	190.0	201.7	195.6	208.3
– industrial served by dist.companies	24.8	25.9	28.4	29.2	31.5
Industry served by Distrigaz	123.8	135.1	139.3	150.8	168.5
– of which firm	85.3	92.4	90.4	98.7	109.3
– of which interruptible	38.5	42.7	48.9	52.1	59.2
Electricity production	62.2	64.5	66.9	70.2	85.2
Exports Luxembourg	18.9	19.9	20.3	20.6	24.0
Total	**422.4**	**435.4**	**456.6**	**466.4**	**517.5**

Source: Ministry of Economic Affairs

Looking back over the last ten years, the electricity sector is the one where the strongest consumption growth has been seen. Total increase in gas deliveries since 1985 has been 45%, but public distribution has increased less rapidly (29.5% over the period) whereas the transmission sector (the segment served directly by Distrigaz) has grown more briskly (total increase of 64.1%, the major part due to increase in gas to power generation). The relatively modest grow in public distribution reflects the fact that the Belgian gas market has reached a high level of penetration: 433 out of 589 municipalities now have gas distribution, which means that about 74 of the Belgian municipalities at least partly have access to gas. The penetration rate, however, is lower: about 65% of the total dwellings in Belgium are located along the gas grid, and about 55% of these are connected. Only 43% of the total number of houses in Belgium use gas for space heating. However, gas now takes about 70% of the new build market where gas is available.

The number of customers in public distribution at the end of 1995 was 2 299171 of which the vast majority were households; industrial and other non-domestic customers in this sector numbering 80236. The number of customers in the transmission (large customer) part is about 300.

In terms of regulation (see below), the gas sector is characterised by a relatively high degree of government intervention and explicit price regulation through the setting of tariffs for most users.

The central player in the Belgian gas sector is Distrigaz, an enterprise which has been in mixed ownership since 1965 but has now been privatised except for a "golden share" still held by the Government. Distrigaz imports, exports, stores, transports and sells gas. It sells gas directly to large industrial customers and power generators (transport sector) and to the regional intermunicipal distribution companies which distribute gas to small customers through their own distribution networks.

Up to the privatisation of Distrigaz in 1994, the state directly and indirectly held 50% of the shares. The ownership structure as of the beginning of 1997 was as follows:

The government:	one share	
Tractebel:	33.25%	
Publigaz:	16.62%	(intermunicipal finance company owned by the intermunicipal distribution companies).
Distrihold:	16.75%	(Distrihold's capital is in the hands of Tractebel and Publigaz - Tractebel holding half of the shares plus one and Publigaz the other half minus one).
Shell Belgium:	16.67%	
Shares quoted on the stock exchange:	16.71%	

Among the consequences of the privatisation of Distrigaz is the modification of several of its by-laws, in compliance with the Royal Decree of 16 June 1994. This decree put an end to the intervention of some public bodies, such as the Commissaire du Gouvernement and representatives of the Flemish- and French-speaking areas, but introduced another form of state control through the acquisition of a "golden share". This specific share, owned by the state, allows the Minister of Economic Affairs to suspend some decisions of the board of directors which would not comply with the state's energy policy and to take other decisions in order to protect national interests (for instance opposition to sales of strategic assets).

The 23 regional distribution companies taking care of all gas distribution in the narrow sense of the word have the legal status of 'intercommunales' and are regulated by the Act of 22 December 1986. Under this act, municipalities are permitted in the public interest to establish an association or intercommunales. This association is a legal entity in public law and may include the participation of other public and private legal entities. They can therefore be divided into pure and mixed intercommunales (for the time being there are 19 mixed intercommunales and 4 pure intercommunales). The mixed companies dominate the market, accounting for close to 90% of total gas distribution. The mixed companies are managed on a day to day basis by the private partner, but public representatives have a majority on the board of directors. The only private company which participates in the mixed intercommunales is Electrabel, the largest electricity generating and distributing company. This company also plays an important role as gas buyer in the Belgian market. Electrabel is the result of a merger of several electricity companies in the past which all held participations in local gas distribution companies. The fact that Electrabel is the only private company in gas distribution may therefore be said to be an accident. Electrabel is partly owned by Tractebel, the group holding the majority of shares in Distrigaz.

Table 2 gives an overview of all the distribution companies in Belgium.

Table 2   Distribution Companies in Belgium

Company Name	1995 Gas Sales TJ	Market Share	Other activities E=electricity distrib., TV=cable television
19 Intermixed Intercommunales			
IDEG	2.702	1.11%	E
IGEHO	2.126	0.87%	E,TV
I.G.H.	20.272	8.34%	—
INTEREST	sales started '96		E,TV
INTERLUX	786	0.32%	E
INTERMOSANE	1.322	0.54%	E,TV
SEDILEC	6.814	2.80%	E
SIMOGEL	1.894	0.77%	E,TV
INTERGA	17.238	7.09%	—
SIBELGAZ	22.470	9.25%	E
GASELWEST	24.758	10.19%	E,TV
IGAO	34.065	14.02%	—
IMEWO	24.978	10.28%	E
INTERGAS	104	0.04%	—
INTERGEM	9.282	3.82%	E,TV
IVEKA	12.094	4.98%	E,TV
IVERLEK I & II	24.127	9.93%	E,TV
PLIGAS	9.669	3.98%	—
IEGA	638	0.26%	E
Inter-Régies Companies **4 Pure Intercommunales**			
Association Liégoise du Gaz (ALG)	18.164	7.47%	—
Intercommunale voor Energie (IVEG)	4.272	1.75%	E
Vlaamse Energie- en Teledistributiemaatschappij (VEM)	1.258	0.51%	E,TV
West-Vlaamsche Elektriciteitsmaatschappij (WVEM)	3.805	1.56%	E,TV
Total in Public Distribution	**242,845**		

Source: Electrabel and Inter-Régies Annual Report 1995

There is a high degree of horizontal integration between gas and electricity distribution in Belgium in that most of the companies that distribute gas also distribute electricity. Even in some of the cases indicated in table 2 where the company distributes only gas (6 of the 23 are pure gas companies), there is co-operation with an electricity distribution company. In fact, only one of the companies in the Intermixt group, Pligas, is a pure gas distribution company. All the companies in the Interregies group except Association Liegoise du Gaz are also combined companies. A large number of the companies are involved in teledistribution. In recent years, some companies have also attempted to become involved in water distribution, but this development has been slow. In general, the distribution companies are not much involved in associated services like sales of appliances. The role of the companies is rather to act as a catalyst in the market by giving advice to consumers.

Gas distribution in Belgium shows a relatively high share of concentration in that the seven biggest companies account for about 70% of the total sales. The biggest one has a market share of 14.2%.

REGULATION

The Energy Department of the Ministry of Economic Affairs has responsibility for the gas sector, while the Government has representatives on the board of Distrigaz. It may intervene by virtue of its golden share in Distrigaz. The terms of this are set out in a royal decree as well as in the company's articles of association.

The Comite de Controle de l'Electricite et du Gaz (CCEG), an autonomous public utility body, is responsible for price regulation. Its main task is to see to it that technical, economic and tariff related policies of the gas and electricity industries are serving the general interest and are integrated into the general energy policy. It is composed of representatives of the two industries (the parties to be controlled), trade unions, industrial representatives and the authorities (central and regional). Its mandate extends beyond price control since it may study any technical, financial or other matters with a bearing on the system for importing, producing, transporting and distributing gas. Regulations are made by means of opinions and recommendations to the industries concerned and, where appropriate, to the relevant authorities. These recommendations do not have the force of law, but the controlled parties have signed an engagement to follow them. The tariff recommendations for residential and small industrial users issued by the CCEG are subject to ministerial decision and are published as a decree. These tariffs are considered as maximum prices, but it would not be illegal for a company to charge lower prices. It would, however, be contrary to the agreement with CCGE. The activity of the CCEG is governed by an agreement between the controlling and the controlled parties which was signed on the 21st of March 1995. The work and the recommendations of the CCEG are to a large extent based on a consensus model.

Pipelines and Storage

These activities are governed by the Act of 12 April 1965 and subsequent laws. The relevant laws distinguish between transport of gas for supply to public distribution companies and transport of gas for other purposes, principally industrial use and the transit of gas across Belgian territory. For the first, a licence is required, known as a "concession de service public" to be granted by the King. The licence may have a term of a minimum of 30 years and a maximum of 50. For the transport of gas for other uses, a permit is required, granted by the Minister of

Economic Affairs. Under the terms of a concession, unlimited in time and conferring exclusive rights under the Law of 29 July 1983, Distrigaz has authority to transport natural gas by pipeline and to store it, irrespective of its place of origin and its source of supply. Nevertheless, transport of gas to public distribution companies is still only possible for Distrigaz if it goes through the procedure of obtaining a gas transport concession in each case, consulting the local authorities and government departments, according to the 1965 law. Transport of gas for industrial purposes, for which a permit is required, is governed by a different procedure. In such cases, Distrigaz has a de facto exclusivity since up to now, permissions have only been granted to this company. Belgium has both underground storages and LNG storage facilities, both owned and operated by Distrigaz.

Distribution

The distribution of gas is treated as an activity of municipal interest and is governed by the Constitution as well as relevant legislation regarding the municipalities. The public service aspect of gas distribution means that distribution companies have an obligation to supply, subject to some economic constraints Such a constraint could for instance be the cost of connecting a household. Practice varies between the companies, but in some cases the distribution company is not obliged to connect a household if the investment cost is more than 30 000 BEF.

The distribution network is monopolistic in character, not by law but by reason of the privileges which the municipalities have. According to the Gas Transportation Act of 12 April 1965, customers are placed in the "transportation" (i.e., transmission) category if they take more than 33 412 GJ of gas annually. This means that they are served directly by Distrigaz. Large industrial consumers, electricity producers and the distribution companies fall into this category. Responsibility for distribution to domestic and non-domestic small customers lies with the local authority, either directly or acting through the public distribution companies, formed by the local authorities themselves or co-owned with the private companies. A distribution company has an exclusive right to distribute gas in its area. Each company enters into a contract with Distrigaz which undertakes to supply the company with the gas it requires, whereas the company undertakes to purchase all the gas it needs from Distrigaz.

There is no concession system as such for distribution. The municipalities may form intermunicipal companies (Intercommunale) which would have a monopoly on distribution in the area covered by the company in question. In all but four of the 23 companies in Belgium, Electrabel undertakes the distribution activity through execution of the decisions of the Conseil d'Administration of the Intercommunale in question. This takes place by virtue of an agreement of association between the Intercommunale and Electrabel. The municipalities always retain the majority on the Conseil d'Administration no matter what the ownership share of the private partner is. Table 3 shows the ownership share of Electrabel in the companies.

The Electrabel ownership varies from around 42% up 99.16%, but in the majority of cases Electrabel holds the majority of the shares. The distribution companies have their individual contracts with Distrigaz, but the sales of the Intermixte companies are accounted for in the Electrabel accounts but without adding any margin for this. The agreement of association

Table 3   Electrabel Ownership Shares in the LDCs

Company Name	Business Activities	Ownership Share Held by Electrobel (Private)	Other Ownership Shares	Return on Equity	Gas Sales as a % of total sales
IDEG		83.44%		18%	%
IGEHO		99.16%		14%	%
I.G.H.		61.60%		11%	%
INTEREST		97.47%		14%	%
INTERLUX		81.48%		16%	%
INTERMOSANE		76.74%		9%	%
SEDILEC	natural gas and electricity distribution	80.47%	23.95% Interenergie – Association Intercommunale Coopérative	12%	23%
SIMOGEL		95.84%		30%	%
INTERGA	natural gas distribution	50.00%	23.95% Interenergie – Association Intercommunale Coopérative	9%	100%
SIBELGAZ	natural gas distribution	50.00%		33%	%
GASELWEST	natural gas distribution cabel television and electricity	86.26%		31%	%
IGAO		53.84%		29%	%
IMEWO		73.13%		32%	%
INTERGAS	natural gas distribution	50.00%	23.95%	not meaningful	%
INTERGEM		75.70%		31%	%
IVEKA KEMPEN CV	cable TV, natural gas and electricity distribution	43,94%		33%	
IVEKA NETE LIER CV		42.14%		35%	%
IVERLEK I		74.84%		9%	%
IVERLEK II		64.91%		31%	%
PLIGAS		49.12%		11%	%

between Electrabel and the companies normally has a duration of 20 to 30 years and are exclusive and can be extended. The tendency in recent years has been to shorten the duration of these agreements. The only possibility for other companies to get involved in gas distribution in Belgium is when an agreement of association expires. At such point in time the municipalities are free to consult other companies. The agreement with the present private partner is normally extended, but in electricity distribution there has been at least one exception to this rule.

The different activities of the Intercommunales have to be accounted for separately. The capital of an Intercommunale has to be divided into as many categories of shares as there are activities. For each activity there are at least two types of shares:

- shares corresponding to the capital brought forward by the private partner (whose share is shown in the table above) and the municipalities.

- shares attributed to the municipalities as a compensation for the exclusive right to serve; the number of shares can be proportional to the population.

The distribution of dividends between the various forms of shares is made according the statutes of the Intercommunale after a decision by the Conseil d'Administration. After allowance for reserves (which should reach 5% of total capital) dividends are distributed between the shareholders but the private shareholder receives less per share than the municipalities. On the other hand, Electrabel does not pay anything corresponding to concession fees often paid by distribution companies in other countries.

Today, Distrigaz has a de facto monopoly on gas supplies to the distribution companies. In the standard contract between Distrigaz and the distribution companies, however, it is stated that if a distribution company can obtain continuous and secure supplies satisfying the necessary quality requirements, from another supplier, the Comite de Controle can allow the distribution company to buy its supplies elsewhere after having given Distrigaz the opportunity to respond by changing its price and /or conditions. According to Distrigaz, there is also a clause in the contract saying that a lower price obtained elsewhere should be to the benefit of all the companies.

Some of the distribution companies in Belgium are thinking in terms of buying their own gas directly from the producer, but realize that most of them are too small to pursue an independent purchase policy. The requirement that a lower price should be to the benefit of all companies could also limit the incentive to look for lower prices.

Imports

Distrigaz has no exclusive right to import gas. In principle, an industrial user can import gas. In 1983, the Law of 29 July removed the exclusive right of import which Distrigaz enjoyed. In practice, however, Distrigaz has remained the only gas importer in Belgium.

CONTRACTUAL ARRANGEMENTS

The distribution companies are obliged to take their gas from Distrigaz and do so under a standard contract with the company which is an "evergreen contract" in the sense that it runs for three years and then is extended. Each of the Intercommunales has an individual contract with Distrigaz but they are identical. The distribution companies carry no volume risk under the contracts in that Distrigaz is obliged simply is required to supply the gas that the companies need. Large industrial users enter into similar standard contracts with Distrigaz with the

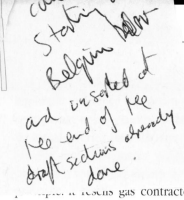

such contracts are not regulated. Industrial contracts do not ge industrial customers pay an annual subscription. Contracts duration of one to five years. Interruptible contracts could be gaz sells gas to the electricity sector, that is Electrabel in ompanies through the Pool des Calories, using a back-to-back gas contracted with gas sellers in Algeria and Norway under long-term contracts (20-25 years). Since import contracts have limited volume flexibility, Electrabel buys storage and modulation services from Distrigaz and pays a fee for this.

PRICING AND TARIFFS

Gas prices in Belgium are regulated through a cost-plus approach, calculated on the basis of a netback system, which gives gas tariffs with little or no explicit reference to other fuels, although gas purchase prices reflect crude oil and petroleum product prices with a lag. In 1994, the average gas purchase cost accounted for 83.9% of the average wholesale gas price and about 57% of the average price of gas to end consumers, which means that gas prices only partially reflect the prices of competing fuels.

Pricing of gas takes place in one or two steps depending on the sector: prices to end consumers in the residential/commercial/small industry sector (the distribution sector) contain two elements – the price paid by the distribution company to Distrigaz and the price paid by the customers beyond this purchase price to cover the cost of distributing the gas and profits. Pricing to large industrial customers and power producers takes place in only one step – Distrigaz determines prices for these customer categories. The gas purchase price paid by Distrigaz is the point of departure for all tariffs charged by the company. To understand pricing through the whole chain it may be useful to consider figure 1 which indicates the way the import price is calculated and how costs in the rest of the chain are added.

The calculation of the average import price (called "G") is in principle very simple: the border price of Dutch, Algerian and Norwegian gas including the direct and indirect costs of importing the gas is calculated by dividing total costs by total volumes. There are, however, some complicating elements:

■ One contract with Algeria is supposed to partly cover gas to power generation, some large industrial customers and fully interruptible customers and is kept outside the calculation of the average price. The volumes under this contract goes to the market "outside G".

■ Within the G market all customer categories pay a contribution F to the coverage of fixed import costs. However, in one special tranche, supplies to the ammonia industry, only 30% of this fixed element is paid. This segment of the market pays a negotiated price to Distrigaz, not following the general industrial tariff described below.

■ Gas buyers in the electricity market (part of the "outside G" market) pay a proportional import price plus a contribution to fixed import costs. Customers in the electricity sector also pay a

Figure 1   Pricing Setting in Belgium

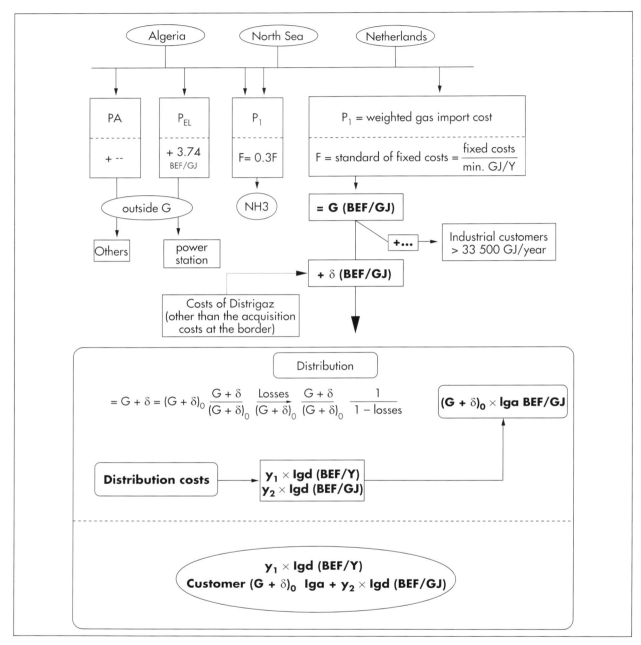

negotiated price for their gas, subject to no published tariff, although conditions are back-to-back with the import contracts.

■ Large industrial customers in the "outside G" are interruptible and contribute to the coverage of fixed import costs only to the extent the competitive situation on the market permits. To the extent that prices to the interruptible market are higher than the proportional import cost, this contributes to lowering the import costs charged to other consumers. Interruptible customers also pay negotiated prices.

As figure 1 shows, the customers within the G market, that is ordinary large industrial customers taking more than 33500 GJ/year and the distribution companies, pays a contribution to Distrigaz to cover transportation costs as a part of their price. In addition, end customers in the public distribution sector pay a contribution to cover distribution costs as a part of their price.

Before cost allocation in transportation and distribution is dealt with, a description of tariffs to large industrial consumers and distribution companies will be given.

Industrial Tariffs

Industrial customers taking more than 33 500 GJ/year are subject to the tariff described in table 4.

Table 4   Industrial Tariffs

Fixed element = $(1-Rh)4.371*RDZ*Sn*K$ [BEF/month]
Proportional element = $1.02*(Gn,m-61.35) + (76.26+6*RDZ*Cne)*P*K$ [BEF/GJ]

where
Rh is an hourly regularity factor defined as $Qa/8760*Sn$ where Qa is annual consumption during the year and Sn is the subscribed volume in GJ/hour.
RDZ is a parameter reflecting the development in cost other than the gas purchase cost. It contains two indices reflecting wages and materials in the electricity and gas industries.
Sn is the sum of subscribed volumes (firm and interruptible) in GJ/hour.
K is a reduction factor which occurs in the fixed element and in a part of the proportional element. Its function is to reduce the unit price with increasing volume. K has the following values for monthly volumes:

- for the first tranche of 41.870 GJ: K=1
- for the second tranche of 41870GJ: K=0.99
- for the third tranche of 41.870 GJ: K=0.98
- for the fourth tranche of 41.870 GJ: K=0.97
- for the fifth tranche of 41.870 GJ: K=0.96
- for volumes beyond 209350 GJ: K=0.95

Gn,m:average border price for all imported gas in BEF/GJ including all direct and indirect charges associated with gas supplies to Belgium.
Cne:Interruptibility coefficient whose value lies between 0 and 1 according to the degree of interruptibility of the supplies. Cn is defined as the firm supplies divided by total supplies, both in GJ/h/year.
P: coefficient reflecting the use that is made of the gas. P has the value 1.1 for thermal uses in which gas is not substitutable. When the supply is fully interruptible, P is reduced by 0.1. For non-specific gas uses P=0.9. When the supplies are firm, P is increased by 0.1. When gas is used as feedstock, P=1.0 for firm supplies and 0.9 for interruptible supplies.

It should be noted that interruptible supplies here means supplies that are interruptible only between 15th November and 15th of March.

Industrial Tariffs (2)

It should be stressed that these tariffs are applicable only to non-interruptible industrial customers of a certain size. All large industrial customers pay a monthly connection fee which is a function of length from the high pressure transmission system and the offtake.

The industrial tariff described above has the following characteristics:

- The tariff consists of two elements, one fixed and one proportional. A low share of total costs is covered through the fixed element.

- The major part of the proportional element is a function of the use which is made of the gas and the volumes taken. The price differentiation taking place between the different uses is an indirect way of competitive pricing – when gas is exposed to competition from other fuels a reduction in price is granted. This reduction is, however, limited to some 10% on the major element of the proportional price element. The volume rebate through K is limited to a maximum of 5% on the same element.

- The transportation element (which in 1995 was about 10 BEF/GJ, that is roughly 10% of the average industrial price) is influenced by volume taken, the use of the gas and the degree to which the deliveries are interruptible. The transportation element is thus lower the higher the volume, the higher the substitutability and the higher the degree of interruptibility.

Tariffs to Public Distribution

The distribution companies buy gas from Distrigaz on standard contracts based on published tariffs. Table 5 describes the tariff used.

Table 5

Tariffs to Public Distribution Companies

All monthly accumulated quantities of the contract year n less than or equal to the reference quantity are invoiced according to the following price formula:

$P_{n,m} = G_{n,m} + \text{delta } n + \Sigma p \text{ from 1 to P of } [K_p{}^*QR_p]/\Sigma p \text{ from 1 to P}[QR_p] + \Sigma p \text{ from 1 to P}[J_p{}^*QR_p]/\Sigma p \text{ from 1 to P}[QR_p] \text{ [BEF/GJ]}$

where

$P_{n,m}$: the price in BEF per GJ supplied to the distributor for month m of the contract year n;

$G_{n,m}$: the average border price for the gas in the relevant month

delta n: share of Distrigaz' transportation cost to be carried by the distribution company

The base value is 28.222 BEF/GJ and it is indexed with the RDZ index.

P : the number of supply points of the distributor

Qrp: reference quantity of supply point p

Kp: impact of the parameter K on the price at supply point equal to 170.275*(K/Km-1) where Km is he weighted average of the K reduction factors of all the supply points in the public distribution network

K: reduction for agreed annual volumes taken at each supply point. K varies as follows:

- 0 to 175.845 GJ/a: K=1
- 175.845 to 351.690 GJ/a: k=0.99
- 351.690 to 527.535 GJ/a: k=0.97
- 527.535 to 703.380 GJ/a: K=0.95
- 703.380 to 879.225 GJ/a: K=0.93
- beyond 879.225 GJ/a: K=0.90.

Jp: impact of the parameter J on supply point p of the distributor equal to $28.222*[(143.423/J)-1]$

J: a parameter expressing the load factor at the delivery point in question. The J values are fixed in the supply contracts. Its calculation refers to the arithmetic mean of the fraction Qa/Qjmax for the five July-June periods of the years 1977/78 to 1981/82 at the relevant supply points. Qa is the quantity taken at the considered supply point during each of the periods defined above, and Qjmax is the maximum daily quantity taken at this supply point during the same periods.

On volumes sold under the ND3 tariff (see below) Distrigaz grants a rebate of 1 BEF/GJ to the distribution company.

For additional volumes beyond the reference quantity the following tariff is applied:

$P=Gn,m+delta\ n[BEF/GJ]$

$P=Gn,m+delta\ n+170.275(K/Km-1)+delta\ n*(143.423/J-1)\ [BEF/GJ]$

where

Gn,m: average border price for the gas in the relevant month

delta n: share of Distrigaz' transportation cost to be carried by the distribution companies. The base value is 28.222 BEF/GJ and it is indexed with the RDZ index.

K: reduction for agreed annual volumes. K varies as follows:

– 0 to 175.845 GJ/a: K=1
– 175.845 to 351.690 GJ/a: k=0.99
– 351.690 to 527.535 GJ/a: k=0.97
– 527.535 to 703.380 GJ/a: K=0.95
– 703.380 to 879.225 GJ/a: K=0.93
– beyond 879.225 GJ/a: K=0.90.

Km: average reduction factor for all delivery points based on volumes for the year 1 July 1993-30 June 1994.

J: a parameter expressing the load factor at the delivery point in question.

On volumes sold under the ND3 tariff (see below) Distrigaz grants a rebate of 1 BEF/GJ to the distribution company.

For additional volumes, the following tariff is applied:

$P=Gn,m+delta\ n[BEF/GJ]$

The tariff to distribution companies has the following characteristics:

■ it is cost plus based and makes all distributors pay a price covering the gas import cost plus a transportation element covering the share of distribution volumes in the total transportation and modulation cost.

■ the transportation element does not distinguish between geographical locations (perequation des tariff). A rebate is, however, given according to the load factor: a high load factor reduces the transportation element.

■ the tariff also provides for a volume rebate which is calculated in relation to the total weighted average rebate obtained in the system.

■ If the distribution company takes volumes beyond the reference quantity, it should pay a price equal to G+delta. If the reference quantity is paid at a lower price, this price will, however, also be valid for the extra quantities.

Allocation of Costs on the Transportation System

The delta between the sales price and the import price in the tariff formulae for industry and public distribution is supposed to cover the cost of transporting the gas from the import point to the end consumer (in the case of industrial consumers) and to city gate in the case of public distribution, including needed storage. Simplifying somewhat, it may be said that deliveries to large industry contribute 6 BEF/GJ (value fixed in 1991, later indexed) to transportation costs whereas distribution companies contribute 33.49 BEF/GJ (since 1 October 1997) to these costs. This cost allocation raises the question about cost drivers in gas transmission and distribution. By relating the two figures to each other, it looks like distribution customers pays 4.77 times more for transportation than industrial customers. According to Distrigaz, the distribution companies account for 50% of total sales, 90% of the storage cost and 67% of the transportation costs.

One obvious reason why industrial customers should pay less for transportation than distribution customers is that it costs less to serve industrial customers. The service of distribution customers is much more demanding in terms of modulation of supplies. Figure 2 shows the variation in monthly sendouts in the two sectors.

In Belgium, daily and seasonal variations in demand are taken care of by using storage, flexibility under import contracts and interruptible contracts. In terms of physical modulation, most of this takes place in the high pressure transmission part of the gas chain. This means that building of storage and physical overdimensioning of the network to take care of swing in demand primarily is a transportation cost. Modulation on the distribution system basically takes place by using line pack. As will be shown below, this, however, does not mean that all consumers pay

Figure 2   Daily Demand for Natural Gas in Belgium

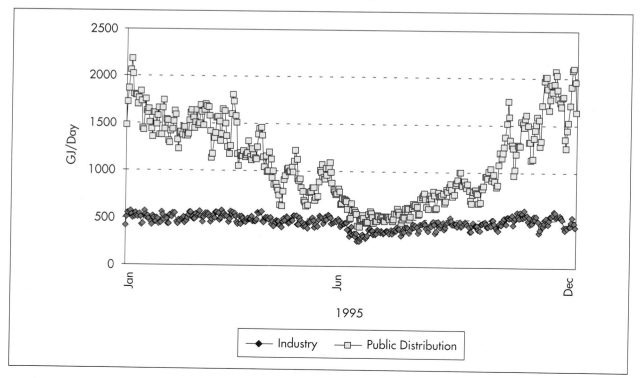

the same share of distribution costs. The methodology used to calculate the delta values for the two sectors is one of the few examples of explicit published thinking in this area and therefore merits some attention.

Methodology for the Determination of Delta

The costs that Distrigaz has to distribute over its various customer categories are the following:

■ joint costs, i.e. overhead costs relating to the total activity;

■ transportation costs, i.e. costs relating to transporting the gas across the country in a narrow sense of the word;

■ peaking and security costs, i.e. costs relating to modulation and to security of supply;

■ profits, i.e. remuneration of equity capital;

The customer categories used in this context are as follows:

■ firm sales to public distribution, to industry and exports to Luxembourg;

■ partly interruptible sales (ventes effacables) - customers that are interruptible for maximum 35 days a year only between 15 November and 15 March;

■ interruptible sales: sales to industry and power stations that are interruptible at any time.

Joint costs are distributed evenly over all volumes except interruptible sales. The reason for this exception is that interruptible sales do not imply any obligation to supply and therefore only costs directly attributable to these sales are allocated to them.

Transportation costs are divided into operating charges and investment charges. The operating costs are essentially linked to the length of the network and are for practical purposes not related to pipeline diameter or regularity in supplies. After deduction for compressor fuel and administration costs pertaining to interruptible sales, the remainder is distributed evenly over total volumes except interruptible ones.

The investment cost necessary to build a transmission network of a given length depends on the maximum transportation capacity needed and thereby on the diameters of the pipes. If the gas flows were constant all over the year, the maximum capacity needed would be equal to the average flow. In reality, the maximum flow, because of the irregularity in demand, will be higher and this implies an overdimensioning of the network and extra costs. Consequently, total investment cost can be divided into:

■ base load cost, i.e. The cost of the network that would have to be constructed to serve a customer base with perfectly regular offtake all through the year;

■ overdimensioning cost, i.e. The extra cost that is incurred to be able to face the irregularity in demand.

The base load cost associated with the theoretical network necessary to transport a constant gas flow (minimum diameter) can be deducted from the following formula:

Total investment cost multiplied by the load factor raised to a power of 0.38 (the formula will not be developed here).

The load factor is given by the relationship between the average flow and the maximum flow, equal to the flow of firm deliveries at -11 degree Celsius (taking into account that partly interruptible and interruptible supplies can be eliminated at peak). Another design criterion is that the system should be able to satisfy demand during a so-called one per cent winter, that is the statistically coldest winter over a hundred year period. The base load transportation investment determined in this way is distributed over total sales except 35/365 times the partly interruptible sales and the regular interruptible sales.

The investment made in overdimensioning of the network, that is total investment costs minus base load investment, is attributed to the distribution sector and the industrial sector according to their responsibility for the difference between maximum hourly offtake and the average hourly offtake of their firm customers.

Peaking and security of supply costs are related to the construction and operation of production and storage facilities with a view to covering seasonal peaks in demand and contributing to security of supply. In this context it has been considered that 20% of the total costs attributable to these functions are related to security. These costs are distributed over all firm and partly interruptible sales. The remainder, that is 80% of total peaking and security costs, is allocated taking into account that necessary storage in the distribution sector amounts to 34% of annual sales and that necessary storage to serve the rest of the firm and partly interruptible customers correspond to 2.74% of annual sales to these customers.

The contribution of Distrigaz' customers to its profits is regulated by the Comite de Controle and has been fixed at 10% of equity after taxes. The principle underlying the way the contribution to profits is allocated is that it should come from all customers in relation to their purchases. Interruptible volumes do, however, not contribute.

The cost of modulation and security of supply is explicitly taken into account in tariff setting in Belgium, but apparently only in the transportation part of the gas chain.

The methods for the calculations of delta and the parameters used are presently under revision.

END USER PRICES IN THE PUBLIC DISTRIBUTION SECTOR

The distribution companies distribute gas to household consumers, commercial customers, small industrial customers and public buildings. Table 6 give the structure of the tariffs to these customers.

Table 6 End User Tariffs in the Public Distribution Sector

Tariff	Fixed element	Variable element (c/MJ)
Household		
A (< 70 GJ/a)	38.49. lgd (BEF/months)	23.8678 . lga + 25.3286 . lgd (<15.474MJ/a)
		23.8678 . lga + 18.0255 . lgd (>15.474MJ/a)
Social tariff		23.8678 . lga + 25.3286 . lgd
B(>70 GJ/a)	2.717. lgd (BEF/year)	23.8678 . lga + 8.0320 . lgd
C	152 lgd (BEF/month connection)	23.8678 . lga + 4.8688 . lgd
Small users		
ND1 (35-527 GJ/a)	5.737 . lgd (BEF/year)	23.8678 . lga + 7.2409 . lgd
ND2 (527-3.517 GJ/a)	14.722 . lgd (BEF/year)	23.8678 . lga + 5.5359 . lgd
	50.732 . lgd (BEF/year)	23.8678 . lga + 1.1382 . lgd <10.550 GJ/a)
ND3 (>3.517 GJ/a)	4.377 . lgd (BEF/max MJ/ d / a)	23.8678 . lga + 1.1382 . lgd - 0.8(>10.550 GJ/a)
Public buildings		
< 174 kW/< 879 GJ/a	38.49 . lgd (BEF/month)	23.8678 . lga + 10.4479 . lgd (< 88 GJ/a)
		23.8678 . lga + 5.3415 . lgd (>88 GJ/a)
> 174 kW/> 879 GJ/a	50.732 . lgd (BEF/year)/	23.8678 . lga + 3.7060 . lgd
	4.377 . lgd (BEF/max MJ/ d / a)	

lga = gas bought index; lgd = gas distribution index
Source: Comité de Contrôle

Tariff A is for household customers using gas mainly for cooking and water heating. In addition to a fixed element valid for all volumes up to 70GJ, there is a variable element with two volume tranches (below and equal to 15.474 MJ/a and above this volume up to 70 GJ/a).

Social tariff: Same as the first tranche of tariff A but without fixed element. This tariff is applicable to certain categories of elderly and handicapped people. In 1995, this tariff would be about 7% lower than tariff A.

Tariff B is typically applied to residential customers using gas for cooking, water heating and space heating.

Tariff C is for centrally heated multidwelling houses with at least ten apartments.

Tariffs for non-domestic customers:

ND1: for customers with annual consumption between 35 and 527 GJ/year.

ND2: for customers with annual consumption between 527 and 3517 GJ/year.

ND3: for customers with annual consumption above 3517 GJ/year.

The ND1 and ND2 tariffs have one fixed and one variable element, the fixed element being a fixed amount per year. The ND3 tariff has two fixed elements, one fixed amount per year and an element differentiated according to the meter capacity. This last element distinguishes between meter capacities below and above 350 m^3 per hour. The variable element is differentiated between two volume tranches, below and above 10550 GJ/a, which means a rebate for volumes above this limit. Tariffs for domestic customers may be applied to other types of customers if they use gas for the same applications as domestic customers. If gas use is mixed (both domestic and non-domestic uses) a domestic tariff is applied if domestic uses are the most important. The distribution companies are generally obliged to supply customers at the most favourable tariffs applicable to the use in question.

Tariffs for Public Buildings

These tariffs are applicable to buildings owned or rented by municipal services. Municipal services having an industrial or commercial character are not considered as municipal from a tariff point of view. Two tariffs are applied: one for loads less than 174 kW or an annual consumption of 879 GJ (243.483 kWh/a) - which is subject to payment of a fixed element and a variable element differentiated according to two different volume tranches (below and above 88 GJ/a) - and one for loads above 174 kW or an annual consumption of 879 GJ. The latter tariff contains the same fixed elements as the ND3 tariff, but has no volume tranches.

General Observations on Tariffs in the Distribution Sector

The end user tariffs in the distribution sector have the following characteristics:

■ The tariffs are all cost plus based and do not contain any direct reference to the prices of other fuels - the only influence from other fuels comes through the gas import price which is linked to oil prices.

■ The tariffs are all, except the social tariff, composed of a fixed and a variable part. When calculating the share of the fixed element in the total price to end consumers, the following shares are found:

A: 7.3%

B1:   24.0%

B2:   9.4%

C: 11.9%

ND1:   8.7%

ND2:   5.1%

ND3: 27.2%

Even after deducting gas acquisition cost it is clear that the fixed element recovers a relatively low share of total costs.

■ The prices contain three elements: gas import cost, gas transportation (including storage and modulation) cost and a gas distribution cost element.

■ The first two elements are all the same for every cubic meter, no matter which customer segment it is sold to. The thinking seems to be that all the cubic meters delivered to the distribution companies all carry the same share of transportation cost including modulation cost. This cost is, as described above, higher for volumes to distribution than for volumes to industry.

■ The gas distribution cost element varies a lot from one customer category to another.

The coefficient of the distribution cost element varies from 25.3286 for the A tariff clients to 1.1382 for the ND3 clients. This implies that the tariff A customers contributes about 22 times as much per cubic meter to the distribution costs through the variable price element than an ND3 customer. A household customer using 35 GJ a year would contribute almost 2.5 times as much to distribution cost than a small industrial customer using the same volume. As modulation cost has already been taken into account at an earlier stage, the difference cannot be explained by modulation costs. Representative of distribution companies admit that the coefficients are a heritage from the past and do not necessarily reflect the cost of serving the various customer categories. The policy of tariff equalisation (perequation des tarifs), the need to be competitive versus other fuels and social reasons are mentioned as explanations for the fact that priced do not reflect costs.

Table 7 shows how tariffs for typical customers in the various segments have developed in recent years.

Table 7   Tariffs to Typical Clients in Public Distribution

Client	1991	1992	1993	1994	1995
A	484.9	473.2	473.1	474.1	477.0
B1	321.8	307.9	304.3	302.9	302.3
B2	276.2	260.4	257.1	255.0	253.4
C	239.6	223.2	219.2	216.6	214.2
ND1	263.2	247.2	243.6	241.4	239.5
ND2	233.5	216.9	212.9	210.2	207.6
ND3	200.6	183.4	178.9	172.6	169.3
For referenceG (import price)	123.6	105.4	100.0	95.7	90.9
Delta n (share of transport cost)	33.4	33.7	33.8	34.4	35.1

Source: Comité de Contrôle

As seen above, the Belgian tariff system to a certain extent takes costs related to modulation and security of supply explicitly into account. The calculation of delta on the transportation system is a clear example. The extent to which tariffs to end users reflect real cost of serving the various customer groups is more unclear. The implicit cost of having a system of interruptible contracts is indicated by the difference between interruptible and firm deliveries to industry (on an average the difference in 1994 was 12.7 BEF/GJ). Whether this reflect the real cost is another matter. There are no indications that an explicit tradeoff is made between further development of storage and extended use of interruptible contracts.

Pricing and Taxation of Gas in Relation to Competing Fuels

Because of the way gas prices are fixed in Belgium there is no strict relationship between oil product prices and gas prices. In the household sector gas prices have consistently been far above light heating oil prices on a heating equivalent basis. Although the difference has been smaller, this has also been the case for heavy fuel oil prices in relation to industrial gas prices. Belgium has among the lowest prices for oil products in Europe, partly because of low taxes on such products. Before the EC Directive on minimum rate for excise duties on mineral oil took effect on 1 January 1993 there were no energy taxes in Belgium on gas oil and HFO. Today the energy tax rates amount to 250 BEF/t for low sulphur HFO and 750 BEF/t for high sulphur HFO.

For gas oil the rate is 210 BEF/1000 litres for heating uses and 750 BEF/1000 litres for industrial uses. In terms of competition between gas and electricity it is clear that gas has been given priority in areas where a gas grid exists.

From 1 August 1993, an energy tax (called "cotisation") was introduced on natural gas, gas oil and electricity sold to small consumers. After the introduction of the "cotisation", the taxes on gas oil and natural gas for residential and commercial uses are equal. There are no environmental components in energy taxes in Belgium. Tax policy is in principle one of neutrality between energies, and gas has thus not benefitted from high excise taxes on for instance oil products as in many other countries. The fact that Belgium does not respect the minimum European excise duties on low sulphur HFO may be characterised as a discrimination against gas. Gas is also discriminated against since it is subject to specific taxes that are not imposed on other fuels:

■ a special withholding tax in income earned by gas distribution companies and payments to the municipalities for the distribution of gas;

■ specific local taxes levied by some municipalities.

The general VAT on energy is 21% except for coal which benefits from a reduced rate of 12%. VAT is recoverable for commercial and industrial purposes.

Cross Subsidies

Through taxation of other fuels and taxation of the gas sector, natural gas is to a certain extent discriminated against and can in no way be said to be subsidised. Inside the gas sector, however, there are some examples of what may be characterised as cross subsidies:

■ Belgium has a system of perequated tariffs which means that everybody in the same customer category pays the same price no matter where he is located. This means that customers far from the import point implicitly are subsidised by customers close to the import point. In a small country like Belgium this is less significant than for instance in a country like France;

■ the tariff system features a social tariff in favour of certain disadvantaged groups. The price reduction obtained by these groups (which is not a large share of the normal price) can be characterised as a cross subsidy;

■ industry in Belgium seems to be very favourably treated in terms of gas prices. Some of the volumes for large industrial customers are kept outside the G calculation and thereby benefits from a special price. In the calculation of delta interruptible customers are exempted from contributing to the transportation costs. Industrial customers in the distribution sector seem to pay a low contribution to distribution cost. All these elements indicate that there are some implicit cross subsidies to industry from other consumer groups.

■ Distrigaz is studying a modification of the gas entry price in order to better take into account the demand characteristics of its clients. As a consequence, prices to customers with regular demand could be decreased.

Price Transparency

Gas prices in Belgium are transparent in that prices to all customer categories except large industrial customers taking interruptible gas and power producers are published. The way prices are calculated is quite transparent in that the border price, transportation cost elements and distribution cost elements are identifiable.

Cost, Value Added and Profits in the Gas Chain

The following table shows the main figures for gas distribution in Belgium for 1994 in terms of value added or gross margins along the chain. The figures are in BEF/GJ.

Table 8   Gross Margins in the Gas Chain

Average sales revenue	234.4	100%
Average gas purchase cost	126.7	54.0% of sales revenue
Gross margin on the retail level	107.7	
Average gas import cost	95.7	40.8% of sales revenue
Gross margin on the wholesale level	31.0	

Source: CCGE

The gross margin at the retail level is the one realised by the distribution companies. They all buy their gas from Distrigaz. Gross margin at the wholesale level is the one realised by Distrigaz. This margin, however, only stems from sales to the distribution companies. The gross margin at the wholesale level is supposed to cover all costs incurred by Distrigaz to serve the distribution sector. As pointed out above it may be questioned whether the respective deltas for industry and distribution reflect the cost of serving these categories. Anyway, a breakdown of the delta used in the tariff formula for distribution companies (originally fixed at 28.28 BEF/GJ) gives an interesting impression of the relative weights of the various types of costs. The breakdown of the delta is as follows:

Joint, overhead costs:	15.5%
Transportation operating costs:	16.2%
Transportation investment cost, base load:	12.3%
Transportation investment cost, overdimensioning:	8.7%
Storage for security reasons:	3.6%
Modulation to meet normal fluctuations in demand:	23.9%
Contribution to profits:	19.8%.

The wholesale gross margin (31.0 BEF/GJ in 1994) is somewhat different from the delta value (33.9 BEF/GJ in 1994), the most important reason being that the delta applied to industry is lower. Table 8 shows the breakdown of the operating result of the distribution sector over the last few years

Table 9   Operating Result for the Gas Distribution Sector (BEF/GJ)

	1990	**1991**	**1992**	**1993**	**1994**
Operating revenue	248.5	258.8	244.3	238.3	236.9
Total operating costs	208.1	217.6	203.9	193.5	196.2
Gas purchase	141.2	153.6	136.5	130.0	126.7
Operating expenses	49.3	46.1	49.7	46.1	50.9
Depreciation	16.9	15.4	16.5	16.7	17.8
Other	0.7	2.4	1.2	0.7	0.8
Gross operating result	40.7	41.4	40.6	45.0	40.8
Financial; expenses	8.8	7.3	7.4	6.3	5.6
Municipal participation	22.7	23.1	25.0	26.0	28.1
Misc.	−1.1	2.9	−0.4	0.6	−4.3
Taxes	2.2	2.5	2.7	2.7	2.9
Net profits	8.1	5.6	5.9	9.4	8.5

Source: Comité de Contrôle

A noticeable feature of this overview is the item called "municipal participation" which amounts to 11.86% of the average operating revenue. This item is primarily a municipal tax on the operating profit of the distribution companies. The private partner in the companies always receives a share of profits which is lower than its equity participation. In a case of 50% ownership, the private partner would typically receive 1/3 of the profit, the remainder accruing to the municipalities. Ordinary, net profits amounts to 3.6% of the average sales revenue. The sharing of profits between Electrabel and the Intercommunales is regulated by the articles of association of each single company.

The value added in distribution in 1994 was as follows (BEF/GJ):

Total	113.7
Salaries,etc.	50.9
Depreciation	17.8
Interests	5.6
Taxes	2.9
Profits	8.5 + 28.1 = 36.6 (including dividends paid to the municipalities)

Value added according to this definition was 89.7% in relation to total average gas acquisition cost.

According to the Comité de Contrôle the net profitability for the whole gas distribution sector in Belgium has been as follows over the last few years:

1990: 6.41%
1991: 5.41%
1992: 5.51%
1993: 7.83%
1994: 5.96%

Profitability is defined as net profits divided by average invested capital. These profitability figures relate to Electrabel's participation in distribution. A rate of return of around 6% corresponds to about 7% before a small tax is paid. The profit rate for the municipalities is around 10%.

To shed some light on costs and value added in the whole chain, a breakdown of the operating result of the transportation part of the chain is also included. The figures given in table 9 relates to Distrigaz' total activity except transit of gas.

Table 10 Operating Result in Gas Transmission 1990-1994 (BEF/GJ)

	1990	**1991**	**1992**	**1993**	**1994**
Operating revenue	124.7	137.3	120.0	116.4	113.1
Operating cost	122.0	133.7	117.7	114.1	111.4
Of which:					
gas purchase	105.7	117.7	101.3	96.0	94.9
Operating expenses	12.5	12.6	11.2	13.1	12.2
Depreciation	3.2	2.8	3.4	3.2	3.4
Miscellaneous	0.6	0.6	1.8	1.9	0.9
Gross operating margin	2.8	3.7	2.5	2.4	1.8
Financial expenses	2.2	1.3	0.9	0.8	1.2
Miscellaneous	-1.3	0.2	-1.3	-1.5	-2.5
Taxes	0	0.1	0.6	1.0	0.9
Net profit	1.9	2.1	2.3	2.1	2.2

Source: Comité de Contrôle

In 1994, profits in relation to operating revenues were 1.94% for Distrigaz and 3.58% for the distribution activity. Because of the high share of debts in Distrigaz' balance sheet, the rate of return on equity is higher than for the distribution activity. According to the Comité de Contrôle, the rate of profitability (net profits on average equity capital) of Distrigaz in recent years has been as follows:

1990: 10.67%; 1991: 11.97%; 1992: 13.11%; 1993: 12.16%; 1994: 12.44%.

FRANCE

Statistical information

Natural Gas Supply/Demand Balance (Mtoe, 1996):

Indigenous production	2.4
Imports	30.3
Exports	-0.7
Stock change	0.7
Total natural gas supply (primary energy)	32.7
Electricity and heat production	0.6
Other transformation and energy use	0.4
Total industry	13.1
Residential	9.1
Commercial	8.6
Other	0.2
Statistical difference	0.7

Natural Gas in the Energy Balance (1996):

Share of TPES	12.9%
Share of electricity output	0.0%
Share in industry	28.7%
Share in residential/commercial sector	29.2%

INDUSTRY STRUCTURE

France produces only about 7% of the natural gas it consumes. Production, centred on the Lacq gas field in the southwest of France, declined steadily through the 1980s and, despite rebounding since 1990, is expected to resume its downward trajectory over the next few years. Preliminary statistics on 1997 show Norway as leading supplying country with a share of 29.2% in total gas supplies, followed by Russia (26.8%), Algeria (25.4%), the Netherlands (12.4%) and indigenous production (6.2%).

Gas consumption has risen steadily in all end-use sectors over the past two decades, reaching 32.7 Mtoe in 1996 - more than twice the level of 1973; natural gas now accounts for 28.7% of TFC in the industrial sector (compared with 10% in 1973) and 29.2% in the residential/commercial sector (10% in 1973). As a proportion of TPES, gas accounted for 12.9%

in 1996, a rise of 5 percentage points since 1973. This share is nonetheless well below the average for IEA Europe for three main reasons: France is relatively sparsely populated, making it uneconomic to connect remote communities; gas use in power generation is negligible because of the dominance of nuclear power; and electricity has a large share of the residential heating market. Unpublished forecasts by Gaz de France (GDF) show a 25% in gas demand over 1995-2005 and a 33% increase over 1995-2010 (see table 1).

Table 1   Gas Demand Projections (Mtoe)

Sector	1995	2005	2010
Residential/commercial	15.8	20.0	21.2
Industrial	12.7	14.0	14.1
Power generation	0.7	3.3	4.2
Other	0.9	0.3	0.4
Total	**30.1**	**37.6**	**39.9**

Source: Gaz de France

The rate of grid expansion has slowed considerably in recent years as most urban districts are connected to the network. Approximately 70% of the population in metropolitan France is now within the gas supply area. Within that area, around 40% of households are connected to the grid, which means that about 28% of the total French population is connected to the natural gas network. The total transmission pipeline network totalled 32000 km at the end of 1996. 90% of the transmission pipeline network is owned by GDF, the remainder mainly by Gaz du Sud Ouest. The total length of the distribution network is 140590 km, i.e. 4.4 times as long as the transmission network. About 95% of the distribution network is owned by GDF; the remainder belongs to local distribution companies that are not part of the GDF system. Some further grid extensions into areas not yet served by gas are planned.

The French gas sector is dominated by the wholly state-owned company, Gaz de France (GDF), which since its creation in 1946 has enjoyed monopoly rights over imports and exports, a large part of the transportation system and the overwhelming bulk of the distribution network. GDF supplies most of the gas to end-users served by the low pressure distribution grid and also supplies through its transportation system a significant proportion of the gas ultimately distributed by the 17 non-GDF distributors. The share of gas not transported on the GDF transmission system is limited to the domestic gas production in southern France which is in decline and accounted for only 7% of total gas supplies in 1996. GDF also operates over 80% of the total storage capacity in France.

The Government controls the gas sector through its ownership of GDF, concession agreements with transporters and distributors and tariff regulation. The 1946 Nationalisation Law provides the basic legal framework for the organisation of the gas sector, encompassing key public service obligations. There are no rights of third party access and there is no gas-to-gas competition. A few local distribution companies not owned by GDF potentially could choose between supplies from GDF or GSO, but in reality there is no competition.

MAIN ACTORS IN GAS DISTRIBUTION

The main players involved in the distribution and sale of gas in France are as follows:

■ GDF, established as the national gas utility under the 1946 Nationalisation Law, is completely owned by the state. It has a legal monopoly of imports and exports and is the sole company entitled to distribute gas in most of France. It operates most of the gas transmission as well as the distribution network. It accounts for around 89% of gas sold to final consumers. GDF is organised according to 11 transmission regions, which operate the high pressure transportation system, and 100 distribution areas.

■ There are 17 other companies involved in gas distribution, most of which are régies -local public corporations with special legal status. These companies, exempted from the 1946 nationalisation of the gas industry, account for around 2.7% of total gas sales in France. Depending on their locations, they buy gas from GDF or from Gaz du Sud-Ouest (GSO). They mostly sell to household, commercial and small industrial customers, though some companies sell gas to larger industrial customers. The largest of the non-GDF distribution companies are in Bordeaux, Strasbourg, Grenoble and Monaco. GDF and Elf both have a 16% ownership stake in Gaz de Bordeaux. Total holds a 49.9% minority participation in Gaz de Strasbourg (half of which it might cede to GDF as part of a larger agreement between both companies). Since beginning of 1998 do distribution companies no longer necessarily have to be at least 50%-owned by the state, or publicly-owned companies or by municipalities, but shall 'only' have at least 30% public capital.

■ Gaz du Sud-Ouest (GSO), owned by Elf (70%) and GDF (30%), operates its own transportation system in southwestern France. It buys gas from Elf's Lacq field and additional supplies from GDF. GSO sells this gas to industrial users connected directly to its network, and to local distributors, but plays no direct part in local distribution.

■ Compagnie Française de Méthane (CFM), owned by GDF (50%), Elf (40%) and Total (10%)[1], leases pipelines in central-western France built, owned and operated by GDF. CFM buys indigenous gas from GSO and imported gas from GDF, and resells its to local distributors and to industrial consumers linked directly to the transportation network. It does not distribute gas directly to end-users through the low-pressure grid.

Table 2 gives an overview of the sources of the gas distributed to end consumers in France. The following observations can be made:

■ The overwhelming share of gas supplies is imported (93%); GDF holds the exclusive right to import natural gas.

■ The domestic production is delivered to GSO and CFM. These two companies also buy imported gas from GDF. ELF, which is the majority owner of GSO, has its own import contracts, but has to sell this gas to GDF at the French border and repurchase it at the inlet of its own pipeline system in southern France. In some cases GDF purchases gas from GSO and CFM for its local distribution companies.

---

1. GDF might cede 5% of its assets in CFM to Total as part of a larger cooperation agreement between both companies of December 1997.

Table 2 Distributed Gas: Detailled Balance 1994 (millions of kWh PCS)

	Natural Gas Production		Gas Industry					TOTAL
	EAP	Others	Gaz de France	Régies	Non-nationalised enterprises	SNGSO	CFM	
Resources								
Natural gas production	34 306[1]	–	1 615	–	–	–	–	35 921
Natural gas imports	–	–	347 205	–	–	–	–	347 205
Other	–	–	1 393	–	–	–	–	1 393
Exports	–	–	–8 234	–	–	–	–	–8 234
Stock variation	–1 516	–	–9 428[2]	–	–	–	–5 816	–16 760
Total (I)	**32 790**	**–**	**332 551**	**–**	**–**	**–**	**–5 816**	**359 525**
Exchanges								
EAP	–29 470	–	–	–	–	+28 370	1 100	–
GDF	–	–	–82 515	+5 810		+620	+76 085	–
Sales to public distribution	–	–	+68 482	+3 951		–18 872	–53 561	–
Total (II)	**–29 470**		**–14 033**	**+9 761**		**+10 118**	**+23 624**	**–**
Total gross available (I) + (II) = (III)	**3 320**		**318 518**	**9 761**		**10 118**	**17 808**	**359 525**
Internal use (compressor station included)	–	–	2 546	48		472		3 066
Non-accounted for gas	–89	–	–6 041	–349		–595		–7 074
Total (IV)	**–89**	**–**	**–3 495**	**–301**		**–123**		**–4 008**
Total net available (III) – (IV) = (V) 363 533	**3 409**	**–**	**322 013**	**6 514**		**3 548**	**10 118**	**17 931**
Individual domestic sales	–	–	102 701	3 590	917	–	–	107 208
Collective and commercial sales	–	–	90 353	2 190	1 472	–	–	94 015
Industrial sales:								
Power station	–	–	309	–	–	–	–	309
Other	3 409	–	128 650	734	1 159	10 116	17 925	161 993
Vehicle fuel	–	–	–	–	–	2	6	8

(1) Lacq Meillon, Ucha 34 185, St MaRcet Auzas, Proupiary 121
(2) Réservoirs et canalisations
Source: Ministère de l'Industrie

■ The non-nationalised distribution companies buy gas from GDF, GSO and CFM.

■ GDF has a market share of about 96% in the residential/commercial sector. The remainder is accounted for by the non-nationalised companies.

■ SNGSO and CFM together account for roughly 20% of the industrial market. The small volumes going into power generation are delivered by GDF.

■ In total, GDF holds 88.5% of the end user markets for gas.

Since GDF holds such a high share of the end user markets for gas and does not, at least externally, distinguish between its main activities like transportation, storage and distribution in accounting terms, the French gas industry has to be characterised as vertically highly integrated. The part of the French gas industry held by Elf is even more integrated in that it also includes production.

Horizontal integration is not very widespread in the French gas industry. In physical terms GDF only distributes natural gas. Some of the independent regional companies, however, combine gas distribution with distribution of electricity and heat and other public services. Marketing of gas and electricity in the residential sector is integrated in the sense that GDF and EDF have joint marketing services here.

LOCAL DISTRIBUTION COMPANY (LDC) SERVICES

GDF is primarily a gas transmission and distribution company, but has interests in gas and oil production within and outside of France and some commercial non-gas activities including technical services for the evaluation of combined heat and power projects, supply of heat equipment and energy management. Under the 1946 Nationalisation Law, GDF is not entitled to market gas appliances and equipment. Regulations introduced in 1994 also limit the extent to which GDF is allowed to diversify outside of its core gas transportation, distribution and marketing activities in France. GDF is not involved in electricity distribution.

In operational, financial and management matters at the distribution level, GDF works closely with EDF; gas and electricity bills, for example, are sent out jointly. Three of GDF's seven corporate divisions and two departments are managed jointly with EDF. In most cases, local distribution centres are run jointly. Around two-thirds of the total staff of GDF/EDF work in common units. The accounts of the two companies are nonetheless entirely separate. Marketing is also carried out separately to ensure a degree of inter-fuel competition.

Of the 17 other gas distribution companies, three only distribute gas (Gaz de Strasbourg, Gaz de Bordeaux and the gas company in Huningue). The first two companies are partly privately-owned. The rest of the distribution companies are diversified, distributing also electricity, and/or water and/or district heat. They are all wholly or majority municipality-owned (either individual municipalities or groups of municipalities).

The general trend in both GDF and the non-nationalised distribution companies is that of offering complete services to the end users, not only natural gas. Complete packages including the fuel, equipment and maintenance are increasingly offered.

MONTHLY GAS FLOWS AND SEASONALITY/SECURITY OF SUPPLY

■ Gas distributors are faced with the task of matching fluctuating demand with supply, which for cost reasons is preferred to be as stable as possible. The fact that demand fluctuates on a daily and seasonal basis makes it necessary to modulate supply to satisfy demand at the right time. For this purpose, the distributor has mainly three instruments at his disposal: flexibility in supply contracts, use of storage and use of interruptible contracts. Figure 2 gives an overview of the monthly gas flows in France in 1995 which clearly illustrates the problem with which GDF and the other gas companies are faced:

■ Total consumption fluctuates considerably over the year - the total consumption in August (the month with the lowest consumption) is only 23 % of what it is in the month with the highest consumption, which happened to be January in 1995.

■ The fluctuations in demand in the residential/commercial sector are considerably larger than in the industrial sector - the relations between the volumes in the highest and lowest months are 6.6 and 1.73 respectively. In terms of swing (highest month in relation to monthly average) these figures mean 195% in the public distribution sector and 124% in the industrial sector. This means that public distribution naturally has to carry a high share of the load factor cost.

■ Both domestic gas production and gas imports shows much less variation over the year than demand, the relations between the highest and lowest month being 1.37 and 1.46 respectively.

■ Fluctuations in deliveries under the various import contracts differ quite widely, the ratio between the highest and the lowest month varying between 3.0 and 1.2. The swing in the volumes taken over the year is bigger in the Algerian and Dutch contracts than in the Russian and the North Sea contracts. The explanations for the swing in monthly off-take over the year are at least twofold: The contracts have different provisions about take or pay volumes i.e. some contracts permit larger flexibility in annual off-take volumes than others. Within the flexibility conditions defined by the contracts, the gas importer will obviously try to minimise total import cost by buying as much as he can of the cheapest gas.

■ The major buffer against seasonal fluctuations in demand in France is storage. The typical pattern is storage fill during the summer months and withdrawal from storage during the winter months. On extremely cold winter days, withdrawal from storage has accounted for up to 58% of total supply to the market.

■ In addition to storage for load factor purposes, GDF also stores gas for strategic reasons. About one third of the total storage capacity (about 10.5 bcm of working gas) is used to store gas to counter supply disruptions. These volumes can, however, also be used to cover demand peaks in winters with below-average temperatures (this has not been the case in the last years due to warmer-than-average winters).

■ About 40% of total sales to the industrial sector are interruptible volumes. This means that supplies to these customers can be interrupted and diverted to firm customers in a shortfall situation or in times in which temperatures fall below normal values.

Figure 1   Yearly Gas Demand Fluctuation, France, 1995

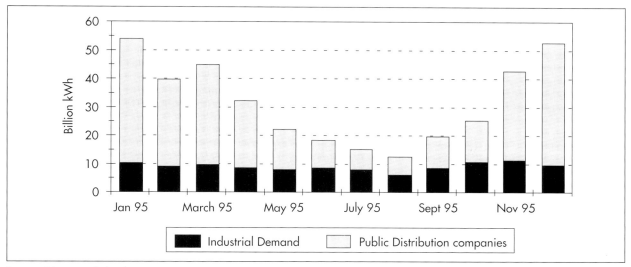

Source: Ministère de l'Industrie

To understand and to have a critical look at gas distribution tariffs in France (and elsewhere) it is necessary to have an idea about the economic mechanisms governing load factor and security of supply measures. It is difficult to calculate precisely the cost of all the instruments used, but an attempt at describing the principles (as generally known in the gas industry - not just in France) will be made:

In gas contract negotiations there will normally be a tradeoff between price and volume flexibility. Seen from the gas seller's point of view it is in principle possible to calculate the price increase he must obtain to justify a higher flexibility in off-take conditions. In most cases the seller has to make large investments in production and transportation facilities, and he has a clear economic interest in using the capacity at a maximum. By making a discounted cashflow analysis he can calculate the tradeoff between price and volume flexibility. The buyer, on his side, may have to evaluate the cost of increased flexibility under his import contracts against the cost of other measures to modulate and assure security of supply.

The cost of storage is a function of the type of reservoir, capacity installed and other physical characteristics. In principle, the cost is relatively easy to calculate. To say what the value of storage in general is, is much more difficult. The value of storage may be defined in terms of possibilities to earn money/reduce costs by speculating on price fluctuations. In a shortfall situation the value of storage may be defined by the price that a customer is willing to pay for gas to satisfy his demand. This value is, however, irrelevant as long as allocation of gas volumes in a shortfall situation is not allowed to take place as a function of willingness to pay but only according to other criteria.

To a certain extent there may be a tradeoff between the flexibility under import contracts and the building of storage since they are both ways of modulating supply.

The French gas system of import contracts, production and storage is designed so that it can continue to supply all of its customers continually during a 2% winter, i.e. The coldest winter that is statistically probable in any 50 year period. This criterion, along with the rule of thumb that

France should be able to sustain an interruption in one of its main supply sources for 12 months, determines whether additional storage capacity is necessary. In addition construction of storage is subject to the general investment criterion of obtaining an internal rate of return of 12%.

Interruptible contracts could in principle be used as a load factor instrument and as a means to enhance security of supply. So far, for reasons of customer care, GDF has tried to avoid as much as possible to use them for load factor purposes. Not making use of the contractual provisions to interrupt a customer at specific, mutually agreed times and conditions is not cost-efficient. This practice produces costs of the size of the reduction in price allowed to the interruptible customer compared to the price paid by firm customers, and which have to be borne by the system, i.e. The firm customers.

In principle, though, interruptible contracts if exercised properly, can be a more cost-effective means for load factoring than building extra storage.

The section on tariff below will look into how load factor and storage costs are taken into account when constructing the tariff system.

REGULATION

Distribution Concessions

The gas industry is regulated primarily by the Ministry of Industry. Under the 1946 Nationalisation Law, the concession for gas distribution is held by GDF in almost all areas except those already served by *Régies* or private companies at the time GDF was created. Today, there are 17 non-GDF distributors. Until 1982, the terms and conditions of concessions for all gas distribution companies were laid down in a standard contract. Within the context of broader moves to decentralise some aspects of government, municipalities and urban districts are now free to negotiate their own terms and conditions when concession agreements come up for renewal, though a model concession agreement – drawn up collectively by all interested parties in 1994 – is typically adopted. This model concession agreement imposes upon the concessionaire a number of rules in terms of obligation to supply, connection and delivery conditions, tariff setting and investment criteria for network extensions. The requirement used for extensions is that discounted profits from the project divided by the discounted investment stream of the project over a deemed duration of 20 years should be equal to or larger than 0.3. This means that new investments are required to give a nominal internal rate of return of 12% using a discount rate of 8%. Concessions normally last 30 years. In any event, concession agreements must contain public service obligations as laid down in the 1946 law.

The 1946 law prevented the non-GDF distributors from expanding their networks outside of their existing supply areas. Thus, any municipalities not already connected to the grid could only be connected by GDF. In some cases, this restriction prevented municipalities from being connected to the grid in those areas where it would have been efficient and profitable for an existing non-GDF distributor to extend their grids but not for GDF to do so (due to reasons of proximity of capacity constraints). There were also instances when non-GDF distributors could have connected new supply areas more efficiently than GDF. These distortions led to Parliament adoption in April 1996 of a change to the 1946 law allowing municipalities situated next to a non-GDF supply area to choose between being connected by GDF or the neighbouring non-

998, another change of the law was adopted, requiring GDF to make
g gas to new municipalities. Municipalities that will not be included
F can approach other companies (with at least 30% public capital)
oncession for the start-up of gas distribution. These changes in the
ly limited effects. Already GDF is considering a list of 200
ear-plan. Since most of the unconnected municipalities which will
small and located in remote areas, it is not clear to what extent
es (minimum 30% publicly owned) that are interested in starting a
nce has about 40000 municipalities. Approximatively 30% of them
epresent about 70% of total population.

Under all concession agreements, gas distributors are subject to the public service obligation to supply any customer within their concession areas. All distributors (including marketers of gas off the high pressure transportation grid) must also sell firm gas according to published, non-discriminatory tariffs, i.e., customers within a particular region with identical load characteristics must be charged the same tariff.

Transport Concessions

The French gas transportation system consists of two parts: an integrated high pressure transmission system where gas flows in all directions, and the antennas linked to this system, on which gas flows in only one direction. The antennas also carry out high pressure transmission but are local branches of the main high pressure system (normally called the "Grand Reseau de Transport" – hereafter called GRT). A concession is needed to build and operate both these two types of high pressure systems, and the requirements imposed upon the concessionaire is defined in the terms of reference for such concessions adopted already in 1952. An interesting point in the terms of reference is that the minister responsible for gas may require the concession holder to transport gas beyond what is defined in his concession if it is judged to be in the public interest and there is spare capacity. Such utilisation may, however, may only take place on a temporary basis.

The terms of reference also contain provisions about rights of way, extensions, assignment of concessions and tariff principles. It is important to note that no tariffs are defined for transportation in itself; the price to be paid is always included in the price for gas delivered at the outlet of the transportation system.

Tariffs

Gas prices in France are regulated. This goes for prices of gas to end users served by the distribution system as well as for gas taken directly from the high pressure transportation system (both the GRT and the antennas). Gas prices are calculated using a cost plus approach which takes into account:

- the cost of building, maintaining and renewing storage, transportation and distribution installations;

- gas purchase cost;

- operating costs for storage, transportation and distribution facilities.

The decree regulating gas prices (decree no. 90-129 of 20 November 1990) states that the invoiced price of gas may include a capacity charge and a commodity charge which may both take into account the size of the subscribed volumes, the volumes effectively used and other parameters like distribution of the customer's off-take over the year.

Within the principle described GDF (and also the other companies selling gas to end users) is allowed to take the market situation into account when tariffs for the various categories of customers are determined. In practice this means that a large share of the gas sold is priced in relation to the main competing fuels, most of the time gasoil and heavy fuel oil through gas import price indexation.

The only exception to the general rule of price or tariff regulation are prices to large industrial customers (deliveries of more than 5 GWh a year) that are connected to the transmission grid and those connected to distributors that apply horizontal tariffs. Prices to this type of customers can be set freely, although the gas supplying company has still to submit them to the authorities. In practice, prices to these customers are set in relation to prices of competing fuel, most of the time distillates or heavy fuel oil. The overwhelming majority of these large customers are supplied from the high pressure transmission system by GDF, GSO and CFM.

Another type of regulation influencing gas prices is the one defined in the Contrat d'Objectifs between GDF and the Government. Average annual tariff adjustments are calculated according to a price cap formula set forward in the contract. The formula implies that half of all productivity gains realised by GDF is to be reflected in the tariffs. For the period 1997-99, GDF is obliged to reduce its annual costs (not including gas purchase costs). 50% of the absolute cost reduction is supposed to be reflected in the tariffs which are supposed to drop on average by a minimum of 1.6%.

Under the 1990 decree on price regulation, all modifications to tariffs (final and transport) must be registered with the Ministry of Finance. After consultation with the Ministry of Industry, the Ministry of Finance may reject these tariffs within a specified time frame on the grounds that they do not sufficiently reflect costs, though this power has never been exercised.

To understand pricing to all end users, one has to start with the gas prices paid at the outlet of the transmission system. Some customers, like large industrial customers take their supplies directly off the GRT. A few distribution companies do the same and are in that case charged the same price as an industrial customer for gas taken on the same conditions. Table 3 shows the tariff for such customers introduced on the 15th of April 1995:

Table 3   Tariffs for Deliveries off of the GRT as of April 15, 1995

Point of Delivery Main Transportation Network	Tariff Parameters	Volumes
Subscription	39552.1 FF	
Fixed Rate	106.0 CF/KWH/day	daily subscribed off-take
Fixed Rate reduced by the summer off-take	50.6 CF/KWH/day	daily subscribed off-take
Winter Proportional Price, first tranche [1]	6.53 CF/KWH	annual consumption in winter [2], first tranche(kWh/year)
Winter Proportional Price, second tranche	6.20 CF/KWH	annual consumption in winter [2], second tranche (kWh/year)
Summer Proportional Price, first tranche [1]	5.26 CF/KWH	annual consumption in summer, first tranche (kWh/year)
Summer Proportional Price, second tranche	4.93 CF/KWH	annual consumption in summer, second tranche (kWh/year)

(1) For annual consumption below 24 millions KWH/year
(2) From November 1 to March 31.
Source: Ministère de l'Industrie, DIGEC

The tariff described gives a price not distinguishing between the gas cost and the transportation fee. The implicit transportation fee can, however, be found by deducting the average purchase price of the gas from the price paid at the outlet of the GRT. The tariff has the following characteristics:

- It comprises a subscription fee, a capacity fee and a commodity fee. In terms of share of total price, the commodity fee is by far the most important; for a large industrial customer it typically accounts for 85 to 90% of the total price.

- The transportation element contained in these tariffs are the same for all buyers. On the GRT everybody pays the same price independent of the location of the off-take point (as pointed out below, this is not the case for the implicit transportation fee paid on the antennas).

- There are two price tranches for industrial customers: one for off-takes below 24 million kWh/year and one for off-takes above this volume. Thus the system give an incentive to increase volumes.

- The fixed capacity charge is multiplied by the daily off-take per year (kWh/day /year). If the need for load factor is low, i.e. if the off-take is evenly spread out over the year, the customer pays a lower unit price than in the case where the off-take is uneven. This means that customers with a high need for load factor pays a higher unit price to contribute to the costs caused by load factor. In this sense there is a link between the cost of load factor and the price paid by the customer.

- The tariff also distinguishes between off-take in the summer and in the winter in that both the capacity charge and commodity charge are lower in the summer. The commodity price is 1.27 c/kWh lower in the summer than in the winter, the difference being the same in both volume tranches. To pay a higher price for gas delivered in the winter is also one way of contributing to the higher cost associated with winter gas since it has to be stored. The price difference is also an incentive to concentrate off-take to the summer period. Relatively this incentive is stronger for the small customers that for the large ones.

The greater part of the distribution companies and some of the industrial customers take their gas from one of the antennas. The implicit tariff for transportation on the antennas is in principle calculated in the same way as the implicit tariffs on GRT but with one major difference: the tariff is not perequated, but calculated in order to allow to recoup specific grid investment costs (depending to a large extent on the length of each connection), in accordance with the volumes transmitted. There are six different zones, reflecting the distance.

There is one exception to the rule of not perequating the cost on the antennas: the Paris region. All buyers in this region pay the same price which is higher than the price paid for direct off-take from the GRT.

Table 4   Tariffs for Deliveries for the Paris Region as of April 15, 1995

Point of Delivery Parisian Region	Tariff Parameters	Volumes
Subscription	39552.1 FF	
Fixed Rate	138.0 CF/KWH/day	daily subscribed off-take
Fixed Rate reduced by the summer off-take	61.7 CF/KWH/day	daily subscribed off-take
Winter Proportional Price, first tranche [1]	6.65 CF/KWH	annual consumption in winter [2], first tranche(kWh/year)
Winter Proportional Price, second tranche	6.32 CF/KWH	annual consumption in winter [2], second tranche (kWh/year)
Summer Proportional Price, first tranche [1]	5.29 CF/KWH	annual consumption in summer, first tranche (kWh/year)
Summer Proportional Price, second tranche	4.96 CF/KWH	annual consumption in summer, second tranche (kWh/year)

(1) For annual consumption below 24 millions KWH/year
(2) From November 1 to March 31
Source: Ministère de l'Industrie

The tariffs in the most of expensive zone on the antennas are identical to those in Paris except for the commodity charges which are higher than in the Paris region.

End User Tariffs in Public Distribution

The tariffs discussed so far are only applicable to large industrial customers and distribution companies. Table 5 shows the tariffs for end users served by GDF.

Each of the tariffs merit some comments:

The Base tariff: This is a tariff chosen by customers who use gas only for cooking. The tariff structure is extremely simple: the same price is paid everywhere in France and consists of a fixed subscription charge and a price per kWh. At a consumption of 1000 kWh per year, the unit price is 38 c/kWh. Within the upper limit of 1000 kWh per year, the unit price declines. This customer segment represents 20% of the total number of customers, but accounts for only about 0.6% of the volumes invoiced.

The B0 tariff: Has exactly the same structure as the base tariff but is geared at customers using gas both for cooking and water heating, thereby consuming bigger volumes. This segment represents more than 20% of the customers but accounts for only 1.6 % of total sales in volume terms.

The B1 tariff: About 55% of the all customers have opted for this tariff and volumes sold under the conditions specified by this tariff account for around 30% of total sales. The typical customer is a household using gas for cooking, water heating and space heating. The only difference in tariff structure compared with the other tariffs described so far is that the commodity charge varies according to geographical zone. The difference in price between the cheapest and the most expensive zone is 2 c/kWh. This feature of the tariff is linked to pricing at the outlet of the antennas: the price paid for gas on the antennas is a function of distance from the GRT system and the size of the pipeline serving the customers on the antenna. The same zone system is used in gas distribution and is reflected in the 6 different price level indicated in the table. The presumption underlying this must be that a distribution system lying far away from the GRT typically is small and costly to develop and therefore also require a high price.

Table 5 Prices to Individual Domestic, Tertiary-sector and Small-scale Industrial Customers (net of taxes as of April 1, 1996)

Tariff	Base	B0	B1	B2I	B2S		TEL	
Tariff Codes	741-841-941	711-811-911	712-812-912	710-810-910	846-856-946		824-834-924	
Annual Consumption	Below 1,000 kWh	Between 1,000 and 6,000 kWh	Between 6,000 and 30,000 kWh	From 30,000 to 150,000 and 350,000 kWh[3]	In excess of 150,000 to 350,000 kWh[3]		In excess of 5,000,000 to 8,000,000,000 kWh[3]	
Examples of Usage	Cooking [1]	Cooking and hot water	Individual heating, hot water and/or cooking	Collective provision of heating and/or hot water	Heating and hot water in boiler heating systems		High powered heating for simple management of high energy consumption	
Subscription	110,88 F /year [2]	172.56 F /year	624.84 F /year	885.00 F /year	3837.96 F /year Price/kWh		35,330.04 F /year Price/kWh	
Price Levels	Price/kWh in centimes	Price/kWh in centimes	Price/kWh in centimes	Price/kWh in centimes	Winter Summer in centimes		Winter Summer in centimes	
1			14.06	13.20			13.10	10.09
2			14.46	13.60	13.10	10.09	13.72	10.26
3			14.86	14.00	13.50	10.49	14.34	10.43
4	26.98	21.32	15.26	14.40	13.90	10.89	14.96	10.60
5			15.66	14.80	14.30	11.29	15.58	10.77
6			16.06	15.20	14.70	11.69	16.20	10.94
					15.10	12.09		
Reduction in second tranche								
Threshold (kWh)							4 mil	2 mill
Amount (CF/KWH)							2.44	3.71

(1) A simplified formula, the kitchen package, is proposed under certain conditions.
(2) The values indicated corresponds respectively to the price of renting and maintaining a gas meter for consumption between 5 and 10 m^3/hour.
(3) According to the usage and the composition of consumption in winter and summer.
(4) Winter season extends from November 1 to March 31. Summer season extends from April 1 to October 31.
Source: industry

The B2I tariff: This tariff is typically used in collective heating systems, small industry and commercial enterprises. The structure is exactly the same as for the B1 tariff.

The B2S tariff: The same type of tariff as the B2I tariff, but for larger customers. The only difference in structure is that it has an element of seasonal pricing: prices in the summer are lower than in the winter. This reflects that load factor costs are higher in the winter than in the summer.

The TEL (Tarif a Enlevements Libres): a tariff for large customers who have a fluctuating demand for climatic reasons and who needs less strict off-take condition than other customers. A typical example would be large collective heating systems. The structure of this tariff is the same as for B2S with one exception: There are two volume tranches, the effect being that higher volumes leads to a reduction in price.

The three last tariffs are applicable to about two thirds of total gas sales through public distribution. Collective heating accounts for about 10% of total distribution volumes, the commercial sector for about 15% and industry for the remainder.

Table 6   Tariffs as of April 1996

Tariff	Capacity Charge (C/KWH)	Commodity Charge (C/KWH)	Unit Gas Price (C/KWH)
BASE	11.09	26.98	38.07
B0	2.88	21.32	24.20
B1	2.08	14.06	16.14
B21	0.59	13.20	13.79
B2S	1.10	13.10	14.20
TEL	0.44	13.10	13.54

Source: industry

Table 7 shows the unit cost of gas according to the various tariffs (calculated using maximum volume for the tariff and assuming winter delivery in the least expensive zone).

Table 7   Average Costs in the French Gas Chain in 1990 (centimes/kWh)

	Total	Industry	Residential/ commercial
Acquisition cost	5.75	5.75	5.75
Transportation cost	0.70	0.70	0.70
Storage cost	0.25	0.10	0.50
Load factor cost	0.30	0.05	0.60
City gate price	7.00	6.60	7.55
Distribution cost	5.30	0.10	8.40
Sales price	12.30	6.70	15.95

Source: DIGEC, Rapport d'activité 1992

As should be expected, small users pay a gas price which is much higher per unit than for a large user, the highest unit price being almost three times as high as the lowest. The unit subscription fee shows an even larger difference, varying from 11 centimes to 0.44 c/kWh. In terms of share of the total price, the capacity charge varies from 29.1% to 3.4%. The prices shown in the table above are all designed to cover fixed costs, and also include the gas acquisition cost. In April 1996 this is estimated to about 4.6 c/kWh. Even after deducting this amount from the commodity charge, the commodity charge accounts for a high share of the total price.

In general, a very high share of total distribution costs are fixed costs. Given this cost structure in gas distribution, the French tariff system implies that a substantial share of the fixed cost are recovered through the commodity charge. One of the arguments in favour of this structure is that the cost of load factor (which to a large extent consists of fixed costs) should be distributed according to the extent to which the cubic meter in question needs load factor. (The philosophy seems to be that load factor cost is recovered through the commodity charge according to the load factor and that the cost of security of supply, that is the cost of strategic storage, implicitly is recovered through a flat rate on every cubic metre sold.)

All customers are concerned by this tariff system. Industrial customers taking more than 5 GWh, and being directly supplied by the transmission grid or via the antennas, have individual contracts with different antenna charges. These are, however, based on the basic tariff system. Only interruptible clients pay prices that are directly related to the competing energy product. These prices are not tariffs as such but are also registered. As normal elsewhere, interruptible customers pay a gas price which is significantly lower than the tariffs for normal industrial customers.

Pricing in Relation to Other Fuels

Given that the French gas import contracts mainly are escalated with crude oil and petroleum product prices and that a cost plus approach to tariff calculation is used, there is a relatively good correlation between gas prices and petroleum product prices. Figure 2 shows the price development in the residential sector and the industrial sector since 1991. All prices are exclusive of taxes.

Figure 2   Prices ex-Taxes to Industrial and Residential Users, France, 1991-1995

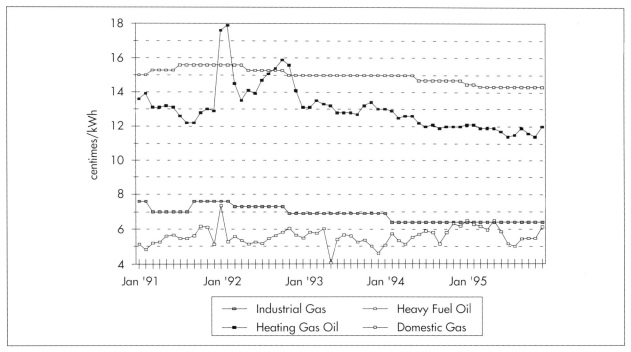

Source: Ministère de l'Industrie, DIGEC

The following observations can be made:

■ Due to the way tariffs are regulated, gas prices shows a more regular development than the prices of the main competing oil products, but both in the residential and the industrial sector there is a good correlation over time between the gas price and the petroleum product price.

■ The prices depicted in figure 2 are, however, not the ones with which end consumers are confronted since they do not include taxes. To get a more realistic comparison between gas and oil product prices, taxes have to be taken into account.

Natural gas is subject to VAT (refundable to commercial and industrial users) Up to the end of 1994, a special rate of 5.5% was applied to the standing charge, while the general rate of 18.6% was applied to the commodity charge. From 1995, all gas charges are subject to the general rate. On 1 August 1995, the general rate was increased from 18.6% to 20.6%. Both GDF and EDF absorbed this increase in VAT.

Figure 3   Industrial and Residential Fuel Prices, France, 1985-1996

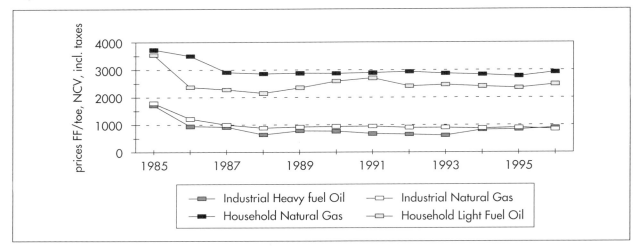

Source: Ministère de l'Industrie, DIGEC

Gas sales to industrial consumers of more than 5 million kWh/year are subject to a special tax of 0.721 centimes/kWh. The tax is paid on the difference between actual consumption and a threshold of 4.8 million kWh. Given that the VAT is refundable for industrial users, this tax on large consumers is in fact the only real gas tax in this sector.

Historically, oil products have been subject to excise taxes and VAT. As of 1 January 1996, the excise taxes are 115 FF/t for low sulphur heavy fuel oil, 154 FF/t for high sulphur fuel oil and 49.50 FF/hl for gasoil. In spite of a much heavier taxation of these oil products, the prices of these fuels after tax have most of the time been lower than those for gas. Figure 4 shows the development in annual prices between 1985 and 1995.

In spite of the fact that the prices of oil products including taxes consistently have been lower than gas prices, gas has been able to increase its market share over the time period. This must be due to the fact that consumers also take into account other factors than the fuel cost when they choose their fuel. Lower investment cost and increased convenience associated with the gas option probably are the main factors explaining this phenomenon.

Cross Subsidies in Gas Distribution

The fairly complicated French end-user tariff system goes some way to reflect the degree to which the servicing of each customer incurs costs. In fact WEFA indicated in a recent study that cross-subsidies in the French tarification system are low by European standards. Still, some form of cross-subsidisation takes place.

The system of tariff perequation on GRT implies an implicit subsidisation of customers located far from the import and production points by customers closer to these points. This probably has some influence on the relative penetration of gas in various geographical areas. The tariffs paid on the antennas of the gas transmission system are not perequated in that they reflect transportation distance. Within the GDF distribution system, the tariffs charged also reflect the physical characteristics of the distribution area in question.

Interruptible customers pay a lower price than firm customers. If such customers are never interrupted, one might argue that the rebate obtained by these customers constitute a kind of implicit subsidy in the order of the price difference minus alternative costs of strategic storage. For cases where such customers are interrupted and have been prepared for this, the argument is void.

Network extensions have to satisfy specific profitability criteria. In some cases, projects that do not fulfill these criteria are allowed to go ahead if they can be subsidised by regional or local authorities. These cases, however, remain marginal; they occur in the context of local economic development projects.

Price Transparency

End user gas prices in France are generally transparent. There are, however two important exceptions to this general observation: prices to industrial customers beyond a certain off-take and prices for gas to power generation. These prices are subject to negotiations and the contracts are not in the public domain.

Costs, Value Added and Profitability in the Gas Chain

Prices, costs and value added in the gas chain are dynamic. It is therefore difficult to get a precise picture of the relative orders of magnitude. Table 8 gives gross estimates of prices and costs in the French gas industry for 1990. These figures still give a good indication of the relative costs in the chain; gas purchase price and sales prices, however, have been somewhat lower over the last few years.

The table merits some comments:

■ In this case the average gas acquisition cost accounts for about 47% of the average sales price, that is less than half of the end user price.

■ The total cost of bringing gas from the import point (or domestic production site) to the end consumer accounts for about 53% of the average sales price in this case. The total cost of bringing gas from the import point can be broken down as follows:

    Transmission: 10.7%
    Storage: 3.8%
    Load factor: 4.6%
    Distribution: 80.9%

The most striking feature in this context is the dominating position of the distribution cost, implying that the greater part of the total costs is incurred downstream of city gate.

In April 1995, the average import price for gas was about 4.6 c/kWh. At the same time, the average price off the high pressure transmission system (GRT) was 6.53 c/kWh (excluding a fixed element which for large customers would amount to about 10% of this amount). For the whole year, the average sales price of GDF was 10.68 c/kWh. As indicated in the tariff examples above, prices to the various customer segments vary a lot. In 1994, the average GDF sales price was 10.91 c/kWh, whereas the average sales prices in the various segments were as follows (in c/kWh):

Individual households:	18.95
Households with collective heating:	12.47
Commercial sector:	13.76
Industry:	12.79
Large industrial uses:	6.13
Sales to other companies:	5.57

In terms of value added, the record of GDF over the last few years is as follows (figures in c/kWh):

	1995	**1994**	**1993**
Average sales price	10.68	10.91	11.18
Average gas acquisition cost	5.06	5.71	5.50
Value added	5.62	5.71	5.68

Making the same calculations for some of the regional companies, the following figures appear (for 1995, in c/kWh):

	Gaz de Strasbourg	**Régie de Bazas**	**Gaz et Electricité de Grenoble**
Average sales price	14.29	14.57	16.76
Average gas acquisition cost	7.01	6.77	6.84
Value added	7.28	7.80	9.92

The figures indicate that value added is higher in the regional companies than in GDF. The difference in gas acquisition cost is due to the fact that the regional companies pay for transportation on the GRT and on the antennas, and this cost is passed on to end consumers. The difference in value added may be due to several factors; one of them could be the fact that GDF has a much higher share of large industrial customers paying relatively low prices. This is reflected in the fact that it has a much lower average sales price than the regional companies.

Looking into the total value added created by GDF in 1995, the following picture emerges:

	Million FF	**Share of total**
Total value added	23501	100%
Salaries	8278	35.2%
Depreciation	6867	29.2%
Interests	3237	13.7%
Taxes	3199	13.6%
Net profits	1920	8.1%

Gaz de Strasbourg, a company majority owned by the municipality of Strasbourg, selling about 3.9 billion kWh annually to about 100 000 customers, realised the following value added in 1995:

	1000 FF	**Share of total**
Total value added	138 250	100%
Salaries	69 311	50.1%
Depreciation	27 929	20.2%
Interests	9 073	6.5%
Taxes	11 773	8.5%
Net Profits	20 164	14.5%

One conspicuous difference between GDF and Gaz de Strasbourg is the difference in the salaries' share of total value added: this is not unexpected since a pure distribution company has a more labour intensive activity than a company present in the whole gas chain. The higher share of taxes in GDF's value added is primarily due to the fact that GDF in 1995 was subject to extraordinary taxation by the Government.

For the analysis of GDF profitability a very crude approach has been taken: net profits and profits before taxes has been related to average total invested capital in the company. This approach gives the following result for the period 1993-1995:

	Net profits as percentage of Average invested capital	**Net profits before taxes as percentage of Average invested capital**
1993	1.36%	2.11%
1994	1.67%	3.25%
1995	2.27%	4.8%

GDF is a fully integrated company and does not, at least externally, account separately for its different activities. This means that the figures above concern GDF's total gas activity, including transmission, storage and distribution. Gaz de Strasbourg, which is a pure gas distribution company realised profits very similar to GDF in 1995: 3.0% after tax and 4.76% before taxes.

Looking at balance sheet of the gas companies in France, it is not obvious what should be considered as equity capital. By deducting total debts from total capital, the following figures have been obtained for profitability in relation to equity (equity thus being defined as total capital minus debts) for GDF:

	Net profits as percentage of equity	**Net profits before taxes as percentage of equity**
1993	2.1%	5.3%
1994	2.4%	4.8%
1995	3.2%	6.8%

The corresponding figures for Gaz de Strasbourg in 1995 were 5.0% and 8.0%, respectively.

State Supervision of GDF

GDF is run as a commercial enterprise, but is under Government influence. Government representatives comprise one-third of the board and a Government commissioner and deputy commissioner sit with the board. The Chairman and the General Director are appointed by the Council of Ministers. Since 1991, the Government has used planning agreements (contrats d'objectifs) as the principal mechanism for state control and supervision of GDF. A second agreement came into effect in 1994. The agreement, which covers 1994-96, sets objectives for operational and financial performance, including the following:

- Secure gas supplies at lowest cost, in part through overseas investment in gas resources and transportation, and increased flexibility in meeting gas demand through the development of storage facilities and a policy of offering supplies under interruptible contracts. In addition to the contract with Norway for additional gas supplies from 1996/7, GDF purchased in 1994 an exhausted gas field in the Marne region for development as a storage facility and is studying a number of potential sites for building additional storage.

- Improved quality of service and relations with regional and local authorities. GDF, in conjunction with EDF, launched in 1995 a programme of service guarantees for small and medium-sized enterprises. GDF is also providing customer information services to promote the use of high-efficiency gas boilers, collaborating with boiler manufacturers on further technical improvements and developing meters to enable customers to better monitor their consumption.

- Grid expansion and increased sales, particularly for space heating and co-generation. The agreement sets a target for new gas-fired co-generation capacity of 600 MW by 1996 and 1 000 MW by 2000, and a goal of connecting 50% of new housing in areas supplied by gas. A target for connecting one million inhabitants in 650 new communes is also set. GDF signed an agreement with DATAR, the French regional development agency, in May 1994 covering part funding by DATAR, the EU Regional Development Fund and local communities.

- Reduction in GDF's debt from FF 13.2 billion at the beginning of 1994 to FF 4.8 billion by the end of 1996. At the end of 1994, debt had fallen to FF 10.6 billion - a slightly slower rate of reduction than planned due to depressed gas sales resulting from mild weather, a weak dollar and lower prices for competing oil products.

- Average reduction in non-gas operating costs per kWh of 3% per year in real terms, compared with 2% under the previous agreement, provided GDF's sales volume grows by at least 3%. Tariffs for domestic customers were reduced by an average 2% in 1994 and 2.5% (before taxes) in 1995.

Infrastructure Financing

GDF is subject to the same rules on investment in new business as other French nationalised industries. A real discount rate of 8% over 25 years is applied to any new project, including extensions to the distribution grid to cover new towns or municipalities and improvements to the transportation system, which must be profitable for them to proceed. If not, the municipality or Departement may provide subsidies up to a specified ceiling to reduce the cost of development to GDF. The company has an obligation to suppl, even if this requires an improvement in the transport or storage infrastructures. The costs of this are not incorporated in the profitability calculations for the supply of the specific customer or customer group.

GERMANY

Statistical information

Natural Gas Supply/Demand Balance (Mtoe, 1996):

Indigenous production	16.1
Imports	62.8
Exports	-2.8
Stock change	-2.6
Total natural gas supply (primary energy)	73.5
Electricity and heat production	13.3
Other transformation and energy use	3.2
Total industry	22.4
Residential	25.3
Commercial	6.7
Other	2.6
Statistical difference	0.0

Natural Gas in the Energy Balance (1996):

Share of TPES	21.0%
Share of electricity output	8.7%
Share in industry	30.7%
Share in residential/commercial sector	31.8%

MARKET AND INDUSTRY STRUCTURE

In 1995/6, Germany is the biggest single natural gas market in Europe and accounts for approx. 23 % of total OECD Europe consumption. Germany has a mature gas market (the eastern parts of the country perhaps still somewhat less), and is a sizeable gas producer (covering around 21% of total supplies in 1995 and 1996). In 1996, imports came from Russia (41.7%), the Netherlands (34.1%), Norway (21.6%), Denmark (1.5%), Great Britain (0.6%) and minor quantities from other sources (0.4%), all through pipelines. Germany is also a very important gas transit country, transporting gas from Russia, Norway and the Netherlands to other European countries.

The structure of the German gas industry is more complex than that of any other country in continental Europe, both because of its sheer size and because of the way it is organised. The reunification of the country has only added to this complexity. The main reason for the complexity is that actors all along the gas chain - from the wellhead or the import point down to the burner tip - sell to end consumers: producers, transmission companies (some of which are also gas importers) and distribution companies.

Table 1 gives an overview of which actors serve which end consumers.

Table 1 Direct Natural Gas Deliveries by Suppliers to Final Consumers, 1996 (million kWh)

	Gas Producing Companies	Supra Regional (mostly transmission) Companies	Regional and Local Distribution Companies	Total
Industrial Sector	15139	172962	177423	365524
Public Electricity Production	2101	23941	47008	72600
Public District Heating	-	11098	41938	53036
Private Households	-	34308	285595	319903
Other	43	11004	106390	117437
Total	**17283**	**252863**	**658354**	**928500**

Source: Gas Report 1996, Bundesministerium fuer Wirtschaft (Federal Ministry of Economics)

The regional and local distribution companies account for about two thirds of deliveries to end consumers, the remainder being taken care of by national transmission companies, and to a more modest degree, by gas producers.

In the following, regional and local distribution companies will in some cases be lumped together and called Local Distribution Companies (LDCs). Table 2 gives an overview of the tasks in which the various types of companies in the German gas industry are involved.

A multitude of companies are involved in gas supply in Germany, including natural gas producers, which in some cases sell gas directly to end consumers, natural gas transmission and distribution companies. Storage takes place at two levels: large volume seasonal storage is mostly undertaken by transmission companies and low volume diurnal storage is undertaken by distribution companies. German gas production is concentrated in that five companies account for about 98% of total production. Most of the producing companies also have stakes elsewhere in the gas chain.

Up until recently, all natural gas supply companies have had the possibility to delimit their respective supply areas from each other by concluding so-called demarcation contracts with each other. This possibility was widely used, but has now been outlawed with entering into force of the new Energy Industry Act on 25.04.1998. In practice the demarcation contract represented an instrument of market sharing, but it could not preclude gas supply companies with whom no such contract had been signed. This explains why it was possible for Wingas, a

Table 2   Tasks Performed by the Actors in the German Gas Chain

Tasks	Producer	National transmission	Supra-regional gas companies	Regional gas companies	Local gas distribution companies
Exploration, field development	x				
Production	x				
Gas conditioing	x				
Gas mixing	x				
Gas import or purchase from domestic producer		x			
Transportation in high pressure network		x	x	(x)	(x)
Seasonal, large volume storage		x	(x)		
Quality assurance		x	(x)		
Supply from national transmission companies at border for demarcation or concession area			x	(x)	(x)
Supply from supraregional companies at border of delivery area				x	x
Supply from regional companies free oncession area					x
Distribution in middle or low pressure network				x	x
Low volume storage (diurnal storage)				x	x
Load factor (Strukturausgleich)	(x)	x	(x)	(x)	(x)
Technical customer advice		x	x	x	x
Marketing		x	x	x	x

x: main activity
(x) : limited involvement
Source: Energiewirtschaftliches Institut an der Universitaet Koeln.

joint venture between Wintershall AG (owned by chemical giant BASF AG) and Gazprom (Russian gas giant), to start offering at the end of the 1980s/beginning 1990s competing gas supplies in large parts of the German natural gas market: not having signed such demarcation agreements, and on the basis of its own transmission network – which it developed across Germany, including through some of Germany's most industrialised regions – , and in some instances on TPA agreements with established gas supply companies, Wingas is able to compete for gas customers (end-users as well as other gas supply companies) in the largest part of Germany. Wingas claims to have captured in 1997 already 12% of market share in Germany.

The transmission companies (Ferngasgesellschaften) buy natural gas from foreign and/or domestic companies in order to sell it to other gas supply companies (transmission and distribution), as well as to large end-consumers. For this purpose, they transport the gas via their systems and provide necessary storage and load factor services. Table 3 lists the German transmission companies.

Natural gas distribution is handled by supply companies which either restrict themselves to local distribution only (most of them Stadtwerke) or which are also involved in gas supply

Table 3   Gas Deliveries from and Revenues for the Transmission Companies (Ferngasgesellschaften) in Germany, 1995

	Deliveries (million kWh)	Revenues (million Deutsche Mark)
Ruhrgas AG	580,428	11,544
Gasversorgung Süddeutschland GmbH	74,029	1.704
Thyssengas GmbH	70,117	1,387
Erdgas-Verkaufs-Gesellschaft mbH	56,855	1,053
Bayerngas GmbH	53,104	1,194
Vereinigte Elektrizitätswerke Westfalen AG (VEW Dortmund)	48,269	8,546
Gas-Union GmbH	42,429	969
Saar-Ferngas AG	41,221	955
Wintershall Gas GmbH	37,663	2,754
EWE Aktiengesellschaft	33,572	2,658
Westfälische Ferngas-AG	30,138	906
Ferngas Nordbayern GmbH	26,624	637
Ferngas Salzgitter GmbH	25,124	593
Deutsche Shell AG	1,178	
BEB Erdgas und Erdöl GmbH	179,841	1,152
Mobil Erdgas-Erdöl GmbH	47,813	
Verbundnetz Gas AG, Leipzig	129,726	2,761
Erdgasversorgungsgesellschaft Thueringen-Sachsen mbH, Erfurt	15,472	

Source: BGW, annual reports

activities on a geographically wider scale (e.g. region). Similar to the large transmission companies, the latter are gas trading companies with their own transportation networks (of mainly middle and low pressure; some high pressure and storage capacity) which buy natural gas (often from the former) and sell it to their customers (Stadtwerke or end users). The local distribution companies distribute gas in the middle and low pressure pipeline grids to end consumers. They buy their gas from transmission or regional gas companies. They have been protected against competition through *exclusive* concession agreements with the municipalities (the exclusivity has now fallen with the new energy reform act).

It is difficult in some cases to distinguish clearly between regional and local distribution companies. According to BGW, there were 673 Orts- und Regionalgasversorgungsunternehmen, or LDCs, in Germany at the end of 1995. As pointed out above, only a minority of these companies are pure gas distribution companies. Table 4 gives a list of the 20 biggest companies in this category.

Table 4   The 20 Biggest Gas-only LDCs Ranked by Volumes Delivered, 1995

City	Company	Gas Deliveries (mill. kWh)	Market Share %
Hamburg	Hamburger Gaswerke GmbH	30,270,955	3.62
Frankfurt/Main	Maingas AG	22,182,950	2.65
Berlin	GASAG	13,135,876	1.57
München	Erdgas Südbayern GmbH	11,708,285	1.40
Halle/Saale	GSA GmbH	9,476,900	1.13
Sarstedt	Sarstedt Landesgasversorgung AG	7,599,201	0.91
Chemnitz	Erdgas Südsachsen GmbH	6,960,636	0.83
Hannover	Gasversorgung Südhannover Nordhessen GmbH	6,947,903	0.83
Augsburg	Erdgas Schwaben GmbH	6,702,348	0.80
Koblenz	Energieversorgung Mittelrhein GmbH	5,879,305	0.70
Erfurt	Gasversorgung Thüringen GmbH	5,585,557	0.67
Dresden	Gasversorgung Sachsen Ost GmbH	5,121,086	0.61
Leipzig	Erdgas West-Sachsen GmbH	4,358,904	0.52
Potsdam	Erdgas Mark Brandenburg GmbH	4,299,089	0.51
Lörrach/Baden	Badische Gas AG	4,205,000	0.50
Schwerin	HGW HanseGas GmbH	4,133,482	0.49
Dortmund	Ortsversorgung Westfälische Ferngas AG	3,564,079	0.43
Bayreuth	Fränkische Gas-Lieferungs-GmbH	3,228,331	0.39
Emmendingen	Emmendingen Gasbetriebe GmbH	3,071,859	0.37
Schönebeck	Erdgas Mittelsachsen GmbH	3,022,235	0.36
Total		**161,453,981**	**19.32**

Source: BGW

Table 5 lists the 20 biggest mixed LDCs in terms of natural gas sales.

Table 5   The 20 Biggest Mixed LDCs Ranked by Gas Volumes Delivered, 1995

City	Company	Gas Deliveries (mill. kWh)	Market Share
Oldenburg	EWE Aktiengesellschaft	29,455,480	3.52
München	München Stadtwerke	17,468,591	2.09
Mainz	Kraftwerke Mainz-Wiesbaden	15,756,380	1.89
Stuttgart	TWS AG	15,331,215	1.83
Dortmund	VEW Energie AG	12,254,159	1.47
Mannheim	MVV GmbH	11,891,119	1.42
Rendsburg	SCHLESWAG AG	10,363,908	1.24
Hannover	Hannover Stadtwerke AG	10,280,864	1.23
Nürnberg	EWAG	10,168,127	1.22
Köln	GEW AG	9,410,510	1.13
Bremen	Bremen Stadtwerke AG	8,672,135	1.04
Köln	Rhenag	7,525,146	0.90
Darmstadt	Südhessische Gas- und Wasser AG	7,502,067	0.90
Dortmund	Dortmund Energie- und Wasserversorgung	6,963,643	0.83
Essen	RWE Energie Aktiengesellschaft	6,710,952	0.80
Düsseldorf	Düsseldorf Stadtwerke AG	5,909,651	0.71
Kassel	Städtische Werke Aktiengesellschaft	5,382,056	0.64
Halle	Halle Energieversorgung GmbH	4,523,227	0.54
Freiburg	Freiburg Energie und Wasserversorgung AG	4,138,555	0.50
Augsburg	Augsburg Stadtwerke	4,137,181	0.50
Total		**191,046,003**	**24.39**

Source: BGW

The level of concentration in German gas distribution may be said to be low as the twenty biggest companies in these two categories (pure and mixed distribution) accounted for around 43% of total deliveries in 1995. The biggest company in the pure gas distribution category held 2.65% of the market and the biggest mixed distribution company 3.6%. However, the biggest companies in both categories represent a gas consumption of 2 to 3 bcm, not much lower than the total consumption in countries like Denmark and Finland, which means that these companies are sizeable actors in the gas market.

In total, the German distribution companies supply more than 15 million customers. Table 6 gives a breakdown per customer category.

Although Germany could be considered a mature gas country, the number of customers is still increasing: the number of customers increased by about 624 000 from 1994 to 1995, out of which about 354000 were added in the new Bundesländer (former East Germany).

Table 6   Total Gas Distribution Companies' Customers in the Various Sectors, 1995

Customer category	Customers in old Bundesländer	Customers in new Bundesländer	Customers in Germany as a whole
Residential	11,014,479	2,133,863	13.148.342
Heat supply	906,058	363,909	1.269.967
Commercial	346,185	8,767	354.952
Public sector	67,639	4,378	72.017
Industry	64,546	7,133	71.679
Electricity companies	309	37	346
Other	80,931	22,617	103.548
Total	**12,480,147**	**2,553,832**	**15.124.399**

Source: BGW

Organisation and Ownership in German Gas Distribution

The 673 gas distribution companies that existed in Germany at the end of 1995 were a heterogenous group in terms of company form. Table 7 shows the different company forms in the old Bundesländer (former West Germany):

Table 7   LDC Company Forms in the Old Bundesländer (1995)

Company form	Number of companies	Share of total	Gas deliveries (million kWh)
1) Zweckverband	4	0.7	4,005
2) Eigenbetrieb	189	34.4	94,708
3) AG or GmbH as Eigengesellschaft	186	33.9	200,321
4) AG or GmbH as public company	35	6.4	48,557
5) AG or GmbH as mixed/public company	105	19.1	241,597
6) AG or GmbH as legally private company	27	4.9	36,327
7) Other private companies	3	0.6	3073
Total	**549**	**100.0**	

Source: BGW

In the cases where the form of "Zweckverband" (Special Purpose Association) has been chosen, the distribution activity is taken care of by an administration set up by several municipalities. The most widespread company form when the municipality itself takes care of the distribution activity, however, is the "Eigenbetrieb" which have to do their own separate accounting but where income streams from several activities go into one purse. In many cases the municipalities have chosen to create an AG or GmbH as "Eigengesellschaft" or as public company. These two categories exclusively have only public shareholders. A widespread form is also an AG or GmbH with mixed public and private ownership. This type of companies typically are created when big cities decide to sell some of their interests in distribution, transforming an "Eigenbetrieb" into an AG or GmbH with some private participation. On an average this type of companies are the biggest ones in terms of gas sales in the old Bundesländer. There are, however, also a relatively small number of AGs and GmbHs that are set up as entirely private companies. It is difficult to estimate the private ownership share in German gas distribution companies, but it seems clear that only categories 5, 6 and 7 in the table have some degree of private ownership, which means less than 25% of the total number of companies. In terms of differences between the AG and the GmbH the most important are that the shares of the GmbH are not publicly traded, and is less easy to privatise. The political influence is stronger in an GmbH than in an AG.

The tendency in recent years has been towards privatisation. A number of large cities have sold part of their shares in the municipal distribution companies to remedy a difficult budget situation. It is expected that with the reforms already decided and with implementation of the EU Gas Directive, the prospect of more competition in the German gas market will give an additional impetus for the municipalities to sell their shares. Concentration in this sector is also widely expected.

As indicated above, the gas industry in the new Bundesländer has gone through a complete transformation after reunification. This transformation process is still going on. The most salient features of this transformation at the distribution level have been the creation of new regional gas companies and Stadtwerke. On a regional level, about 25 new companies have been created, typically with mixed ownership, leaving at least 49% of the ownership to the municipalities, in several cases 51%. The private partners are typically west German regional gas companies and foreign companies (for instance Gaz de France). A number of cities (for instance Dresden and Leipzig) have created their own Stadtwerke, often with private participation. Table 8 gives the distribution of the distribution companies in the new Bundesländer by company form. Close to two thirds of the distribution companies in the new Bundesländer have private interests.

Table 8   Natural Gas Distribution Companies in the New Bundesländer by Company Category

Company category	Number of companies	Share of total	Gas deliveries (million kWh)
1) Zweckverband	2	1.6	251
2) Eigenbetrieb	1	0.8	32
4) AG or GmbH as Eigengesellschaft	31	25.0	14,466
5) AG or GmbH as public company	11	8.9	2,998
6) AG or GmbH as mixed public/private company	66	53.2	61,873
7) AG or GmbH as private company	13	10.5	20,631
Total	**124**	**100.0**	

Source: BGW

Degree of Horizontal and Vertical Integration

In the old Bundesländer, the large majority of companies distributing gas are involved in other activities like distribution of electricity, water and heat. Quite a few have traditionally also been involved in public transportation and public baths. Table 9 gives an overview of their different activities:.

Table 9   Gas Distribution Companies and their Degree of Horizontal Integration
in Terms of Other Distribution Activities · Old Bundesländer

Type of activity	Number of LDCs	Share of total	Gas sales (million kWh)	Gas sales per company (million kWh)
Pure gas companies	89	16.2	162,949	1831
Gas and water	116	21.1	94,966	819
Gas and electricity	49	8.9	46,937	958
Gas, water and electricity	295	53.8	323,735	1097
Total	**549**	**100.0**		

Source: BGW

About 40% of the total number of companies also provide district heating services. Only 16.2% of the companies are pure gas distribution companies, but these companies by far have the highest gas sales per company. More than half of the total number of companies distribute both gas, electricity and water. As table 10 shows, pure gas distribution companies are more widespread in the new Bundesländer (41.9% of the companies). On a regional level, most of the companies are pure gas companies, whereas the new Stadtwerke often also have other activities.

Table 10   The Gas Distribution Companies and their Degree of Horizontal
Integration in Terms of Other Distribution Activities – New Bundesländer (1995)

Type of activity	Number of companies	Share of total	Gas sales	Gas sales per company (mill. kWh)
Pure gas companies	52	41.9	67,194	1,292
Gas and water	16	12.9	2,338	146
Gas and electricity	31	25	15,663	505
Gas, water and electricity	25	20.2	15,053	602
Total	**124**	**100.0**		

Source: BGW

As mentioned above, the transmission companies deliver a considerable share of the gas supplied to end users. This in itself may be said to be a kind of vertical integration in the gas chain. Quite a few of these companies are also vertically integrated in that they have ownership shares in distribution companies (which in this context comprise both regional and local distribution companies). Table 11 gives some examples of transmission company ownership in distribution companies. Apart from noting that such ownership is on the increase, it is difficult to estimate the share of distribution company ownership held by companies higher up the gas chain.

Table 11   Downstream Ownership by Transmission Companies

Company	Downstream Ownership (status 1998)	
Ruhrgas Energie Beteiligungsgesellschaft	Dresden Gas	19.50%
	Stadtwerke Bremen	12.50%
	Stadtwerke Hannover	12.00%
	GASAG Berliner Gaswerke AG	11.95%
	Stadtwerke Chemnitz	10.00%
	Thüga AG	10.00%
Vereinigte Elektrizitätswerke Westfalen AG (VEW Dortmund) (also through VEW Energie AG)	AVU Gevelsberg	50.00%
	Stadtwerke Dülmen	50.00%
	Stadtwerke Merseburg	49.00%
	Dortmunder Energie- und Wasser Versorgung	44.50%
	Energieversorgung Halle	36.70%
	Gelsenwasser AG	27.90%
	Erdgas WestSachsen	25.50%
	Erdgas Mark Brandenburg	24.90%
	Stadtwerke Borna	24.50%
	Stadtwerke Bernburg	22.50%
	Erdgasversorgung Halle	20.00%
	Städtische Werke Magdeburg	17.00%
	Stadtwerke Brandenburg	6.25%
Gas-Union GmbH	Erdgas Westthüringen Beteiligungen (through which they own):	34.00%
	Werragas Bad Salzungen	49.00%
	Ohra-Hoersel Gas, Waltershausen	49.00%
	Eisenacher Versorgungsbetriebe	23.90%
Saar-Ferngas AG	Pfalzgas Frankenthal	100.00%
	Südwestgas Saarbrücken	100.00%
	SpreeGas Cottbus	50.01%
	Stadtwerke Bad Kreuznach	24.52%
EWE AG, Oldenburg	Märkische Erdgas 100% (through which they own):	
	Stadtwerke Finsterwalde	22.50%
	Stadtwerke Ludwigsfelde	20.00%
	Städtische Betriebswerke Ludwigsw.	20.00%
	Stadtwerke Schwedt	10.00%
Ferngas Salzgitter	Erdgas Mittelsachsen Schönebeck	26.40%
	Stadtwerke Zerbst	24.50%
	Stadtwerke Halberstadt	20.00%
Verbundnetz Gas AG	Gasversorgung Delitzsch	49.00%
	Havell. Stadtwerke Werder	36.80%
	Erdgas Westsachsen	25.50%
	Spreegas Cottbus	24.80%
	Stadtwerke Borna	24.50%
	Stadtwerke Schkeuditz	24.50%
	GV Sachsen Anhalt, Halle	24.00%
	Stadtwerke Güstrow	10.00%
Westfälische Ferngas AG (WFG)	Erdgas Mark Brandenburg	24.90%
	Stadtwerke Oranienburg	24.50%
	Stadtwerke Paderborn	10.00%
	GV Sachsen Anhalt, Halle	10.00%
	Städtische Werke Brandenburg	6.13%

Source: 1998, BMWi

Contractual Arrangements

The merchant companies or transmission companies in Germany buy gas on long-term (20-25 year) take-or-pay contracts from producers abroad and on depletion contracts from producers in Germany. This implies that the merchant companies carry the volume risk under these contracts. These companies consequently try to reflect the conditions in their supply contracts in the contracts with their buyers. The contracts that the merchant companies have with the distribution companies are typically long-term and can run up to 15/20 years, but they normally do not have take or pay clauses. Up to a few years ago the contracts between merchant companies and distribution companies typically were supply contracts with an obligation for the latter to take all present and future gas volumes under the contract with the merchant companies. In recent years, new contracts are typically limited to specified volumes, leaving some freedom for the distribution companies to take gas from alternative suppliers.

Contracts between merchant companies and large industrial users often have a duration of 10 to 15 years, but shorter term contracts also exist. Large industrial contracts normally have a take or pay clause. Interruptible industrial contracts exist both between the merchant companies and large industrial users, and between LDC and large industrial customers. The contracts between merchant companies and power producers and between LDCs and power producers, to a large extent have the same structure as contracts with the industrial sector.

The distribution companies distinguish between tariff customers and contract customers. The tariff customers do not have contracts that are limited in time – most of the time they are "evergreen" or standing agreements to supply gas at specified conditions. Contract customers typically have contracts with a three to five year duration, but contracts with up to 10 year duration also exist.

Demand Fluctuations and Security of Supply

Annual gas demand fluctuation in Germany is typical such that demand in the month with the highest demand would be roughly three times as high as demand in the month with the lowest demand.

The fluctuations are of course even more violent on a daily basis. This is illustrated in the following table showing the monthly highs and lows in daily send-out of gas from Ruhrgas which should be a good indication of the situation in Germany as a whole given that close to 70% of the gas volumes in Germany pass through the Ruhrgas network.

Table 12   Gas Send-out of Ruhrgas AG, Daily Highs and Lows, 1995 (million kWh)

Month	High	Low
January	2644	1882
February	2182	1761
March	2238	1697
April	1970	1125
May	1679	846
June	1377	879
July	1040	715
August	1338	706
September	1641	943
October	1690	954
November	2267	1599
December	2634	1789

Source: Ruhrgas AG annual report 1995.

On the coldest day in January, gas send-out was almost four times as high as send-out on the warmest day in August. These fluctuations are met by using flexibility under the supply contracts, storage and, in extreme cases, interruptible contracts.

Germany has a strong tradition of leaving security of supply to the gas companies. Historically, German authorities have played little more than a supervisory role in this area. The merchant/transmission companies play the key role in terms of security of supply. The contracts between merchant companies and distribution companies place a strong responsibility for security of supply on the merchant companies. At the same time the distribution companies are under a strong obligation to serve: German legislation is one of the few that specifies punishment in cases where the obligation to serve is not observed. On a physical level, load factor in Germany takes place on two levels, at the transmission company level and at the regional retail level. This reflects the fact that at both levels companies own storage facilities. This is different from most other gas consuming countries. Given the market value approach taken to gas pricing in Germany, no attempts to calculate and allocate costs pertaining to security of supply are made.

REGULATION

General Framework

The legal framework for regulation is largely contained in two laws:

- the Energy Act of 13 December 1935 (Energiewirtschaftsgesetz), applicable to the gas (and electricity) industry in all Länder of the country. The act authorises the government to perform a supervisory role over gas and electricity supply and obliges the energy companies to provide information, obtain licences and permits, and to notify the state when they wish to enter a market or make investments.

- the Act against Restraints of Competition 1957 (Gesetz gegen Wettbewerbsbeschränkungen) which contains articles with special reference to the gas industry and is an additional instrument of control over the industry. It was amended by the Fifth competition Act Amendment 1989. This law contains provisions directly affecting the energy industries. The amendment came into force in January 1990.

Both Acts have been reformed just recently.

The objective of the Energy Act is to provide a secure and adequately priced supply of energy. Through this law a degree of government control is exercised over the gas and electricity supply industries. Any company wishing to commence supplying gas has to apply for a licence. The act imposes extensive obligations on energy supply companies including the general connection and supply obligations. When a company supplies a specific area, it is obliged to announce publicly its general conditions and tariffs and to connect everyone, on the basis of these conditions and tariffs, to its grid and to supply them.

In principle any company may engage in gas supply operations. Market entry for new companies is regulated under section 5 of the energy Act which states that they require a licence. A licence is usually granted to a company for the whole of Germany. The state authority then has to notify authorities in other states that the licence has been awarded.

The Energy Policy Department of the Federal ministry of Economic affairs is responsible for drafting new laws and energy policy. Detailed control of the gas utilities in their area is the legal responsibility of the Länder. Each Land has an energy control department (Energieaufsicht) within its ministry of economic affairs which grants licences to gas companies.

Until its reform, the Competition Act permitted exceptions for gas and electricity suppliers from the general competition rules, notably the so-called demarcation agreements and the exclusive concession contracts (both now no longer permissible). These were private agreements and had the effect of creating protected service areas, excluding competition. There were two principal kinds of agreement:

- The demarcation contract - in which the parties agreed to refrain from competing in the supply of gas by pipes in an area defined by the contract. Such contracts were made between the gas utilities themselves or between a gas utility and a municipality and affected only the parties to the contract. The permitted maximum length of these contracts was 20 years. Contracts could be renewed under certain conditions. The Federal Competition Office (FCO) could declare an agreement wholly or partially invalid when an application was made for an extension or the agreement was made only under specific and strictly defined circumstances. The contracts did not prevent the entry of new market entrants, as is illustrated by the expansion of Wingas's operations in Germany. Nor did they prevent inter-fuel competition. However, they have had an inhibiting effect on competition at the final consumer level.

- The (exclusive) concession agreement – concluded between a utility company and a municipality under which the latter commits itself to grant the former the exclusive right of constructing and maintaining a network for the public gas supply to end-consumers within a specific area. Even after the reform, concession agreements may still be signed but they are no longer allowed to be exclusive. They must be notified to the responsible competition offices.

In January 1992 a regulation was issued (Konzessionsabgabenverordnung) which laid down rules on the admissibility and assessment of royalty payments by energy supply companies to communities and municipalities. Royalties are payment for the simple or exclusive right to use public highways for the laying and operation of pipelines for supplying gas or electricity to end-users located within the boundaries of a community. The Royalty regulation sets maximum royalty rates and establishes specific exemptions.

The FCO has primary responsibility for the administration and enforcement of the Competition Act at the federal level, where more than one Land is affected. The Land competition offices are responsible for all other matters, in particular matters relating to horizontal and vertical restraints of trade or restrictive conduct not extending beyond the territory of the respective land.

Imports

There is no legislation on imports or exports of natural gas. The relevant laws are contained in the law referring to foreign trade of 6 October 1980- and its implementing regulations. Their effect is to liberalise the imports and exports of natural gas entirely and free it from quantitative restrictions. It may be noted, however, that any contract specifying deliveries for more than 24 months requires authorisation by the federal government.

Pipelines

The laws relevant here are the Energy Act, the External Trade Act and the regulations which implement them. The Mining Law applies to offshore pipelines. According to section 4 of the Energy Act, projects for the construction, renewal, extension or decommissioning of an energy system must be notified to the competent supervisory authorities of the Länder. Natural gas pipelines are subject to these regulations, as is any system for the manufacture, transmission, distribution and delivery of gas. The authorities have the power to object to these projects in the public interest, even to the extent of preventing the projects to go ahead. If they have no objections, a declaration of no objection is issued. Strictly speaking, the requirement to notify and to obtain a declaration of non-objection is not the same as requirement for government approval. Obtaining a declaration of non-objection is not required by law. However, it has become standard practice to obtain the declaration which is designed to clarify the legal position. A utility is almost certain not to go ahead with its plans before obtaining a declaration of non-objection since the declaration acts as a guarantee that the authorities will not take any action under section 4 of the Energy Act.

Apart from the provision in the Act on Competition, there is no obligation in federal or Land law for a gas pipeline owner to grant a competitor company access to its pipeline nor for a gas pipeline owner to transport gas on behalf of a competitor. There is also no provision in the law for tariffs and conditions for the transport of gas. The Ministry of Economic Affairs is considering changes in the law which could shift the burden of proof on access issues so that a pipeline owner refusing access would have to justify that decision and there would be a presumption in favour of access for the shipper (introduced for electricity with the recent reform).

Distribution

Before an energy distribution company may be established, a licence has to be sought from the Land minister of economic affairs, according to section 5 of the Energy Act. Any company which supplies a third party with either gas or electricity is defined as an energy distribution company. Once licence is secured, it is not necessary to obtain a new licence should the company wish to extend its working area or to integrate horizontally or vertically. The licence does not grant a monopoly or exclusive rights to the distribution company. Exclusivity derives from the concession, a private law agreement between the municipality and the utility. The concessions are identical for all companies. They grant the utility an exclusive right to construct and maintain, within the territorial limits of the municipality, a pipeline network for the purpose of supplying the end user with gas as a matter of public service. The distribution companies pay a fee for this concession.

Distribution companies are not entrusted with any duty in the public interest apart from observance of he general aim in the preamble of the Energy Act to provide the safest and cheapest supply of energy possible. However, this does imply a general obligation to provide connections and supplies incumbent upon distribution companies as set out in the "Verordnung über Allgemeine Bedingungen für die Gasversorgung von Tarifkunden (AVBGasV), which is based on section 6 in the energy Act. In the AVBGasV there is a provision for payment of damages to the customer by the distribution company in the event of interruptions or irregularities of supply. The draft energy law mentioned above would imply a ban on concessions as far as the exclusive right of way is concerned.

Storage

All gas storage facilities above and below ground are energy installations within the meaning of the energy Act 1935 and therefore have to be declared in accordance with section 4 of that law. Consent to projects may be refused by the Land authorities, if that is deemed to be in the public

...orage facilities without tanks are subject to the provisions of the Federal ...erground exploration undertaken to determine suitability for such facilities.

PRICING AND TARIFFS

In general, public authorities do not play a predominant role in the fixing of gas prices. There are no specific controls over transmission and wholesale gas prices. Tariffs on the pipelines are entirely a matter for the pipeline owners, subject only to competition law and EU law, although there is a requirement in the Energy Act to publish tariffs and conditions for the supply of gas. The only legal instrument for intervention in the area of pricing by the government is the Federal Regulation on Gas Tariffs (Bundestarifordnung Gas). It distinguishes tariff from non-tariff customers, basically a distinction between domestic and non-domestic customers. The regulation concerns small consumers only. It requires the distribution companies to offer at least two tariffs to small consumers, defined as those with annual consumption of less than 1200 cubic metres. There is a small consumers tariff comprising a metering charge and a commodity charge. The consumer is offered a standard supply contract. Alternatively, there is a basic tariff, comprising a demand charge and a commodity charge valid for all uses and all kinds of consumers. These customers are not offered a standard contract, although some are offered similar contract terms. They are therefore called 'normalised customers'. According to section 4 of the regulation, the consumption-based fraction of the tariff offered may not exceed 60% of the unit consumption rate for the small-scale consumer. Apart from this restriction, rates may be freely set within these tariffs.

In practice, many distribution companies offer three or more tariffs to domestic consumers, with the first two applying to natural gas used for cooking, hot water, or both, and the third applying to gas used in central heating. Price reductions based on volume and load factors are applied, depending upon the terms on which the distribution companies have purchased the gas from their suppliers. The general principle is that of market value pricing for each customer, with the distribution company setting the gas price according to the next closest alternative.

This flexibility in pricing is subject to two conditions. Firstly, it is subject to the provisions in the Competition Act which in strictly defined cases allow the competent authority to take steps to correct subsequently the prices of a company should it consider them to be unjustifiably high. Secondly, the large municipal shareholders in the distribution companies ensure an element of indirect government influence on their market behaviour. When the company is entirely owned by the municipality, the latter will determine the tariffs. The degree of intervention by the municipality will vary from case to case and over time. With municipal elections coming up, all opportunities for price rises justified by escalation clauses in the contracts may not be exploited.

Consumer organisations have regular discussions with the associations of gas suppliers on matters concerning general terms and conditions. There is, however, no legal obligation to consult consumers on prices and conditions of supply.

While the federal regulation sets the conditions regarding supply and tariff structure, the regional ministers of economic affairs have competence to supervise the prices. There are examples that the competition office of a Land has ordered gas suppliers to reduce their prices. The large industrial users negotiate their contracts individually with the transmission or distribution company, with the option of a firm or interruptible contract. The firm contract customers are offered a standard contract called Sondervertrag. Contracts for interruptible supplies are also negotiated individually with heavy fuel oil parity. Prices and supply conditions for non-tariff customers can come under the control of the Bundeskartellamt. It is regarded as an abuse if the prices and conditions of a gas distribution company are less favourable for its customers than those of another gas distribution company. However, there is no abuse if the company can prove that the difference is caused by circumstances for which the company cannot be blamed.

Prices to Distribution Companies

In line with the general approach to gas contracting and pricing, the distribution companies are in principle free to enter into whatever contract they want with their suppliers. In spite of the fact that the LDC pursue contract negotiations individually, their supply contracts are in practice standardised to a very large extent. Part of the reason for this is that the LDCs tend to create joint purchasing consortia to face their supplier, most of the time a national or a regional transmission company. The use of standardised contracts has not changed with the arrival of Wintershall on the market. Established suppliers rather responded to the competition by offering up front payments to keep old customers. Until recently, the supply contracts for LDCs were normally intended to cover their entire need for gas. Now the parties still negotiate a specific volume which they are obliged to deliver and receive but the LDC is no longer obliged to take new volumes from the same supplier as before. This means that the LDCs are free to choose their supplier for additional volumes.

Since the middle of the 1980s, LDC supply contracts are typically based on a two part price formula, one fixed price element and one proportional element, with an additive price escalation formula. The two elements are described in the following table:

Capacity charge:

$LPt=[LP0+yl*(Lt-L0)]/Bd$ [Pf/max kWh/d/a]

where:

LPt: actual capacity charge

LP0: base price (equal to 140,484 DEM at the time when the formula was constructed)

yl: absolute salary element

Lt: actual salary according to municipal salary agreements

L0: base salary

Bd: days of utilisation

The weights are 70% for the base price and 30% for the actual salary level.

The commodity charge:

$Apt+AP0+yHEL*zHEL*(HELt-HEL0)$ [Pf/kWh] where:

APt: actual commodity charge (proportional unit price)

AP0: base price which at a HEL price of 64.39 DEM/hl was fixed at 3.8 Pf/kWh.

yHEL: absolute HEL link equal to 93%

zHEL: heat equivalent conversion factor

HELt: Reference HEL price (Statistisches Bundesamt quotation) Adaptation through a 6-3-3 system.

HEL0: Basis HEL price 1 April 1984: 64,39 DEM/hl (6-3-3).

Source: von Donath

With a base price of 3.8 Pf/kWh, the LDCs obtained a commercial margin of about 2 PF/kWh since the heat equivalent HEL price was about 5.85 Pf/kWh at the time when this formula was adopted. Deducting the capacity charge of about 0.9 Pf/kWh (at 150 utilisation days) at that time, one arrives at a gross margin of about 1 PF/kWh. This gross margin diminishes over time with the development in the salary element which has a weight of 30% the in capacity charge. The absolute oil link in the commodity element means that in a period of increasing oil prices, the gas price will lag somewhat behind and will thereby give an extra margin that could be used to compensate for increases in the LDCs' own costs. In a period with declining oil prices the gas purchase price will decline less than the HEL price. The price escalation formula is therefore in many cases replaced by a substitute formula which makes the link to the HEL price weaker in times of low oil product prices. This means that the gas price becomes more cost oriented at low oil prices. There is, however, no corresponding mechanism to create a ceiling for the gas price when oil prices are high. The type of pricing mechanism that has been described is also used in contracts between national and regional transmission companies.

Prices to Large Industrial Customers

Prices to large industrial customer are negotiated according to the market value principle. The following shows typical price formulae for industrial customers taking about 100 GWh or more.

One part price system with additive escalation:

$Apt = Ap0 + yHS*0.00792(HSt-HS0) + Z[Pf/kWh]$
With multiplicative escalation:
$Apt = yHS*0.00792*HSt + Z[Pf/kWh]$

One part price system (additive escalation) plus cost element:

$APt = AP0 + yHS*0.00792(HSt-HS0) + yHEL*0.09098(HELt-HEL0)$
$+ yI,L*[(0.5*It/I0)* (0.5*Lt/L0)] + Z[Pf/kWh]$
where:
AP = base price
HEL (light heating oil) = Stat. Bundesamt, Fachserie 17, Reihe 2 HEL quotation
HS (heavy fuel oil) = 1% sulphur HS quotation from same source
y = absolute link
I,L = cost elements, I = investment good index, L = salary index
0.00792;0.09098 = heat equivalent conversion factors
Z = gas price premium (because of lower emissions)

Large industrial users tend to get contracts only with a commodity element, reflecting that their load charge is high, often around 70%. The main rule is that the base price is based on HFO being the alternative fuel, but relatively small customers in this category often prefer to have the gas price escalated also with light heating oil. For such customers, the sellers in some cases also include a cost element to pass through to the customers part of his costs. Most of the time the formulae also include an element reflecting an environmental premium for gas.

In addition to the price escalation clause, industrial contracts will normally have clauses whose effect it is to pass through to the buyer price increases and taxes on the seller's side. These clauses reduce the risk seen from the seller's point of view. Some contracts also contain renegotiation clauses that are triggered when the actual gas price reaches a certain proportion of the agreed base price. Volume rebates are given according to a tranche system which could also take into account deliveries to several sites within the same demarcation area.

Industrial gas contracts often have a duration of ten to fifteen years to ensure the amortisation of investments necessary to link the customer up to the distribution system. The contracts normally have take or pay clauses, the usual level being 70% take-or-pay obligation.

Customers having dual or multi-fired equipment can also buy gas on an interruptible basis. Interruptible contracts typically foresee an interruption in supplies triggered by a specified temperature and could specify a minimum interruptibility of 500 to 1000 hours a year. The price reduction given to compensate for interruptibility takes into account the difference in costs between a dual or multi-fired facility and a gas-only facility and the price of substitution fuels. Seen from the seller's side an alternative to an interruptible contract is gas storage, and it could be expected that the rebate for interruptible supplies is fixed in some relationship to storage cost. A rebate of 0.8 Pf/kWh for customers that are interruptible at 6 hours' notice has been mentioned as usual (for comparison: the average sales price to industry for the whole German gas industry in 1995 was 2.55/kWh).

Gas Prices to Electricity Production

The pricing principles for gas sales to electricity production (including cogeneration) are the same as for sales to industry. The market value principle is used; HFO, LFO and coal being the alternatives. There are, however, practically no gas contracts to power generation where the gas price is linked to coal. To avoid paying a fixed element some power producers are willing to be interruptible, which implies that they use LFO when gas supplies are interrupted.

Prices to End Users in the Distribution Sector

Customers in the distribution sector are households, enterprises in the tertiary sector and industrial customers. In terms of prices there are two main types of customers: tariff customers (Tarifkunden) and contract customers (Sondervertragskunden). Table 13 gives an overview of the tariff and price systems in these markets.

Small consumers are offered two types of tariffs: compulsory tariffs (Pflichttarife) and choice tariff (Wahltarife). The compulsory tariffs are aimed at small consumers and there are two categories: the small consumer tariff (Kleinverbrauchertarif – consumption up to 266 cum/year-typically cooking) and the base price tariff (Grundpreistarif – consumption beyond 266 cum/year - typically cooking and water heating).

The name of the choice tariffs points to the fact that the consumers are free to choose among the tariffs in this group according to their consumption. Typical customers are residential heating/all purpose customers. The small consumer tariff and the choice tariffs have in common that they consist of a fixed and a variable element. In volume terms the small consumer tariff accounts for a very low share of total residential gas use (about 5% of that market). The gas

Table 13   Overview of the Tariff and Price Systems in the German Distribution Sector

Contract type	Tariff customers	Tariff customers	Contract customers	Contract customers
Consumer group	Households and small consumers	Households and small consumers	Small consumers/ small industrial customers	Industrial customers
Price charged according to	Compulsory tariff	Chosen tariff		Guided price or individual price
Publication	Compulsory	Mostly voluntary	Unpublished	Unpublished
Purpose	Cooking/ hot water	Heating/ all purposes	Heating/ (process heat)	Process heat/ (heating)
Consumption in kWh/a	< 8000 households < 100.000 enterprises	10.000– 30.000/30.000– 100.000	100.000– 2.5 million.	2.5 million.– 50 million./ 50-100 million.
Pricing system	Two part price	Two part price	One part/two parts/ zoning	One part

Source : Donath

distributor is obliged to charge the customer according to the most favourable tariff. The gas distributors are obliged to publish the small user tariffs, but not the choice tariffs. Most of the time, however, also the latter are published.

Contract customers are supplied according to normalised or individual contracts. As the choice tariff customers, customers with normalised contracts typically use gas for heating and/or all other purposes, for instance in residential blocks. In addition to a zoning system, the price to these consumers will consist of two parts where the fixed element is charged according to the load ordered. Contract customers account for about 55% of the consumption in the household and small consumer segments. The deliveries to these customers are not governed by the "AVBGAsV". The municipalities also receive less concession fees for these customers.

The border line between normalised contracts and individual contracts is not clear and varies between the LDCs. The share of gas for process purposes is an important criterion for distinction as far as industrial customers are concerned. The prices to contract customers are not published, but in most cases are available on demand.

Tariff Customer Price Structure

Both the small consumer tariff (Kleinverbrauchertarif) and the base price tariff (Grundpreistarif) should contain a fixed and a variable element. The variable element of the base price tariff is limited to 60% of the small consumer tariff; consequently the fixed element of this tariff is relatively higher than that of the small consumer tariff: often the fixed element in the base price tariff is two to four times as high as that in the small consumer tariff. The LDCs have considerable discretion in setting their prices: in a survey of about 40 LDCs in Northern Germany it was found that the price to small consumers could vary by a factor of two.

For the choice tariffs there are no regulations concerning price structures and basis for measurement. Two part pricing systems justified by cost and market arguments are common.

In principle, the fixed costs in gas distribution should be covered by fixed payments. The high fixed share of costs in gas distribution (typically 40 to 50% of total costs including gas acquisition cost) would , however, lead to a high fixed element. Such a high fixed element would give prices that are strongly digressive with volumes, and the principle is rarely followed in practice. In the residential/commercial sector the standing charge normally accounts for 20-25% of the total price and the commodity charge for 75-80%. A higher fixed element would also not give a strong incentive to energy saving as a lower volume taken would have to carry a higher fixed cost per unit.

The choice tariffs also contain a fixed and a variable element. They typically provide two or three combinations of a fixed element and a variable element, the fixed element increasing with volume; the variable element decreasing. Basically, the consumer has to find the optimal combination of fixed and variable price, but in most cases the LDCs would charge the price that is the most favourable to the consumer. There are no provisions as to how the fixed element should be calculated. A majority of the LDCs, however, offer a fixed price element dependent on the load factor from a specified level. This is then combined with a general or zoned commodity price.

Since all the LDCs in Germany to a large extent are free to set their prices, it is very difficult to put forward prices that could be claimed to be representative.

Prices to Contract Customers

Contract customers are a very heterogenous group of customers: they could be everything from fully gas supplied multi-dwelling houses via small industrial and craftmenship enterprises to large industrial complexes. The transition from normalised or standardised contracts to individual contracts is not sharply defined and varies from on LDC to another. An approximate volume limit of 2 to 5 million kWh can be indicated. Customers up to this limit are typically small industrial enterprises and public services. Customers taking volumes beyond this up to 20 to 50 million kWh, typically industrial customers, are often supplied according to guided prices. For even larger industrial customers, individual prices are negotiated. Typical pricing systems and price escalation clauses for contract customers are shown in the following table.

In standardised contracts to small industrial customers and public sector gas consumers both zone price systems and commodity charge/standing charge combinations are used. Commodity charge/demand charge combinations are typically used for customers above 2 million kWh since it is only for such volumes that devices to measure hourly and daily flows are economic.

In spite of the fact that each LDC has its own set of pricing systems for contract customers, it is relatively easy to get an idea about the prices charged to these categories through the price surveys that are published. Since 1973, an organisation called Bundesverband der Energie-Abnehmer e.V. has published prices for 45 LDCs spread over the old Bundesländer, and in recent years also for 12 LDCs in the new Bundesländer. This survey gives prices for supplies from 0.5 million kWh a year up to 50 million kWh year, specifying also the load factor. The survey shows that there are substantial price differences between the companies. One example is that on 1 April 1996, the highest price charged to customers taking 10 million kWh a year was 4.81 Pf/kWh, whereas the lowest price was 2.81 Pf/kWh.

Table 14 Typical Pricing Systems and Price Escalation Clauses for Contract Customers in Germany

Zone price system: Volume zones with a specified price in Pf/kWh for each of the following zones: First 500000: ZP 1

Next 1.000.000: ZP 2

Next 5.000.000: ZP 3

Next 10.000.000: ZP 4

For all further 35.000.000: ZP 5 with the price escalation formula:

$ZPt=0.09098(HELt-HEL0)+0.000079(L0-Lt)$ [Pf/kWh]

Commodity charge/demand charge combination:

Commodity charge: $Apt=AP0+yHEL*0.09098(HELt-HEL0)$ [Pf/kWh] with demand charge based on daily off-take :

$Lpt=Lp0+yL(Lt-L0)$ [DEM/max kWh/d/a]

or on hourly off-take:

$Lpt=LP0+yL(Lt-L0)$ [DEM/max kWh/h/a]

Commodity charge/standing charge combination:

Commodity charge: $APt=AP0+yHEL*0.09098(HELt-HEL0)$ [Pf/kWh]

Standing charge: $Gpt=GPt*kW$ [DEM/a]

where: AP=commodity charge, LP =demand charge, GP =standing charge

oil price: HEL 40-50 hl, Statistisches Bundesamt, Fachserie 17, Reihe 2

Cost element L: salaries

y: absolute link

0.09098: conversion factor

Source: Donath

Price Transparency

Gas prices in Germany, especially for small customers, are transparent in the sense that all the LDCs publish tariff schedules. These schedules are in many cases quite instructive in that they make potential customers understand what they will have to pay for their gas. Publication of prices in the contract market is less systematic, but customers have no problems in obtaining a precise price quotation from the suppliers. Prices to large industrial users and power producers are generally not published since they are subject to individual negotiations.

Price transparency when it comes to understanding the prices obtained for the gas sold and paid for the gas purchased by the LDCs leaves a lot to desired. Especially for the companies that have distribution activity besides gas it is in general difficult to get this kind of information from published annual reports.

Taxes

The LDCs in Germany have to pay concession fees to the municipalities in which they have a concession by virtue of the Konzessionsabgabenverordnung dated 9 January 1992. These fees are not taxes as such; they are supposed to be payment for the use of rights of way, but for end consumers they have the same effects as taxes. In 1995, the total amount paid in concession fees on gas was about 1.5 billion DEM. The concession fees are specified in Pf/kWh of gas delivered and vary with the type of contracts under which they are delivered, the use of the gas and the population of the municipality. They are specified as maximum amounts that can be charged, as indicated in the following table:

Number of inhabitants in the concession area	Max. fee on gas for cooking and water heating (Pf/kWh)	Max. fee on gas for other tariff deliveries (Pf/kWh)
Up to 25000	1.01	0.44
Up to 100 000	1.21	0.53
Up to 500 000	1.52	0.66
Beyond 500 000	1.82	0.79

For contract customers, the fee is limited to 0.06 Pf/kWh, independent of the number of inhabitants. Some LDCs indicate in their published tariff schedules the share of the gas price accounted for by the concession fees. For small users, the fee could constitute about 20% of the commodity charge. The progressivity of the concession fees as a function of the number of inhabitants may have a certain logic in that the fee is a payment for using the ground in the area in question; a municipality of 500 000 inhabitants would normally have to grant rights of way to larger areas than a municipality with few inhabitants. In terms of developing new areas to be served one may, however, question the incentive effect of the fee since it would be more difficult to develop a gas network in a large city with high population density than a small town with low population density. One might also question the marked difference in fees paid by tariff customers and contract customers which would probably not be justified on cost grounds.

The gas industry also has to pay royalties to the Länder governments. In 1995, these amounted to 185.9 million DEM which corresponds to a fee of 14% on the price ex borehole. When an excise tax on natural gas was introduced in 1989, the political aim was that the tax burden on natural gas should have a neutral impact on competition, i.e. it should not affect the competitiveness of gas in relation to the oil products. To comply with this criterion, the tax on natural gas was determined as the weighted average of the increase in tax on gas oil and heavy fuel oil.

The published reference prices for gas oil and heavy fuel oil include excise taxes, so indirectly these taxes also influences the gas price. The gas price determined using these reference prices is increased by the natural gas tax which is currently 0.36 Pf/kWh. Where the fact that both the oil tax and natural gas tax have been allowed for in determining the gas price leads to a double burden on the gas price and thus to a competitive disadvantage for natural gas, the customer is granted a discount to counteract this double burden.

The following table gives an overview of the taxes in Germany by sector and fuel.

Table 15   Taxation in Germany by Sector and Fuel as of 1 January 1996 (in ECU/GJ (NCV)

	Residential	Commercial	Power generation	Industry
HFO	-	0.34	0.62	0.34
GO	1.04	1.04	-	1.04
LPG	0.50	0.50	-	0.50
Natural gas	0.51	0.51	0.51	0.51
Coal	-	-	-	-
Electricity	-	-	-	-
Non-deductible VAT (%)	15	0	0	0

Source: Eurogas, 1998

Price of Gas in Relation to Other Fuels

The main message presumably is that the gas price, given the market value principle is closely linked to the oil product prices with a lag.

Costs, Value Added and Profits

The German gas industry is more complex than that of any other country in Europe. One of the consequences is that it is more difficult to get a detailed overview of costs, value added and profits in the industry and the distribution of these items than in most other countries.

In 1996, the German gas industry sold 928.5 billion kWh of natural gas, bringing a revenue of 33.015.8 Million DEM, which means 3.556 Pf/kWh. The same year, the average gas import price was 1.24 Pf/kWh. The difference between the average sales price and the average import price (plus the natural gas tax of 0.36 Pf/kWh) gives an indication of the gross margin in the German gas industry (in fact it underestimates it since the average price of gas bought from domestic producers is lower than the import price). Over the last few years this margin has developed as follows (in current Pf/kWh, not taking account of the gas tax of 0.36 Pf/kWh):

Pf/kWh	1992	1993	1994	1995	1996
Average sales price	3.66	3.65	3.58	3.52	3.56
Average import price	1.39	1.43	1.33	1.32	1.24
Gas tax	0.36	0.36	0.36	0.36	0.36
Gross margin	1.91	1.86	1.89	1.84	1.96
Gross margin/average import price	1.1	1.0	1.1	1.1	1.2

Source: Bundesministerium fuer Wirtschaft

The average sales price over the period has been going down both in nominal and real terms. Most of this decrease can probably be explained by declining import prices, but part of it could also be due to gas to gas competition that has taken effect over this period. The average import price has fluctuated somewhat over the period and explains some of the variation in gross margin. In 1995, the average gross margin was 1.67% of the average import price.

In the residential sector, the highest paying segment of the gas market, gross margin developed as follows over the period:

Pf/kWh	1992	1993	1994	1995
Average sales price to private households	5.36	5.26	5.30	5.11
Average import price	1.39	1.43	1.33	1.32
Gas tax	0.36	0.36	0.36	0.36
Gross margin	3.61	3.47	3.61	3.43

Source: Bundesministerium fuer Wirtschaft

The prices to industry have consistently been going down over the period from 1991 to 1995. This is probably due to the gas to gas competition that has developed in some parts of the market in recent years. In real terms there has been some decline in gross margins:

Pf/kWh	1992	1993	1994	1995
Average sales price to industrial customers	2.69	2.66	2.58	2.55
Average import price	1.39	1.43	1.33	1.32
Gas tax	0.36	0.36	0.36	0.36
Gross margin	0.94	0.87	0.89	0.87

Source: Bundesministerium fuer Wirtschaft

The evolution in the average sales price to electricity production has not been consistent; the same has been the case with the average gross margin:

Pf/kWh	1992	1993	1994	1995
Average sales price to power generation	2.28	2.16	2.04	2.15
Average import price	1.39	1.43	1.33	1.32
Gas tax	0.36	0.36	0.36	0.36
Gross margin	0.53	0.37	0.35	0.47

Source: Bundesministerium fuer Wirtschaft

The trend is not entirely clear for the category called "other customers" either (this category comprising mostly small customers outside the residential sector):

Pf/kWh	1992	1993	1994	1995
Average sales price to other customers	4.18	4.13	4.18	4.00
Average import price	1.39	1.43	1.33	1.32
Gas tax	0.36	0.36	0.36	0.36
Gross margin	2.43	2.34	2.49	2.32

Source: Bundesministerium fuer Wirtschaft

Given the complex structure of the German gas industry it is difficult to get a precise picture of how the gross margin is distributed throughout the chain, one of the problems being that companies often are active in more than one part of it. Therefore it is for instance difficult to get a clear picture of the total gross margin and its distribution in the distribution part of the chain. By looking into some examples of companies present in the chain, however, an attempt will be made to shed some light on their economics.

Ruhrgas AG is the leading transmission company in Germany. In 1995, 69.1% of the total gas volumes sold on the German market passed through its hands. Looking at its gross margin in 1995, this amounted to around 21% of the total gross margin of the German gas industry. The company is basically an intermediary, importing gas, transporting it and reselling to other companies. It sells the greater part of its gas to other national and regional transmission companies, but also sells substantial volumes directly to industry, power generation and LDCs. Table 14 shows how its average sales price and volumes the various customer segments have developed over time.

Table 16   Ruhrgas Average Sales Price (Pf/kWh) and Sales Volumes
(billion kWh) 1986 to 1995

	1986	1987	1988	1989	1990	1991	1992	1993	1994	1995
Average sales price	3.18	1.85	1.67	1.85	2.11	2.48	2.25	2.17	2.12	1.98
Sales merchant companies	240.8	278.1	268.0	292.6	319.4	348.9	361.0	361.2	355.9	373.1
Sales LDCs	98.7	112.0	105.3	111.2	119.2	130.5	129.2	137.4	135.6	144.5
Sales industry	69.7	69.3	69.1	71.7	72.1	69.3	65.1	63.6	64.2	62.8
Total sales	**409.2**	**459.4**	**442.4**	**475.5**	**510.7**	**548.4**	**555.3**	**562.2**	**555.7**	**580.4**

Source: Ruhrgas annual report 1995

Using the average gas import price into Germany as an indicator of the company's gas acquisition cost, the gross margin has developed as follows over the last few years (Pf/kWh):

Pf/kWh	1992	1993	1994	1995
Average sales price	2.25	2.17	2.12	1.98
Gas acquisition cost	1.39	1.43	1.33	1.32
Gas tax	0.36	0.36	0.36	0.36
Gross margin	0.5	0.38	0.43	0.3

Sources: Ruhrgas annual report 1995, Bundesministerium fuer Wirtschaft

In all parts of the gas chain, it is difficult to find companies that could be claimed to be representative because each of them has a different market structure. The following example gives the main figures for a medium sized regional distribution company in the eastern part of Germany which both sells gas to end consumers and to other LDCs (Stadtwerke):

Unit: Pf/kWh

Average sales price	3.01	Tariff customers: 5.53 Contract customers: 2.51 Stadtwerke: 2.45
Gas acquisition cost	2.19	
Gross margin	0.82	Tariff customers: 3.34 Contract customers: 0.32 Stadtwerke: 0.26

Source: Industry sources

This company buys its gas from on of the Ferngasgesellschaften which in turn buys its gas from an importer. The next example involves a longer supply chain:

Average sales price Stadtwerke	4.07
Average gas acquisition price	2.07
Average gross margin Stadtwerke level	2.00
Average sales price regional transmission company	2.07
Average gas acquisition cost regional transmission company	1.72
Gross margin regional gas transmission company	0.35
Average gas sales price national transmission company to regional transmission company	1.72
Average gas acquisition price national transmission company (=import price including gas tax)	1.68
Gross margin national transmission company	0.04

Source: Industry sources

Based on data for all the transmission companies and all the distribution companies in Germany collected by BGW, Rheinisch-Westfälisches Institut fur Wirtschaftsforschung (RWI) has given an overview of costs in the transportation part of the gas industry and the distribution part of it. One of the problems encountered is that quite a few of the companies are active in both parts of the industry. The practical solution to this problem has been to place the companies in the category where they have their main activity. The following table shows the cost structure of transportation part of the gas chain.

Table 17 Breakdown of Costs in Gas Transmission (per cent)

	1985	1990	1991	1992	1993	1994	1995
Gas cost	91.6	66.7	67.1	60.4	60.8	59.4	59.2
Non-gas cost	8.4	33.4	32.9	39.6	39.2	40.6	40.8
Capital cost	4.9	11.7	10.4	13.6	13.7	14.9	14.9
Labour cost	2.5	5.7	5.7	7.0	6.5	6.8	6.7
Other cost	0.9	16.0	16.8	18.9	18.5	19.0	19.2
Total cost	**100**	**100**	**100**	**100**	**100**	**100**	**100**

Source: RWI

The share of gas costs has gone down over time. The increase in the share of capital cost can be explained by the fact that the transmission part of the industry has invested quite heavily in recent years. The share of labour cost has been the most stable over the past five years. The substantial increase in other costs is to a large extent due to the introduction of the gas tax which is collected on the transmission level.

Table 18 Breakdown of Costs in Gas Distribution (Pf/m^3)

	1985	1990	1991	1992	1993	1994	1995
Gas cost	42.9	22.2	26.8	24.6	23.6	22.8	21.1
Non-gas cost	11.9	14.2	13.5	15.2	15.1	15.5	15.0
Capital cost	7.3	8.4	7.8	8.6	8.6	9.0	8.7
Labour cost	3.4	3.8	3.8	4.5	4.4	4.5	4.4
Other cost	1.2	2.0	1.9	2.2	2.1	2.0	1.9
Total cost	**54.8**	**36.4**	**40.3**	**39.8**	**38.7**	**38.3**	**36.1**

Source: RWI

Table 19 Breakdown of Costs in Gas Distribution (per cent)

	1985	1990	1991	1992	1993	1994	1995
Gas cost	78.6	62.5	65.8	60.7	60.8	59.4	58.5
Non-gas cost	21.4	37.5	34.2	38.7	39.2	40.6	41.5
Capital cost	12.9	22.5	19.7	22.0	22.1	23.7	24.2
Labour cost	6.0	10.0	9.8	11.5	11.3	11.6	12.2
Other cost	0.2	5.0	4.7	5.8	5.3	5.3	5.1
Total cost	**100**	**100**	**100**	**100**	**100**	**100**	**100**

The share of non-gas costs in distribution is just slightly higher in distribution than in transportation, but the composition and the absolute figures are quite different. Capital cost takes a higher share, reflecting the fact that distribution networks are more expensive to build per kilometre than transmission pipelines. In recent years, there has been a lot investment in distribution, partly as a consequence of the German reunification. The share of labour cost in distribution is almost the double of that in transmission, reflecting the higher extent of labour intensive contact with customers. One explanation for the lower share of "other costs" in distribution is that no gas tax is collected at this level except for the concession fees. It is impossible to find one particular distribution company that could be claimed to be representative because their market structure and supply costs differ. To give an idea of the orders of magnitude in terms of revenues and costs in the LDC, an example will be given.

The company in question is a mixed company, distributing gas, electricity, heat and operating public baths. In terms of gas deliveries it is among the hundred largest LDCs. The fact that it is a mixed company is reflected in the fact that the gas activity has to carry a share of the overhead costs of the company as such. Table 19 gives an overview of total revenues and costs in absolute terms and in Pf/kWh.

Table 20   Revenues and Costs for a Medium Sized German Distribution Company, 1995

	Thousand Deutsche Mark	**Pfennig/kWh**
Gas Supply Costs	33,936	2.24
Konzessionsabgabe	577	0.04
External Contract Services	2,009	0.13
Personnel Costs	1,583	0.10
Other Expenses	876	0.06
Capital Costs	4,815	0.31
Overhead Costs	4,291	0.28
Operating Profit	9,853	0.65
Total	**57,940**	**3.83**

ITALY

Statistical information

Natural Gas Supply/Demand Balance (Mtoe, 1996):

Indigenous production	16.5
Imports	30.6
Exports	0.0
Stock change	-0.7
Total natural gas supply (primary energy)	46.4
Electricity and heat production	9.9
Other transformation and energy use	0.3
Total industry	15.0
Residential and commercial	19.8
Other	1.3
Statistical difference	0.1

Natural Gas in the Energy Balance (1996):

Share of TPES	27.0%
Share of electricity output	24.4%
Share in industry	41.6%
Share in residential/commercial sector	52.5%

MARKET AND INDUSTRY STRUCTURE

Italy accounts for 13% of gas consumption in OECD Europe and is a relatively mature gas market. In 1996, domestic gas production covered 32% of total consumption. The remainder was imported as pipeline gas from Algeria (35%), Russia (25%), the Netherlands (8%) and as LNG (0.1%).

The Italian gas market consists of three main parts: firstly, the market served by the distribution companies through urban networks i.e. The residential/commercial market including small industrial users; secondly, the large industrial customer market served directly from the high pressure transmission network, including autoproducers of electricity; thirdly the power generation market, which is also served directly from the high pressure transmission system. The size of the direct delivery market (large industrial and power generation segments) and the market served by the distribution companies are about equal in size.

Table 1 gives an overview of the structure of the Italian gas market over the last couple of years.

Table 1 Natural Gas Sales in Italy 1994-1997*

	1997	%	1996	%	1995	%	1994	%
Residential	21.794	41.0	23.013	43.2	22.084	42.1	19.904	42.0
Commercial	547	1.0	549	1.0	516	1.0	464	1.0
Residential & commercial	22.341	42.0	23.562	44.3	22.600	43.1	20.368	43.0
Industrial	23.369	44.0	21.161	39.8	20.966	40.0	19.278	40.6
Industrial firm supplies	18.741	35.3	16.499	30.8	15.780	30.1	14.095	29.7
Industrial interruptible supplies	4.628	8.7	4.662	8.8	5.186	9.9	5.183	10.9
Chemical processing	722	1.4	693	1.3	773	1.5	777	1.6
Thermo-electrical	6.259	11.8	7.265	13.6	7.747	14.8	6.706	14.2
Automotive	325	0.6	308	0.6	283	0.5	263	0.6
Others	129	0.2	245	0.5	141	0.3	–	–
Total Italy	**53.145**	**100.0**	**53.234**	**100.0**	**52.510**	**100.0**	**47392**	**100.0**

* sales by Snam only
Source: Snam

The distribution companies serve more the 14 million customers. The following table shows how they were distributed by category at the beginning of 1995 and what their gas consumption was in 1994.

Table 2 Number of Customers Served by Distribution Companies*and their Gas Consumption in 1995/1996 (in thousand of cubic metres)

	1996		1995	
	Number of customers	Gas consumption	Number of customers	Gas consumption
Domestic cooking and hot water	4,171,408	934,537,887	4,242,875	945,166
Domestic individual heating	9,523,272	12,912,217,770	8,824,904	11,336,178
Domestic centralised heating	238,117	3,860,712,766	242,555	3,409,746
Industrial uses -large firms	5,081	4,371,201,668	5,572	3,669,054
Industrial uses -small firms	64,883	928,673,385	58,693	704,894
Artisans	198,995	698,070,171	182,403	814,724
Commercial customers	619,335	2,956,417,029	562,891	2,575,160
Hospital complexes	478	478,451,598	446	391,527
Total	**14,817,569**	**27,140,252,272**	**14,121,339**	**23,846,452**

* sample of distribution companies covering 98% of total customers
Source: Statistiche Metano

In 1995, 3.929 large industrial customers were served directly from the high pressure transmission system.

Main Actors

ENI, a fully integrated oil and gas company engaged in all aspects of the petroleum industry, is the most important player in the Italian gas industry. Its operations are conducted through six principal operating subsidiaries, among which Agip and Snam are the most important in its gas activity, the first as the leading producer of gas in Italy, and Snam as the dominating gas company.

ENI's natural gas supply, transmission and distribution activities commenced in the 1940's when it began the commercial sale of natural gas to industrial users in northern Italy. Through Snam, ENI is the primary supplier of natural gas in Italy and Europe's third largest natural gas company based on volume of natural gas sold to the domestic market. In 1995 ENI's sales of natural gas totalled 52.5 bcm, representing 98% of the gas consumed in Italy.

Snam owns and operates a domestic natural gas network of some 26000 kilometres and imports natural gas from several sources under long-term contracts through a network of international pipelines. It sells gas directly to large industrial customers (about 3900 customers) and to electricity producers and serves the residential and commercial sector through sales to distribution companies. In addition it holds 40.91% of the shares in It algas, the largest retail gas distribution company in Italy.

In 1997, It algas sold about 6.4 bcm of gas to more than 4.2 million customers, thereby covering more about 27% of the total retail gas market in Italy in volume terms. The company serves its customer base through among others 1262 gas distribution concession contracts.

In 1996, around 5000 municipalities received natural gas from Snam. In total, 14.8 million gas users were served by the distribution companies.

In 1997, there were 714 LDCs in Italy. Table 1 shows some of the characteristics of these companies.

Table 3   Distribution Companies in Italy in 1997

	Total end 1997	Municipal companies	Private companies	Public companies
Companies	714	292	302	120
Municipalities served	5140	404	3907	829
Users	15,000,000	1,039,000	9,224,000	4,737,000
Sales	21,917 mcm	1,806 mcm	12,404 mcm	7,707 mcm

Source: Snam

In the municipal companies, gas distribution is undertaken by the municipality directly. In the public companies, the service is undertaken by a consortium of municipalities, by a special municipal company or a company with mixed capital (public and private). The private companies operate the gas distribution service by virtue of a concession agreement with the municipalities.

The municipal companies are the most numerous but are the smallest in terms of sales per company and the number of customers per company. On an average, the public companies are the largest both in terms of sales per company and number of customers. The private companies, however, account for close to 60% of total LDC sales and serve more than 75% of the total number of municipalities.

Table 4 shows an overview of the 30 largest LDCs in Italy in 1995.

In 1995, the 30 companies accounted for about 58% of the total gas sales through LDCs and held about 64% of the customers. Italgas accounts for a relatively high share of total LDC sales; the second company holds less than 5% of the market. In 1995, there were only four companies selling more than 0.5 bcm a year. The same year, only 25 companies out of 809 sold more than 100 million cubic metres of gas.

The LDCs in Italy are all members of one of the three associations that have been created to look after their interests: Federgasacqua, which have municipal companies as members; ANIG (Associazione Nazionale Industriali Gas) which regroups the large, private distribution companies; and Assogas which regroups the small distribution companies. ANIG has only 136 member companies but these companies account for almost half of the total gas sales and serves 7 million customers. The companies have 12000 employees.

Through its 40.91% ownership share in Italgas, SNAM is present also in the distribution part of the gas chain. Considering that SNAM is responsible both for the quasi-totality of gas supplies and for gas transmission, one may say that there is a considerably degree of vertical integration in Italian gas distribution. As far as the remaining part of gas distribution is concerned, there are no examples of vertical integration.

The degree of horizontal integration between gas distribution and other sectors is low in Italy, but changing. Today, some of the municipal companies are also distributing electricity. Involvement in water distribution and related services is somewhat more widespread. Italgas, in addition to 1442 municipal concessions for gas distribution also held 324 concessions for water distribution, 59 concessions for water cleaning and 27 concessions for sewerage at the end of 1995. Market liberalisation, however, is expected to lead to an increase in horizontal integration (as well as in LDCs concentration).

REGULATION

There is no single framework law applying to the gas sector in Italy. The rules governing the sector are found in various statutes, in the ENI charter and in recently adopted legislation. The legislative framework reflects that the Italian gas industry is in a state of transition; privatisation of ENI and introduction of a new regulatory authority for the gas and electricity industries being important aspects of this.

Although there are significant private elements in the Italian gas sector, the Government has historically played an important role both as owner of the ENI conglomerate of companies and as

Table 4

	cubic metres	customers	employees	cm/ customer	cm/ employee	ownership
IT ALGAS SPA	4,274,271	3671177	6469	1.16	660.73	private
CAMUZZI GAZOMETRI SPA	860,697	669483	935	1.29	920.53	private
AZ ENERGETICA MUNICIPALE	667,330	776813	1212	0.86	550.60	public
ACOSER - AZ Consorziale Servizi Reno	508,370	302646	609	1.68	834.76	public
FIORENTINA GAS SPA	334,844	269779	490	1.24	683.36	private
AZ MEDITERRANEA GAS ACQUA SPA	281,706	312215		0.90		public
ITALCOGIM SPA	275,268	243384	323	1.13	852.22	private
AZ CONSORZIALE GAS ACQUA	267,640	165476	276	1.62	969.71	public
NAPOLETANA GAS SPA	246,602	478260	797	0.52	309.41	private
VENEZIANA GAS SPA	243,357	148757	265	1.64	918.33	private
AZ MUNICIPALIZZATA PUBBLICI SERVIZI	204,447	11012205	141	1.86	1449.98	public
AZ MUNICIPALIZZATA GAS ACQUA	203,456	112693	157	1.81	1295.90	public
SOC. GAS RIMINI SPA	191,420	109849	67	1.74	2857.02	private
AZ CONSORTILE GAS ENERGIA SERVIZI	185,226	119204	150	1.55	1234.84	public
CONS.BIM PIAVE DI TREVISO	180,217	95792	101	1.88	1784.32	public
AZ GENERALE SERVIZI MUNICIPALIZZATI	178,600	116750		1.53		public
AZ SERVIZI MUNICIPALIZZATI	167,423	127307	215	1.32	778.71	public
CONS. INTERCOMUNALE ACQUA-GAS E	163,817	122351	235	1.34	697.10	public
AZ MUNICIPALIZZATA	139,513	78254	180	1.78	775.07	public
SOGEGAS SPA - SOC. GENERALE GAS	131,867	85876	111	1.54	1188.00	private
TIRRENIA GAS SPA	124,567	131596	230	0.95	541.59	private
AIM-AZIENDE INDUSTRIALI MUNICIPALI	116,436	66726	90	1.74	1293.73	public
AREA-AZ. RAVENNATE ENERGIA AMBIENTE	113,223	64331	127	1.76	891.52	public
SIT SPA - SOC. INDUSTRIALE TRENTINA	112,049	55498	115	2.02	974.34	private
AZ COMUNALE ELETTRICITA GAS ACQUA	106,740	110508	173	0.97	617.00	public
AZ GAS ENERGIA AMBIENTE	98,843	62787	71	1.57	1392.16	public
SAG SPA-SOC. ADRIATICA PER IL GAS	97,397	49675	84	1.96	1159.49	private
CIS-CONS. INTERCOMUNALE SERVIZI	94,250	53223	48	1.77	1963.55	public
SICILIANA GAS SPA	87,141	121995	225	0.71	387.29	private
BERGAMO AMBIENTE E SERVIZI	84,988	59254	92	1.43	923.78	public
	10,741,705	8891764	13988	1.21	767.92	

regulator. Recently, however, the role of the Government has undergone changes: the privatisation of ENI presumably means less Government influence in the company. The establishment of an energy regulatory authority makes the regulatory intervention in the gas sector less direct.

Before July 1992, ENI was a public statutory body owned by the State. Since that date, it has been a joint stock company fully owned by the Treasury up to the stagewise privatisation that started in 1996. Since July 1998, the Government holds about 35% of ENI capital. Law no. 474 adopted on July 30, 1994 (the "Privatisation Law") sets forth the procedures to be followed in connection with the privatisation of companies directly or indirectly owned by the state. Pursuant to this law, special provisions have been adopted in ENI's by-laws to limit the voting rights of certain shareholders. These provisions make material acquisition of shares subject to approval by the authorities and give the Treasury veto power over major changes. The law changing the status of ENI in 1992 provided that all special mining rights formerly attributed or granted to ENI are to be exercised by the new company in the form of concessions. The new legislation did not change Snam's position.

On matters of pricing, there has for a long time been a measure of regulation in the distribution sector carried out via CIP (Comitato Interministeriale Prezzi). This system changed in 1994, however, and the CIP's functions were taken over by the Ministry of Industry, Commerce and Crafts. The establishment of the "Autorita per l'energia elettrica e il gas", operative since April 1997, constituted a further change. The Ministry of Industry, Commerce and Crafts together with the Interministerial Committee for Economic planning (CIPE) are responsible for establishing the energy policy guidelines. The "Autorita per l'energia elettrica e il gas" is an independent regulatory body which competences are defined by law.

It has its legal basis in Law number 481 adopted on the 14th of November 1995 which is a general law on regulation of public utility services setting out the framework for regulation in the areas of gas, electricity and telecommunications. This law defines the tasks of the regulator in each field, which for gas is those of monitoring and control, promotion and development of competition, definition of quality standards for the services offered, and regulation of tariffs. More specifically, the competences of the Authority in the gas area have been set as follows:

■ definition of services that should be subject to concession or authorisation and modifications of such concessions;

■ control that access to services takes place under competition and transparency;

■ fix and update base tariffs;

■ publish directives on separate accounting for the various services offered and verify the cost of each service;

■ see to it that supply conditions are widely published;

■ deal with complaints from consumers;

■ ensure that all distribution companies render services according to a directive concerning a public service charter which was adopted in January 1994 (see below).

Under the law, all competence in the above indicated areas has been transferred to the new authority.

Exploration and Production

In the following, regulation in the gas chain will be dealt with only to the extent that it is relevant for gas distribution.

Historically, the ENI group has benefitted from a number of privileges in this area. Since 1991, there is no obligation to offer all gas produced offshore to ENI.

Pipelines

No concession is required to lay or operate a gas pipeline. This does not, however, apply to Sicily where a concession is required.

Under Law No. 9 of 9 January 1991, a limited form of third party access has been introduced which compels pipeline owners to provide carriage of natural gas produced in Italy and used in producers' own plants, in plants either owned by producers' controlling companies or their affiliates, or to be delivered to ENEL or to other companies falling within the scope of existing legislation. The term used in the law is "vettoriamento", which means carriage but without the obligation to redeliver the same gas. The conditions and the fee for the service of vettoriamento will be agreed between the parties taking into account an adequate remuneration of the investments involved, the operating costs, the criteria followed in the European gas markets for determining the fees for vettoriamento and the relevant levels, as well as the evolution of the energy market. Failing an agreement between the parties, the conditions and the fee will be established by CIP after hearing the parties.

All companies owning gas pipelines in Italy (Snam, Edison, SGM and SPI) are subject to the carriage obligation in the law. In practice, this means Snam and the local distribution companies. There appears to be no requirement that owners increase transport capacity, and carriage is therefore confined to the system's marginal capacity.

In December 1994, Snam reached agreement with the Unione Petrolifera (UP) and the Assocazione Mineraria Italiana (Assomin) on terms and conditions for contract carriage of natural gas produced in Italy. The essential features of the agreement are:

- all costs are incremental and must be paid by the shipper;

- Snam is to indemnified against all taxes and duties which may be levied on gas transported;

- the gas is to be transported by the shortest and most economical route.

The agreement emphasises that the carriage of gas for third parties takes second place to Snam's own transport requirements. It applies only to gas produced in Italy and sold either to ENEL or to the producers' own affiliates. In 1997, natural gas produced in Italy and carried for third parties amounted to about 4 bcm.

Storage

The right to store natural gas in depleted hydrocarbon fields is vested in the state but is exercised by the grant of concession to others. The rules governing underground storage of natural gas are to be found in Law No. 170 of 26 April 1974, revised by the Decree No. 625 of

25 November 1996, which among other things, abolished the exclusive right of ENI in the Po Valley. There is no legal barrier to prevent any natural or legal entity to store natural gas, providing that the necessary technical competence is displayed. The storage concession is granted by the Ministry of Industry, Commerce and Crafts for a period of 20 years and it is renewable for a subsequent 10-year period.

Distribution

SNAM has a dominant position in gas distribution in Italy in a wide sense of the term since it accounts for close to all distribution to large customers and through It algas holds a large share of the residential/commercial market. Its position, however, is not based in law – there is no legal prohibition on distribution of natural gas by other parties in Italy, neither for imported gas or gas produced (onshore and offshore) in Italy. In practice, only SNAM has the facilities to organise the import of natural gas from abroad. Most of the local and regional distribution companies purchase their gas from SNAM. Some LDCs purchase their gas from other pipeline owners.

To engage in gas distribution in the small customer segments of the market, an administrative permit or concession is needed. The 8101 municipalities in Italy are entitled to exercise public services such as local gas distribution, but they can also issue concessions for gas distribution to either public and/or private companies. These concessions are normally subject to tendering procedures; in the future they will be subject to international tendering procedures. Once obtained, the concession gives an exclusive right to distribute gas (normally for 30 years, but the duration could be shorter, for instance 20 years).

The conditions for such concessions are negotiated individually between the municipality and the company involved. Their character can, thus, vary from one municipally to the other. Normally the concessions can be renewed. The tendering procedure are based on criteria like pace of development more than on prices to end consumers or quality of service offered. One gas supply company can hold several concessions.

In accordance with Law 273 of July 1993 a legal decree was published in September 1995 containing a general reference scheme for a service charter in the gas sector. The scope of the scheme is to establish and guarantee the rights of gas distribution customers. The legislation requires the LDCs to establish their own service charters along the lines of the reference scheme. The scheme imposes a series of service quality indicators, some of them quite specific. A principle of equal and impartial treatment of customers is laid down. It states that ensuring continuous and regular service is an obligation on the LDC. Furthermore, it defines a number of maximum time delays that the LDCs has to observe in its relation with its customers, for instance for connection and disconnection. The companies are also obliged to indicate measures to facilitate the access of certain customer categories, for instance disabled persons. A concession holder normally has an obligation to supply all customers located within a certain distance (usually 10 metres) from the mains.

Natural gas prices to industrial customers are calculated in accordance with the agreements made between Snam and the most representative industry associations (Confindustria and Confapi), and they apply nationally to all industry sectors. Consumption volumes above 200.000 cm/year may benefit from this industrial tariff system. Annual consumptions below this threshold are treated under a pricing methodology specified by the Italian public administration.

...er be directly supplied by Snam or by LDCs. Final price levels and
the kind of supply network to which the customer is connected.
...les as to the choice of grid to which to connect. They choice
...ilities and economic preferences of the parties.

PRICING AND TARIFFS

Gas prices in Italy are generally market value based but there are also elements of a cost plus approach and rate of return regulation with price capping involved in price setting in some parts of the system, so the case is not clear cut. Tariffs are supposed to be the same all over the country for the various user categories, but there are exceptions to this.

Below, an overview of price determination in the following cases will be given:
• sales from SNAM to urban distribution companies;
• sales from urban distribution companies (LDCs) to final domestic users;
• sales from SNAM to industrial customers;
• sales from SNAM to power generators.

The prices applied in these transactions are all subject to some kind of price control: the price of gas sold by SNAM to LDCs and industrial users are supervised or monitored, while tariffs applied by LDCs to final residential and commercial users are subject to an "administrative" system.

Before looking into price determination, some observations on contractual arrangements are appropriate. Natural gas supply conditions to LDCs are agreed on a national level between SNAM and the association representing Italian municipalities (ANCI), the public distributors (Federgasacqua) and the private ones (ANIG and Assogas). The present agreement between Snam and the LDCs' associations runs until 30 June 2002 with a possibility for prolongation up to 30 June 2005. Individual contracts between Snam and LDCs have the same duration.

Generally, the contracts between the LDCs and their residential customers do not have specified durations.

Contracts for gas sales to industry are based on agreements between SNAM and the most representative associations of Italian industrial enterprises (Confindustria and Confapi). The large industrial users (> 50 million cm/year) can negotiate particular tariff conditions with SNAM provided they accept take-or-pay provisions under contracts with poly-yearly durations. Standard contracts typically have a duration of one year, but often run on as "evergreen" contracts until they are brought to an end.

Electricity producers, whether selling their production to the public distribution system or to final users, can buy gas under special agreements between SNAM and UNAPACE (union of electricity producers). The present agreement in force was signed on 1 June 1998. The individual contracts normally have a duration of fifteen years for producers selling their electricity to the public distribution system, and for three years for other producers, and contain take or pay clauses (take or pay level 50% of agreed annual quantity or agreed daily quantity multiplied by 160, whichever is the higher).

ENEL has a separate contract with SNAM for deliveries of 6 bcm of gas annually, and running until 2003. At present this contract is under renegotiation. ENEL also buys about 4 bcm directly from Algeria. The gas is delivered at the Algerian-Tunisian border and then transported by SNAM to Italy.

Prices to LDCs

The supply tariff has a binomial structure made up of a fixed charge and a commodity charge. The fixed charge is a flat rate updated annually according to inflation indices (at present, it stands at 58.5 Lit/cm). The commodity charge is differentiated according to the average annual consumption of the final consumers supplied by the LDC, and is updated every two months on the basis on heating gas oil price variations. The commodity charge fixed by the decree of the Ministry of Industry, Commerce and Crafts of 16 November 1996 (published in Gazzetta Ufficiale on 23 December 1996):

Per capita consumption (Mcal/year)	LIT/m^3
<= 4.000	146.0
=10.000	277.0
>=16.000	286.5

Values of the commodity price associated to per capita consumption between 4000 and 16000 are obtained by linear interpolation of above given values.

Commodity price formula: $Cpt=CP0+0.538(G0-Gt)$ [LIT].
CP0 is the base price, Cpt the current price, G0 is the basis heating gas oil price and Gt the current heating gas oil price.
Since May 1998, gas oil price = CIF Med quotations published by the Platt's Oilgram Report, expressed in USD/t and converted into Lit/kg using the exchange rate between Lit and USD.
The commodity charge is adapted to the gas oil price through a 6-1-2 system (i.e., the commodity charge applied in January and February is related to the average gas oil prices calculated over the period of six months between July and December of the previous year).

Tariffs to Customers Served by the LDCs

Since 1975, a particular procedure or method for determining and revising tariffs for gas distributed by LDCs has been used (hereafter referred to as the Method). In formulating the Method, the Public Administration aimed to standardise tariffs throughout Italy, minimise costs to the consumer and keep the correct balance between the LDCs' revenue and cost. The principles used are akin to the cost of service approach to gas pricing used in North America: total revenue must equal the costs allowed (which include a rate of return on investment) and costs are controlled to ensure that all expenses are strictly necessary to guarantee a safe and efficient gas distribution service.

In addition to the cost of gas that the LDCs are allowed to pass on, the Method determines the distribution cost, consisting of two components, operating costs and investment costs. The cost determined by the Method is then used as a basis for the tariffs to be paid by the various customer categories according to specified regulations and guidelines. In the following, the three elements will be dealt with in more detail:

Gas Cost

> The gas cost element is calculated according to the following formula:
> $Qm = (cm+iqf)/(9.2*cnc)$ [LIT/Mcal] where
> cm = gas supply price for the LDC in question (function of per capita consumption)
> iqf = fixed price element
> 9.2 = conversion factor Mcal into m^3
> cnc = correction factor to take non accounted for gas quantities into account
> This cost element is escalated according to the agreement with SNAM described above.

Operating Costs

> These cost comprise staff costs, operational costs and general expenses. In 1993, a price-cap formula was introduced for this part of the costs.
> The formula for the price cap is:
> $Qg = Qgprec*(1+(I-cp))$ [LIT/Mcal] where
> Qgprec = previous value of Qg
> I = coefficient for development in inflation since the last tariff adaptation
> cp = productivity coefficient

In accordance with the Miniterial Decree of 19 November 1996

- cp varies from 0.25 to 3.0 accoridng to Qgprec*K.
- inflation rate is 3.4%.

Investment Costs

This part of the costs is supposed to ensure sufficient revenues to cover normal investment costs and investment costs to further develop the network of the LDCs. It should cover technical depreciation of plant, financing costs and give an adequate remuneration reflecting the risk involved. The investment cost is individualised in that it reflects certain characteristics of the company in question. The formula used is as follows:

> The investment cost is set in relation to the per capita consumption in the LDCs in question:
> $Qi = (i*Is*a)/K$ [LIT/Mcal] where
> i = rate of return on debt and equity, for LDCs with equity share of 45% or more i was fixed at 9% (1993), for all other LDCs i = 8.6% .
> Is is 1.140.000 Lit/customer when K < = 9000 Mcal per user per year, and 1.260.000 Lit/customer when K > 9000 Mcal per user per year.
> A = factor that has the effect of smoothing the investments. It varies with the Ip factor which expresses the relationship between the investment of the LDC over the last four years and the standard investment. A varies with Ip as follows:
> When Ip is less than or equal to 2%, then A = 0.6
> When Ip is 4%, then A = 1.0
> When Ip is 10% or more, then A = 1.6.

The incentive effects of the investment element are not easily recognisable, but apparently they are as follows:

- the allowed rate of return on invested capital is higher for companies with high equity share than for companies with low equity share. This could have the effect of strengthening companies that already have a high equity share compared to companies with low equity;

- the allowed standard investment costs are higher for companies with high K (consumption per user per year) than for companies with low K. This should give companies with a high degree of development a stronger incentive to invest than companies with low degree of development. To the extent that a high allowed investment cost leads to a higher gas tariff, this should, however, give some incentive to energy saving;

- the A factor is intended to have at least two effects on investment: firstly, to give companies an incentive to develop their network - if the investment in a year is less than 4% higher than in the three previous years, the company is penalised by an A factor less than 1. If on the other hand investment is more than 10% higher than standardised investment cost over the last three years, the company is penalised by being able to recover less than its full investment cost. For A values between 1.0 and 1.6, the LDCs are allowed to recuperate about 90% of their investment cost in one year over the first three years.

The total cost defined by the formulae above is seen in relation to last years supply, and a standard cost per cubic metre emerges. This standard cost is equal to the average sales price that the LDC is allowed to charge. Declining standard costs, for instance through increase in sales, has to be passed on to consumers. Therefore LDCs can only increase profits beyond the level implied by the above formula by reducing operating costs below the price cap level. The fact that LDCs vary a lot in their characteristics has led to the 40 different tariff basins, i.e. Areas where tariffs are homogenous.

For the transformation of standard costs into tariffs, four different gas uses are distinguished:

- Cooking and hot water (tariff 1);

- Individual heating (tariff 2);

- Central heating and other uses (tariff 3);

- Small artisanal and industrial customers with annual consumption between 100.000 and 200.000 cubic metres a year (tariff 4).

All the four tariffs consist of a standing charge and a commodity charge.

The way standard costs are reflected in the applied tariffs are to a large extent determined by regulation. The standing charges for all tariffs and the commodity charges for tariffs 1 and 4 are determined by the authorities for the whole of Italy. The standing charge is differentiated according to use and is 3000 LIT per month for tariff 1 and 5000 LIT/month for tariff 2. For tariff 3 and 4 there is consumption dependent fixed price of 41.4 LIT/ cubic metre.

The commodity charge for individual heating (tariff 2) and for central heating and other uses (tariff 3) are set by the LDCs according to precise guidelines. These guidelines prescribe prices declining with volume for both tariffs and the offer of at least two tariffs for central heating and other uses. Compliance with these guidelines are monitored by 89 provincial price commissions. The commodity charges are adapted bi-monthly by changes in gas oil prices of more than 11 LIT/kg.

The T1 tariff is fixed by regulation, but this does not mean that it is the same for all LDCs. In the same way as gas purchase cost is fixed as a function of the K factor (consumption per user), this tariff is also a function of this factor. It varies as follows:

K-factor (consumption in Mcal/a per user)	Tariff 1 (LIT/Mcal)
0-5000	75
5.001-7.000	70
7.001-9.000	65
>9.000	60

Source: Donath

The tariffs for cooking and hot water are adjusted by the authorities. The commodity charge in tariff 4 is volume dependent and was 393.3 LIT /cubic metre up to 100.000 cubic metres per year and 365.7 LIT/cubic metre for off-takes above that volume. It is adapted bi-monthly like the commodity charges of tariffs 2 and 3.

Gas taxation will be dealt with more in detail below but already in a tariff context it is important to point out a peculiarity of Italian gas taxation: the fact that gas for heating purposes are so heavily taxed. Both excise taxes, regional taxes and value added taxes are lower for gas for cooking and hot water than for heating purposes. This creates a price structure where prices increases with volumes for small consumers when going from one category of gas use to another.

Industrial Gas Prices

Industrial gas customers taking up to 200.000 cubic metres (2 million kWh) can choose among three industrial tariffs: a "high load factor" tariff and a "low load factor tariff" – both consisting of a connection charge, a standing charge and a commodity charge –, and a one part tariff consisting of a connection charge and a commodity charge (which has build into it a fixed element).

The "high load factor" tariff is designed for customers with a load factor of more than 50%.

The "low load factor tariff", with a standing charge of 60% lower than that of the "high load factor" tariff and a higher commodity charge, is designed for customers with load factor of less than 50%.

The one part tariff, usually chosen by customers with low consumption, is designed for users with irregular off-takes (load factor under 40%).

A discount factor H applying to the commodity charges has been introduced by the new agreement of 11 November 1997 between Snam and the most representative industry associations. This discount amounts to 7 lit/cm from January 1997 to September 1999. From October 1999 onwards, it will be reduced by 1 lit/cm every month until April 2000.

In the following, these tariffs will be described in more detail.

The High Load Factor Tariff

The connection charge under this tariff is LIT 500.000 a month.

The fixed charge is calculated according to the following formula:

Standing charge $= Ca*I$ [LIT/month per max m^3/day] where
Ca = price for ordered load, updated bi-annually. Ca is defined as follows: since July 1998
$Ca = 1.412,7(0.57*SO/SOo)+(0.38*PPI/PPIo)+0.05$ [LIT/month] where SO is an inflation index of prices applied to industry.
I = agreed maximum daily off-take (m^3/day)

A peculiarity of the standing charge is that it allows the maximum daily off-take volumes to be surpassed without punishment. The tolerated excess volumes decrease when the ordered quantities increase. Up to 8.000 m^3/day the ordered quantity can be exceeded by a factor of 1.15., between 8.000 and 12.000 m^3/day the factor is 1.13 and beyond 12.000 m^3/day it is 1.11. When these tolerance limits are exceeded, the extra price paid amounts to 1/7 of the standing charge Ca. The intended effect of this system is to give the customers an incentive to foresee the daily peaks as precisely as possible. The commodity element is differentiated according to nine volume zones, linked to oil product prices and follows a 12-1-1 lag system:

Commodity charge = summa 1 to 9 $Vi*Bi$ [LIT] where
Vi = annual volume (m^3/a)
$Bi = 0.845*(ICI+K)$ [LIT/m^3/a] where $ICI = GI*G+bi*BTZ+ai*ATZ$ and G = gas oil price, BTZ = HFO 1% and ATZ = HFO 3.5%. For the calculation, arithmetic mean quotations for oil product prices CIF Italy and FOB Rotterdam barge for the last 12 months from Platt's Oilgram are used.
K = constant with value = 52 LIT from which the product ($n*Si$) has to be deducted. N has a value of 14 from January to October 1994, then decreased by one unit every two months to disappear completely by 31 December 1996. The value for Si follows the same quantity intervals as for the oil price weights (see below) and declines from 2.83 for the first volume zone to 0.88 in the last volume zone.

The weights of the oil prices in this additive price formula vary over the volume zones as indicated in the following table:

Table 5 Oil Price Weights in the Italian High Load Factor Industrial Tariff

1000 m³/month	gi(GO)	bi(HFO 1%)	ai(HFO 3.5%)
0-100	0.68	0.26	0.06
100-300	0.48	0.46	0.06
300-500	0.40	0.54	0.06
500-700	0.38	0.38	0.24
700-1000	0.34	0.32	0.34
1000-2000	0.30	0.32	0.38
2000-3000	0.26	0.30	0.44
3000-4000	0.24	0.30	0.46
Beyond 4000	0.20	0.28	0.52

Source: Donath

The reasoning behind this variation in weights is that the gas price should reflect the competition with oil products. Large industrial users tend to have HFO as the most interesting alternative, and this is reflected in the weights.

The Low Load Factor Tariff

Consumers with irregular gas consumption can choose between a low load factor tariff and one part tariff (described below). The low load factor tariff is applied to industrial customers with an load factor of about 50%.

In the low load factor tariff a standing charge is added to the commodity price element. This standing charge Cb amounts to one third of the standing charge under the high load factor tariff and only accounts for about 0.04% of the tariff. The commodity price looks like this:

Commodity price = Summa 1 to 9 Vbi*Bbi [LIT] where
Bbi = Bi+0.043836*Cb [LIT/m³]. Here Cb = 1/3*Ca (defined above).
Punishment for exceeding the agreed daily volumes is equal to 1/4*Cb.

The One Part Tariff

This tariff consists of a commodity charge plus a connection charge equal to 500.000 Lit/month (the same paid under high and low load factor tariff), but this commodity charge has built into it two fixed elements:

Commodity price = 0.94*P+F [LIT/m³] where
P = Bm+0.131508*Cb and Bm = 0.845*(0.90*G+0.10*BTZ+K) and F = Ca (defined above).

To this are applied the same discounts of the high and low load factor tariffs, except the seasonal rate.

The amount of 6% of seasonal rebate is applied during the period April-September. An additional discount of 2% is granted at the end of the year according to the weight of off-takes from July to September inclusive compared to total annual off-takes, provided it is higher than 19%.

Customers taking gas at more than one site, each of which has an annual gas off-take of more than 10 million cm, provided the sites are located in the same area and are connected to the same gas pipe and provided they accept take-or-pay clauses under contracts with a poly-yearly duration, in addition to standard discounts, can negotiate particular price conditions with Snam.

The rebate function of the load factor is applied to the whole price.

A rebate equal to 1.5% of the amount of the total invoice is obtained on prompt payment.

The discount given to customers who use gas for power generation will expire starting from January 2000. At present it amounts to 1 lit/kWh of electricity produced with gas; for producers using electricity for own use, this discount was increased to 3 lit/kWh from January to December 1997 and to 2 lit/kWh from January to December 1998.

Interruptible Supplies

The conditions for interruptible supplies are demonstration of ability to use another fuel as substitute for gas, minimum off-take of 1 million m3/year and 5.000 m3/day. As interruptibility period, 4,8,12 or 16 weeks may be chosen. The price reduction compared to firm deliveries increases with the length of the interruptibility period. The one part gas tariff is based on a multiplicative formula linking the gas price to the HFO price through a 1-0-1 lag system. Since January 1997 the formula is:

The gas price $p = [0.875 * (BTZcif + M - I * T - Sloc) * (1 + Pr/1200) * Sm * Kstag - Y] * Reg$ [LIT/m3]

M = parameter for variable oil product transportation cost (from refinery to end user) and HFO taxes.

= 103.5

Pr = prime rate

Y = discount equal to 3.5 Lit/cm, which will be reduced by 0.5 lit/cm each month starting from October 1999

I = a rebate term = 8.8

T increases with the interruptibility period: for 4 weeks t = 0; for 8 weeks t = 1; for 12 weeks t = 2; for 16 weeks t = 3.

Sloc = regionally differentiated rebate, for instance 12 LIT in Napoli and Pisa, 8 LIT in Rome, etc.

Pr = Prime rate (inter-bank interest rate) to take account of the time span between delivery and payment.

Kstag: Seasonal rebate of 2.5% for off-take between April and September inclusive

Sm: volume discount according to the following monthly off-take (million m^3):

0-1: Sm = 1

1-2: Sm = 0.98

2-3: Sm = 0.96

>3: Sm = 0.94

Reg: discount for contract compliance (regular payment): 1.25%. An additional discount is granted when the customer has more than one site: 0.5% discount when he has two to four sites and 1% when he has more than four sites.

Discounts for power generation: it was equal to 3 lit/kWh of electricity produced with gas in 1997; in 1998 it is 2 Lit/kWh; in 1999 it will be 1 Lit/kWh, and in 2000 it will be abolished.

In addition to firm and interruptible contracts, SNAM also proposes supplies with planned off-take suspensions, that is contracts that can be considered a mix between these two types of contracts. SNAM and the customer agree on one or two calendar months, from November to March inclusive, in which off-take is totally interrupted or partially reduced. For the signature of such a contract, the following conditions have to be met:

■ contract period of at least 12 months;

■ daily requirement higher than 7.000 cm/d;

■ minimum reduction of the offtakes amounting to 7.000 cm and in any case higher of equal to 60% of the daily requirement.

The price paid under such contracts has a structure similar to the one paid for firm supplies, but during total suspension no standing charge is paid and during partial reductions in off-takes, the standing charge is reduced proportionally. The base value of the commodity charge is calculated according to the formula used for firm supplies, but a variable discount is granted which varies according to each month agreed for the suspension:

November and March: 8% discount
December and January: 10% discount
February: 12% discount.

The discount granted to customers that have regular off-takes amounts to 0.3%, while all the other discounts and price reductions established for firm supplies apply to the supplies with planned interruptions.

Tariffs for Power Generation

Enterprises buying gas for autogeneration of electricity have to buy their gas based on ordinary industrial tariffs, whether they are firm or interruptible. On purchase of gas for power generation, however, a discount is given until end 1999.

For gas deliveries to independent power producers, i.e. generators who sell at least some of their production to third parties, an agreement was signed between SNAM and UNAPACE on 1 June 1998. This agreement expires on December 31, 2006. The tariff under this agreement consists of a standing charge and a commodity charge. The standing charge is related to the daily requirement according to the following formula:

Standing charge = C*I where
I = daily requirement
C = demand charge for the capacity made available on each day - denominated in lira per months per cubic metre. Its base value, fixed in 1998, is updated annually in accordance with a wage index (weight of 45%) and an inflation index of prices applied to industries (weight of 30%). 25% of the value is not indexed.

The commodity charge has the following structure:

> Monthly commodity charge=Summa one to 7 [(P - Di)*Vi] where
> P = base price calculated according to the following formula:
> P = 195 * (0.3 * Gas oil/269,957 + 0.3 * Crude oil / 224,601 + 0.3 * LSFO/178.034 + 0.1)
> Vi=volume of gas within each offtake bracket, of which there are seven
> Di = discount for each off-take bracket varying as follows:

Off-take bracket (million cm/month)	Di (Lit/cm)
0 - 1	0
1 - 2.5	7
2.5 - 4.5	10
4.5 - 13.5	12
13.5 - 27	14.5
27 - 50	23
> 50	35

GAS OIL=average of the three months preceding the month of supply of the cif Med Basis Italy quotations published by the Platt's Oilgram Report, expressed in USD/tonne and converted into LIT/kg suing the exchange rate between LIT and USD of the same period.

CRUDE OIL=average of the three months preceding the month of supply of the fob break-even quotations of a group of eight kinds of crude oils published by Platt's Oilgram Report, expressed in USD/bbl and converted into LIT/kg using the conversion factor of 7.4 bb/t and the exchange rate between LIT and USD of the same period.

LSFO=average of the three months preceding the month of supply of the cif Med Basis Italy quotations published by the Platt's Oilgram Report, expressed in USD/t and converted into LIT/kg using the exchange rate between LIT and USD of the same period.

The coefficients of the oil products and the crude oils with which the commodity charge is indexed vary over the offtake brackets. They are as follows:

Offtake brackets: million m³/month	gi	oi	bi
from 0.0 to 4.5	0.38	0.30	0.32
from 4.5 to 13.5	0.35	0.30	0.35
beyond 13.5	0.32	0.30	0.38

The link to crude oil remains constant, but the weights of gasoil and heavy fuel oil are changed slightly with increasing offtake volumes. The link to crude oil can be explained by the fact that the gas price in Italian import contracts with Algeria is linked to crude oil.

The agreement with the independent power producers provides which sell their production to the public distribution grid provides for a discount on the value of P which amounts to 6 LIT/m3 In 1997, SNAM sold about 8.8 bcm for power generation. The major part of this went to ENEL which has a separate contract with SNAM.

DEMAND FLUCTUATIONS, SECURITY OF SUPPLY CONSIDERATIONS AND TARIFFS

As in most mature gas markets with a considerable share of deliveries going to residential and commercial customers, gas demand in Italy fluctuates heavily over the year. Demand on a cold winter day could be five or six times higher than on a summer day. These demand fluctuations are met by swing in domestic production and in import contracts, by using storage and interruptible contracts. Italy is well endowed with storage and uses its total available volume both for seasonal storage and for strategic storage.

The prices obtained by SNAM are to a large extent the result of negotiations with the various customer groups. The tariffs applied reflect that gas is sold in competition with other fuels. There are few indications that there is explicit thinking about security of supply cost and allocation of these costs to the various customer categories according to who causes these costs. The tariffs for interruptible supplies, however, offer an interesting example of how the tariff structure can offer economic incentives for customers to take their gas in a way that enhances security of supply.

TAXATION AND PRICES OF NATURAL GAS IN RELATION TO OTHER FUELS

The tax system applicable to natural gas is quite complex: taxation differs both between sectors and between geographical locations. The lowest taxes are found in the industrial sector where an excise tax amounting to 20 LIT/m^3 (0.30 ECU/GJ) plus a regional tax of 10 LIT/m^3 (0.15 ECU/GJ) are applied. Gas used for power generation, in refining processes and gas used as raw material in chemical plants are exempted.

Gas used in the domestic sector is subject to different excise taxes, to regional taxes and to value added tax. The level of these taxes differs among uses and regions.

Excise tax amounts to 332 Lit/cm (238 Lit/cm in the South) for residential and commercial customer (with the exception of hotels) while small and large industrial customers pay the same taxes.

Throughout Italy there are different regional taxes whose level can vary between 10 and 60 LIT/m^3 (0.15 and 0.89 ECU/GJ).

The VAT regime on natural gas is differentiated by sectors (and used to be by regions):

Industry	Power gen.	Cooking and water heating	Heating	Commercial	NGV
10%	20%	10%	20%	20%	20%

Source: Eurogas 1998, Snam

The various taxes on natural gas add a considerable element to the price before taxes. The following table shows the composition of the final price of natural gas in Italy on 1 November 1995. The price used is an average of the four tariffs applied by the LDCs:

	LIT/m³	%
Price before taxes	492	51%
Excise taxes	295	31%
Regional taxes	32	3%
Value added tax	147	15%
Total taxes	474	49%
Final price	966	100%

Source: Italgas

Total taxes amount to almost half of the end user price, which is exceptionally high in a European context.

This table shows how natural gas is taxed compared to other energies on a heat equivalent basis (unit: ECU/GJ (NCV):

	Residential	Commercial	Industry	Power generation
HSFO	1.15	1.15	1.15	0.36
LSFO	0.56	0.56	0.56	0.36
Gas Oil	10.78	10.78	10.78	0.34
LPG	4.01	4.01	4.01	-
Natural gas*	5.46	5.46	0.45	-
Coal	-	-	-	-
Electricity	6.44	6.44	6.44	6.44

* = north and central Italy (includes a median of regional gas tax levels)
Source: Eurogas, 1998

In spite of the high level of natural gas taxation, natural gas is favourably treated in relation to gas oil, its main competitor in the residential/commercial sector. The very high level of taxes on gas oil has no doubt facilitated natural gas penetration in this sector. The tax burden on gas in industry and power generation is relatively modest (in fact none in power generation), but gas penetration is assured by higher taxes on HFO and Gas oil.

CROSS-SUBSIDIES

The Italian gas tariff system features some examples of indirect cross-subsidies between customer categories and geographical regions that are quite evident:

- consumers using gas for cooking and hot water pay a much lower price for the gas than consumers using the gas for heating;

- LDCs with a low degree of development in terms of consumption per user, pay a lower price for their gas than companies with a high degree of development. The allowed return of LDCs with a high equity share is higher than for companies with low equity share;

- residential customers in the south of Italy pay lower excise taxes than consumers in the north;

- in the residential/commercial sector, consumers pay high taxes on gas, but on the other hand may be said to benefit from the fact that taxes on heating gas oil are even higher.

COSTS, VALUE ADDED AND PROFITS IN THE GAS CHAIN

SNAM in practice accounts for the entire gas wholesale activity in Italy. Over the period 1993-1995 its gross margin developed as follows (LIT/m^3)*:

	1995	**1994**	**1993**
Average sales revenue	254	243	230
Average gas purchase cost	143	133	133
Average gross margin	111	110	97

* gross margins given for Snam include:
– a significant difference between FOB and CIF import costs since distances between delivery points and the national border is higher than those of other European countries
– a higher internal transportation cost compared to Distrigaz, Gasunie and Ruhrgas due to longer average distances between entry points and customers, and to a less favourable country orography
– a lower load factor in the transportation system caused by the concentration of winter demand peaks in a very short period.

Source: Own calculations based on SNAM annual reports

Italgas buys its gas mainly from SNAM and realised the following gross margins in 1995-1994 (LIT/m^3):

	1995	**1994**
Average sales revenue	460	509
Average gas purchase cost	285	277
Gross margin	175	232

Source: Own calculations based on Italgas 1995 report.

One of the reasons why the SNAM average sales revenue differs from the gas acquisition cost of Italgas, is that SNAM also sells gas to other customers. Italgas profits per cubic metre gas sold in 1995 were about 13 LIT/m^3.

The following table gives an indication of the value added in the gas distribution chain in Italy as it was towards the end of 1995 for gas sold through LDCs:

	LIT/m^3	**In percent of end user price incl. taxes**
Sales price to end consumer incl. taxes	934	100
taxes	443	47
End user price excl. taxes	491	53
LDC gas purchase cost	310	33
Value added in gas distribution	181	19
Gas acquisition cost wholesale level	143	15
Value added wholesale level	167	18

Source: Italgas, own calculations

It should be stressed that these figures represent a snapshot and may not be representative for the whole gas distribution industry since it only comprises the Italgas group.

THE NETHERLANDS

Statistical information

Natural Gas Supply/Demand Balance (Mtoe, 1996):

Indigenous production	68.3
Imports	4.1
Exports	-35.0
Stock change	0.0
Total natural gas supply (primary energy)	37.5
Electricity and heat production	9.7
Other transformation and energy use	1.2
Total industry	8.6
Residential	10.1
Commercial	—
Other	8.3
Statistical difference	-0.4

Natural Gas in the Energy Balance (1996):

Share of TPES	49.4%
Share of electricity output	55.6%
Share in industry	46.3%
Share in residential/commercial sector	67.7%

MARKET AND INDUSTRY STRUCTURE

The Netherlands account for 12.1% of total gas consumption in OECD Europe and is the country in the world with the highest level of gas penetration. It is the biggest producer and exporter of natural gas in Western Europe and is more than self-sufficient. In spite of this it is an importer of natural gas from Norway and the UK and will import gas from Russia in the future. The country also serves as a transit country for gas from Norway to Belgium and France, and gas from the UK to Germany (total transit volumes in 1995: 7.7 bcm). Gas exports in 1995 were 38.4 bcm (62% to Germany, 13% to Belgium, 13% to France, 10% to Italy, and 2% to Switzerland). Slightly more than half of total production is sold on the domestic market. Table 1 shows how domestic sales have developed over the last few years.

Table 1   Gas Sales in the Netherlands 1991-1995 (bcm)

	1991	**1992**	**1993**	**1994**	**1995**
Small consumers*	16.3	15.7	16.4	15.3	15.9
Large consumers	3.3	3.3	3.4	3.5	3.6
Green-house growers	4.4	3.9	4.6	4.4	4.4
Power stations	8.6	8.8	8.7	8.1	7.5
Industrial users**	12.4	11.9	11.9	12.5	13.3
Total	**45.0**	**43.6**	**45.0**	**43.8**	**44.7**

* Consumption < 170.000 m³ a year
** Including Gasunie's own consumption
Source: Gasunie

The overall tendency in gas consumption over the last few years has been that of stability, which reflects the high level of maturity of the Dutch gas market in terms of gas penetration. In the residential sector, around 97% of all households are connected to the gas grid. Due to energy efficiency improvements consumption per household is declining. In 1995, the average household used 32% less gas than in 1980 (2130 cubic metres compared to 3145 fifteen years earlier). In recent years, gas consumption in co-generation has increased, to a certain extent to the detriment of traditional power generation. The share of natural gas in the various sectors is very high: 98% in the household sector, 70% in the commercial sector, 100% in greenhouses, 65% in industry and 55% in power production.

The natural gas market in the Netherlands may be divided into two distinctive parts: the one served by the distribution companies and the one served by N.V. Nederlandse Gasunie (hereafter Gasunie). Figure 1 shows the organisation of the Dutch natural gas market.

Figure 1   Organisation of the Dutch Natural Gas Market

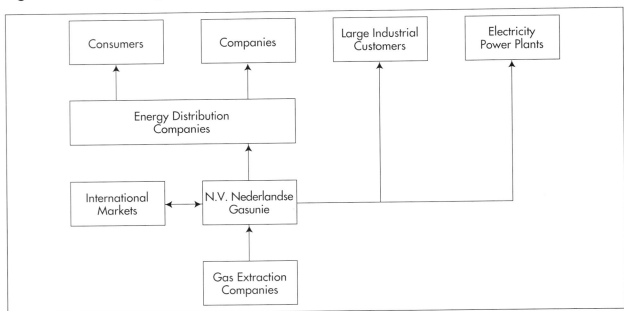

Gasunie is the most important player on the Dutch gas scene. It buys and sells almost all gas produced in the Netherlands.

Following the so-called SEP/Statoil contract - an import contract for gas into power generation - new imports were recently concluded by third parties for direct sales in the Netherlands. The gas will be delivered through the Interconnector.

Gasunie sells gas to the distribution companies as well as to large industrial customers and to power producers. The 34 distribution companies serve residential and commercial consumers, industrial consumers taking less than 10 million cubic metres a year and greenhouse growers. Historically, the border line between customers served by a distribution company or Gasunie has never been precise. Normally customers taking 2 to 5 million cubic metres a year are supplied by Gasunie. For new customers the limit is 10 million cubic metres a year.

In recent years, distribution companies gas sales have developed as follows:

Table 2   Gas Sales by the Distribution Companies 1986-1995 (million cubic metres)

	1986	**1991**	**1992**	**1993**	**1994**	**1995**
Small consumers	14664	15336	14330	15670	14746	15352
Block/district heating	1227	993	929	974	924	917
Green-house growers	3157	4345	4309	4558	4312	4348
Large consumers	3138	3338	3321	3571	3546	3700
Total	**22186**	**24012**	**22889**	**24773**	**23528**	**24317**

Source: EnergieNed

Given the high level of gas penetration in the Netherlands, demand is not expected to increase dramatically over the next few years. Within the planning period adopted by Gasunie, i.e., the period up to 2020, domestic gas demand is foreseen to increase from about 47 bcm in 1996 to 55 bcm in 2020. The strongest growth is expected in the industrial sector where demand is seen increasing from around 16 bcm today to around 22 bcm in 2020. The other sectors are expected to see a relatively modest growth.

THE MAIN PLAYERS IN THE DUTCH GAS INDUSTRY

Gasunie is owned jointly by the state and private companies. The ownership is as follows:

Energie Beheer Nederland B.V.:   40%

ESSO Holding Company Holland, Inc.:   25%

Shell Nederland B.V.:   25%

State of the Netherlands: 10%.

State ownership takes place both directly and through EBN B.V. whose shares are owned by the Government. EBN is a company linked to the chemical group DSM. EBN is (apart from its role as a shareholder in Gasunie) the state's participation vehicle in concessions and licences for the

Table 3   Gas Distribution Companies in the Netherlands, 1995.

Company	Number of Employees	Number of customers			Sales			
		Electricity	Gas	Heat	Electricity ('000 kWh)	Gas ('000 m³)	Share in Gas Sales (%)	Heat (GJ)
1 NV Energie Distributiemaatschappij voor Oost- en Noord Nederland EDON	3,136	849,739	675,848	8,689	9,751,536	2,522,528	10.37	546,679
2 NV ENECO, Capelle a.d. Ijssel	3,435	989,440	852,307	41,971	9,345,169	2,479,751	10.2	3,844,400
3 Energie Noord West NV	3,774	1,122,804	821,871	25,271	11,169,226	2,390,166	9.83	572,000
4 NV MEGA Limburg (Maatschappij voor Elektriciteit en Gas Limburg)	1,675	385,472	351,140	0	6,274,853	1,602,867	6.59	0
5 Nutsbedrijf Westland NV	198	44,603	45,456	0	729,938	1,502,679	6.18	0
6 NV NUON Energie-Onderneming voor Gelderland, Friesland en Flevoland	2,959	1,078,066	406,227	35,054	10,538,200	1,480,889	6.09	1,495,446
7 NV Provinciale Noordbrabantse Energie-Maatschappij (PNEM)	2,448	871,274	431,269	30,303	10,846,266	1,480,246	6.09	2,886,479
8 NV GAMOG Gasmaatschappij Gelderland	408	0	322,502	0	0	1,125,056	4.63	0
9 NV Energie- en Watervoorziening Rijnland (NV EWR)	816	232,303	222,810	4,402	1,791,876	930,393	3.83	480,195
10 Energie Delfland NV	370	108,610	101,566	0	1,020,000	857,000	3.52	0
11 Maatschappij voor Intercommunale Gasvoorziening n Oost-Brabant (OBRAGAS NV)	167	0	175,171	0	0	759,283	3.12	0
12 Gasbedrijf Centraal Nederland NV (GCN)	204	0	189,625	0	0	631,552	2.60	0
13 NV Nutsbedrijf Regio Eindhoven	550	93,111	175,449	0	748,605	584,896	2.41	0
14 NV Maatschappij tot Gasvoorziening Gelders Rivierengebied (GGR-Gas)	154	0	97,739	0	0	533,647	2.19	0
15 NV Regionale Energiemaatschappij Utrecht (REMU)	1,354	461,063	192,729	32,333	3,964,839	506,944	2.08	2,671,000

Company	Number of Employees	Number of customers			Sales			
		Electricity	Gas	Heat	Electricity ('000 kWh)	Gas ('000 m³)	Share in Gas Sales (%)	Heat (GJ)
16 Maatschappij voor Intercommunale gasdistributie Intergas NV	151	0	132,155	0	0	489,563	2.01	0
17 NV Delta Nutsbedrijven	904	180,247	169,851	0	5,161,491	475,341	1.95	0
18 Centraal Overijsselse Nutsbedrijven NV (COGAS)	249	47,372	119,300	851	484,755	451,249	1.85	25,254
19 NV FRIGEM	233	44,401	114,877	969	314,500	425,259	1.75	32,000
20 Nutsbedrijf Amstelland NV	154	0	59,341	0	0	412,901	1.7	0
21 NV Regionaal Energiebedrijf Gooi en Vechstreek (REGEV)	207	38,667	108,860	0	177,412	359,062	1.48	0
22 Energiebedrijf Midden-Holland NV	254	82,193	75,985	0	630,317	342,177	1.41	0
23 NV RENDO Regionaal Nutsbedrijf voor Zuid-Drenthe en Noord-Overijssel	149	27,533	90,846	0	223,875	323,664	1.33	0
24 NV Nutsbedrijf Haarlemmermeer	79	0	42,306	0	0	232,093	0.95	0
25 Gasdistributie Zeist en Omstreken	84	0	58,166	0	0	206,190	0.85	0
26 NV Gasbedrijf Noord-Oost Friesland	92	0	54,147	0	0	183,016	0.75	0
27 NV Intercommunaal Gasbedrijf Westergo	92	0	56,788	0	0	176,912	0.73	0
28 NV Energiebedrijf Rijswijk-Leidschendam (ERL)	100	40,668	40,117	0	369,700	164,900	0.68	0
29 NV Nutsbedrijven Maastricht	197	43,032	45,513	184	391,246	152,342	0.63	11,943
30 Gasbedrijf Midden Kennemerland NV	67	0	49,014	0	0	141,052	0.58	0
31 NV Nutsbedrijf Heerlen	163	45,695	44,016	1,876	399,407	134,179	0.55	136,988
32 NV Energiebedrijf Zuid-Kennemerland	67	11,441	28,496	0	64,262	102,556	0.42	0
33 NV Openbaar Nutsbedrijf Schiedam	222	36,631	42,456	0	259,507	91,079	0.37	0
34 Nutsbedrijven Weert	66	18,797	17,480	0	179,365	65,978	0.27	0
TOTAL	**25,177**	**6,853,162**	**6,411,420**	**181,903**	**74,836,345**	**24,317,410**		**12,702,384**

Source: EnergieNed

production of gas. It was set up after the privatisation of DSM 1989. The state's energy interests were incorporated in the new company, EBN, and in return for management and the performance of other functions, the state pays DSM a fee. Established as a public limited company in 1963, Gasunie's relationship to the state is regulated through a private law agreement. Strictly speaking, the company is a trading and distribution company entitled to a margin to cover costs and a fixed profit. It needs Government approval for the prices it applies, for major investments, export/import contracts and for the 25 year marketing plan. Gasunie is a private law "vennootschap" which is important in terms of the sharing of its profits.

SEP, a co-ordinating body of the four regional electricity generating companies in the Netherlands is a major player in the Dutch gas markets through its responsibility for fuel purchases for the power sector including gas. In addition to being a major gas customer, SEP is also an importer of gas from Norway by virtue of a contract signed in 1989. This contract is special in that the price of the gas is linked to the coal price and inflation and the volumes dedicated to specific combined cycle power stations.

The 34 distribution companies in the Netherlands serve the segments of the gas market normally associated with gas distribution.

Only 11 of the 34 distribution companies are pure gas companies. These eleven companies account for only 20% of total distribution company gas sales and 19% of the total number of gas customers. The biggest gas distributors are found among the horizontally integrated companies, i.e., companies distributing also electricity and/or heat. The seven largest ones, which all distribute more gas than any of the pure gas companies, account for about 55% of total gas distribution.

18 of the Energiened companies, most of them medium and small companies are members also of a loose association called Enercom, an organisation created to increase the influence of small and medium sized distribution companies. Its members generally want to increase competition in the gas market more slowly than some of the biggest companies.

The present structure of the distribution sector is the result of a dramatic restructuring process that has taken place over the last ten years. In 1985, the number of companies was 158, which means that about three fourths of the companies at the time have now disappeared. Consequently, the average size of the companies is four times larger now than ten years ago. Furthermore, there has been a sharp increase in the degree of horizontal integration (see below).

Before 1985, distribution companies were normally under direct municipal authority. Today all of them except two are exploited as public limited companies (plc's) but the shares of these companies are still in the hands of municipal or provincial authorities. Thus, while the legal form is indeed privatised, authority has remained with local/regional government. The companies are organised in the form of Naamloze Venootschap or NV and are as such regulated by the provisions of the Dutch Civil Code. As a shareholder in an NV, local governments have considerably less influence on the company than when they were under direct municipal authority. Being run as private companies, the enterprises have become more customer and market oriented.

There is in principle no obstacles to private shareholding in the distribution companies, but so far this is almost non-existent (although it is quite widespread in some of the affiliates of the companies, for instance in new activities taken up by the distribution companies). There will probably be an increasing trend towards privatisation, but what the Government hopes is that this will take place only when the market has been opened up and the necessary regulation put in place.

LDC SERVICES

The LDCs in the Netherlands, in spite of recent restructuring and rationalisation, go on delivering the energy they were originally established to supply. The pure gas distribution companies still deliver gas but have also taken up other services. The LDCs that distribute electricity now all also distribute at least one more form of energy. A lot of the companies have in fact grown into conglomerates of different activities: in addition to distributing energy, many of them have also got involved in telecommunication (for instance distribution of TV signals and telephone services), waste treatment and water distribution in addition to activities closer to their core business like rental of appliances and energy management consulting services. Some of the companies have also taken up international activities through investments abroad.

The tendency towards restructuring and increased horizontal integration is not a very recent phenomenon: it started in the electricity supply sector shortly after the first oil crisis when it was realised that electricity prices in the Netherlands had to come down through rationalisation of the sector, including separation of generation and distribution. In this process the large electricity companies which provided both production and distribution services were obliged to abandon their production companies. The considerable financial reserves which had been built up, remained in the hands of the electricity distribution companies, enabling them to compensate for their loss of production facilities through purchases of small gas and/or electricity distribution companies.

Meanwhile, the Minister of Economic Affairs, supported by Parliament, had come to the conclusion that increase in scale and, above all, horizontal integration were desirable goals for energy distribution companies. Legislation to enforce this policy was put forward by Parliament which obliged energy distribution companies to become horizontally integrated and serve a minimum number of customers. The arguments put forward by the government in favour of horizontal integration were those of higher convenience and efficiency for the consumer (relationship to only one company, one meter reading, one invoice, etc.) and stimulation of a more balanced evaluation of energy options by the consumer. As observed by one commentator within the industry, however, the drawbacks of such integration were given remarkably little attention. One of the arguments against was that reinforcement of the monopoly position of the distribution companies would lead to the disappearance of competition between gas and electricity. This would put a brake on innovation, besides entailing an evaluation of energy options subject to no external control as well as the possibility of obscuring costs and of see-sawing between gas and electricity tariffs.

Based on government recommendations in favour of horizontal integration and minimum size, the three associations for gas, electricity and heat companies were urged to elaborate a detailed plan for reorganisation. In 1987, regional plans were consolidated into a national plan under which 56 companies should still remain. To foster implementation of the necessary mergers, the three organisations set up a special guidance committee. Thanks to the clear visions on the part of the Government, a co-ordinated effort on the part of the distribution company associations and to the fact that the larger companies were ready to pay a good price for smaller, municipal companies, a much smaller number of companies has emerged. There are, however, a lot of signs that the present situation with 36 distribution companies is still not stable. Once a stable regulatory framework has been put into place, both privatisation and mergers between companies are probable. Under a new competitive regime the influence of local governments will generally decrease, and some of them will probably consider selling their shares.

REGULATION

The Government exercises an influence on all aspects of the gas industry directly and indirectly through three organisations: EBN, Gasunie and the energy directorate of the Ministry of Economic Affairs. The Government bases its intervention on several agreements made with the principal actors in the gas business. Firstly, there is a joint venture agreement of 1963 ("Maatschap") between DSM (now EBN, state owned company), Shell, Exxon and NAM (owned by Shell and Esso). The agreement established Gasunie as the central entity for marketing and transporting gas in the Netherlands and NAM as the body in charge of production of Groningen gas. The principle was set out that all gas produced in the Groningen field was to be sold to Gasunie.

Secondly, an agreement was concluded between the Government and Gasunie which was thought necessary in the absence of specific legislation on the gas sector. The agreement obliges Gasunie to draw up a plan for the sale of gas for a period of 25 years. This plan is drafted each year and submitted to the Minister of Economic Affairs for approval. The plan may be seen as the source of an obligation on Gasunie to supply gas to the Dutch market and especially the household sector. A statutory obligation to supply gas may seem unnecessary since a guarantee to supply is already imposed on Gasunie through the long term supply contracts concluded between Gasunie and the distribution companies.

Imports

In principle, there are no legal barriers to the import of natural gas from outside the Netherlands. The electricity sector currently imports gas from Norway. UK-Interconnector gas has been contracted by various parties. Under the terms of agreement between Gasunie and the State, Gasunie needs prior consent of the Minister of Economic Affairs for any import contract.

For the distribution companies, the scope for imports of gas is limited by the terms of the contract they have with Gasunie, the so-called Standard Agreement, to which all the distribution companies are committed. The agreement requires Gasunie to supply unlimited quantities of gas and the companies to purchase their gas for a determined period of time (see below).

Gasunie no longer has a first right of refusal for the purchase of gas produced in the Netherlands. So far, however, for economic reasons producers seldom sold gas to another buyer.

Transportation and Storage

In principle, the construction of pipelines is not subject to a legal restriction such as a concession requirement and there is no exclusive statutory right to transport. A concession was granted to Gasunie on 12 December 1963 but the Government expressly reserves the right to grant similar concessions to other parties.

The main pipeline network is owned by Gasunie. In the south of the country, new investment in a high pressure grid is being done by other parties. In recent years, there have been demands from distribution companies and large industrial users for open access to the transportation system. This issue is under consideration in the context of the reform work going on, and it clearly is the intention of the Government to facilitate third party access to the transportation grid, even at the distribution level. Gasunie is already offering negotiated access to its transportation system. In anticipation of further market opening Gasunie recently announced the introduction of a new pricing system for transportation and related services for large consumers.

Some of the main offshore gas pipelines, owned by oil companies, already have third party access.

At present, there are no legal provisions applicable to underground storage. However, it is generally accepted that underground storage in mines, which includes empty gas fields and caverns in salt domes, should be governed primarily by mining legislation. There are plans to further develop this view by means of a royal decree based on the above law. It seems that a licensee (for production) will become entitled to build and operate storage facilities, but that a system of permits will be introduced in those cases where the mine and the goods to be stored differ, for instance the storage of gas in salt domes.

Distribution

The "Wet Aardgasprijzen" (1974) is aimed at regulating the price of gas throughout the gas chian in special circumstances, and has only been used a few times during the 1980s. There is also the "Wet Energiedistributie" (1997) that sets general rules for distribution companies but does not regulate natural gas distribution in detail. Apart from these laws, there is no legislation expressly aimed at regulating the sale and distribution of natural gas. Some of the LDCs have concessions, but the concession system has not been consistently developed all over the country. The "Wet Energiedistributie" rules out the possibility for local authorities to give concessions. Therefore, the existing concessions lost their significance. Today, the practice is for the gas distribution companies to negotiate their gas purchases exclusively with Gasunie. They do so within the framework of the Standard Agreement referred to above. The price clause in the contract is renegotiated once every three years. It requires the LDCs to purchase their gas exclusively through Gasunie, and Gasunie to supply any demand generated by the LDCs. The LDCs presently have a de facto exclusive right to sell gas in their area, but this will probably also change in the future. There is no legal obligation for gas connection, but the LDCs have pledged to connect any customer within 25 metres from the mains.

For small consumers, gas is supplied on the basis of a standard contract called Model Algemene Voorwaarden Gas of 1994. This is drawn up under the auspices of two consumer committees of the SER (Social Economische Raad). In case of dispute, there is a special arbitration body set up in the Hague. For large consumers a similar arrangement is under preparation.

In the Netherlands, there are eight LDC selling more than one billion cubic metres a year, the three biggest ones sell about 7.5 bcm altogether, thus being sizeable actors in the market. Some of the LDCs have already, others are interested in arranging for alternative supplies. There is presently a rolling contract between EnergieNed and Gasunie which obliges the LDCs to take industrial volumes from Gasunie for 5 years and volumes for small users for 10 years.

Future Trend in Natural Gas Regulation

In the Third White Paper on Energy Policy published by the Dutch Government in early 1996 promotion of liberalised markets in electricity and gas is one of the focal points. More market and less central planning seem to be the main catchwords, in particular concerning electricity. Concerning gas, the paper states that the European Directive on the internal market in natural gas will determine the form and introduction of TPA in the Netherlands.

The following table summarises the proposed changes to the statutory framework for the gas industry:

Natural Gas	Situation today	In the future
production	free (based on permit)	free (based on permit)
transmission	not regulated	free, non-discriminatory grid access, independent monitoring
trade	producers free, no legal impediments for consumers	– producers free, non-captives free, captives protected (coverage plans, maximum tariffs)
new distribution grids	not regulated	regional decision based on nationwide criteria

Source: Third White Paper on Energy Policy, 1996.

The Government proposals have raised some concerns in the LDCs. According to EnergieNed, the major issues seen from the LDCs point of view are:

■ separation of natural monopolies into natural monopoly functions and contestable functions;

■ definition of the distribution network management function to avoid inefficient networks;

■ introduction of a spot market with equal access for all;

■ free choice for consumers in stages;

■ prevention of abuse of dominant market position;

■ import reciprocity;

■ the establishment of a supervisory body to undertake light handed regulation.

The LDCs are aware that the Government's proposals imply third party access to their distribution networks. In general, they see their risk increasing as a consequence of the new proposals, although it has to be admitted that their present risk is very low (no volume risk in the contract with Gasunie, security of supply assured by Gasunie, generous margins in a monopoly environment, etc.). Some of the new risks as perceived by the LDCs are:

- New entrants - lower barriers to entry

- Regulatory risk

- Competition on services

- Possible change in ownership

- Price risk - no longer more or less guaranteed margins

- Risk of predatory pricing

- Increased supply risk.

Present Contractual Arrangements

The most important single contract as far as gas distribution activity is concerned is obviously the standard contract between Gasunie and EnergieNed. In practice, however, there are individual contracts between Gasunie and each of the LDCs. The contracts provide for different duration for industrial volumes (five years) and for volumes to small customers (fifteen years). In the future, contract duration will probably be reduced. The contract is an example of "streaming" of gas in that LDCs take four different streams of gas (small consumers, large consumers, green houses, combined heat and power) at four different prices at citygate. There is no take-or-pay obligation on the LDCs - in fact they have unlimited volume flexibility at no extra cost. Prices are renegotiated every three years.

Industrial customers taking more than 10 million cubic metres a year (CHP more than 50 million cubic metres per year) are served directly by Gasunie. The LDCs do not, however, lose a customer when his demand exceeds this volume. The general rule for industrial customers is a contract duration of five years, but a few have ten year contracts. More relevant, however, is the notification period after the original contract period. By far, most contracts are "old " contracts. The notification period is then usually one year; in some cases 3 years. Industrial customers are not faced with any take-or-pay obligation but are obliged to take a relatively small volume to cover connection costs.

Gasunie has several contracts for gas to power generation with SEP. The main contract expires in 1999. The buyer also has total volume flexibility under this contract.

Fluctuations in Demand and Security of Supply Considerations

As illustrated in figure 2, fluctuations in gas demand over the year in the Netherlands follow the usual pattern of any country which has a high share of gas consumption in the residential/commercial sectors.

The approach to tackle load factor and security of supply issues, however, has been somewhat special in the Netherlands compared to most other countries. To understand how costs related to these issues have been dealt with, some background is needed.

Up to very recently, the Netherlands have needed very little storage because of the inherent flexibility in the production from the Groningen field and also in the production from the other onshore and offshore fields. Necessary load factor is taken care of by using this flexibility and

Figure 2   Monthly Natural Gas Flows, Netherlands, 1984-1995

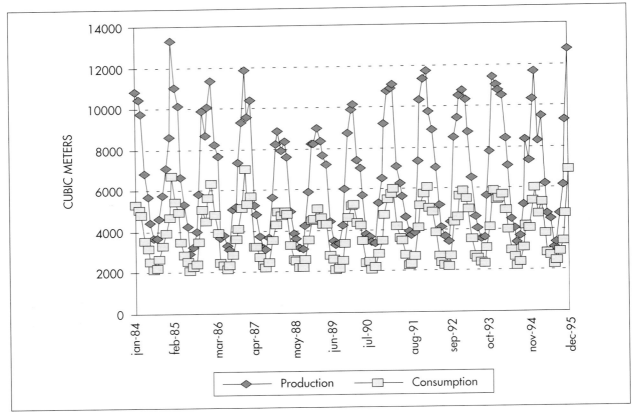

operational linepack in the pipeline system. This means that no extra storage facilities have been necessary. Since the country is more than self-sufficient in gas there has been no need to build storage for strategic security reasons. One might say that the only extra load factor and security of supply cost the gas industry in the Netherlands has incurred is that of dimensioning the pipeline system in relation to a situation where deliveries over the year are even. The pipeline system is designed to be able to satisfy demand in situations where the outdoor temperature goes down to minus 17 degrees C on a daily average. The gas companies do not offer interruptible contracts, but there is an agreement with the electricity sector that part of the deliveries to this sector can be interrupted when the temperature goes below a certain level. Under certain circumstances in the past, it has also been possible to interrupt supplies to some large industrial customers. It is clear that since residential customers requires more load factor, they are in principle more expensive to serve than for instance industrial customers. In the Netherlands, these two customers categories effectively pay different prices. This is, however, not dictated by different costs, but by the market value principle which leads to higher prices for residential consumers because the alternative mostly is gasoil, which has a higher price than heavy fuel oil, the alternative in the industrial sector. As will be pointed out below, no attempt is made to allocate transportation and distribution costs according to the share of costs caused by the various customer categories.

The situation described above is changing: the pressure in the Groningen field is decreasing, and Gasunie is building new storage capacity to ensure sufficient load factor capacity at a considerable cost (three different projects will add a storage capacity of 4 to 5 bcm). This along with possible unbundling of the services offered by Gasunie (for instance requiring Gasunie to offer transportation services to third parties based on published tariffs) may lead to a more explicit thinking in terms of cost estimation and allocation.

Another special feature of the gas supply system in the Netherlands is the mix of different gas qualities: the gas from the Groningen field is low calorific gas, and the gas from North Sea fields as well as from other on-shore fields is high calorific gas. This necessitates two separate pipeline systems and blending of gas. The cost of gas blending is relatively modest (the cost of adding nitrogen to high calorific gas to make low calorific gas, the type of gas distributed by LDCs) is 0.5 to 1 cent NLG per cubic metre.

PRICE REGULATION AND TARIFFS

Under an agreement of 1963 between Gasunie and the Dutch Government, the Minister has the power to approve the selling prices and destinations of natural gas charged by Gasunie to its customers. Under the Natural Gas Prices Law (Wet Aardgasprijzen) of 1974, the Minister is authorised to lay down minimum prices if prices agreed upon by the parties are considered not to reflect the market value for producers' natural gas supplies and for other supplies to distribution companies for the home market and abroad (although this power has not been used since 1985). The Minister also has the power to make binding recommendations to individual gas companies on their tariffs, having recourse to Wet Aardgasprijzen, art. 5, but only if the tariffs are "contrary to the interest of a good energy supply". So far, this article has never been used by the Minister. Apart from this, there are no specific legal instruments regarding prices applied by the LDCs.

Pricing in the Dutch gas sector is primarily based on the market value principle, which means that the price of gas reflects the market value of alternative fuels, which most of the time means petroleum products. Use of the market value principle has been a way of optimizing the state revenues from the gas activity. The price paid by the distribution companies for their gas was until recently negotiated between Gasunie and EnergieNed. The price paid to Gasunie is set as a function of the end user market values for the various end uses (although it is claimed that the full potential in terms of gas prices is not exploited). From the end user prices two elements supposed to cover operating costs and profits are deducted to arrive at the price paid to Gasunie by the distribution companies (the city gate price). In this way, one may say that the distribution companies work on a cost plus basis since this system allows them to cover their costs.

Price to Distribution Companies

The distribution companies pay a price to Gasunie for their gas which depends on the type of customer that is to offtake the gas. The price is a function of the size of the customer to be served, independent of the use of the gas (except for the volumes for greenhouses and CHP). It consists only of a proportional element (commodity charge) and is linked to either the GO price (for small consumers) or the HFO price (for large consumers) through a multiplicative price formula as described in table 4.

Table 4   Supply Price Formulae for Distribution Companies (1996)

Zone	Cubic metres/year	Commodity charge (NLG cent/cubic metre)
a	0-170.000	$(G/CVa)=CVb-Ma$
b	170.000-3 million	$(P/CVa)=CVc-Mb$
c	3 Mio.-10 million	$(P/CVa)=CVd-Mc$

Source: EnergieNed

G: price of GO

P: price of HFO

CVa = first calorific value element

CVb = second calorific value element for zone a

CVc = second calorific value element for zone b

CVd = second calorific value element for zone c

Ma = margin for zone a

Mb = margin for zone b

Mc = margin for zone c

For the small consumers, market prices are adjusted twice a year, in January and July. A dampening factor is built in to reduce the effect of excessive fluctuations in oil prices, and the maximum permissible price change is limited. For the large consumers, the price is adjusted every calendar quarter. The gas price charged follows the oil product prices with a lag.

Since there is a separate price for each type of customer in volume terms ("streaming"), the average gas acquisition price of an LDC will depend on the composition of its customer base. At the end of every year the LDCs have to declare the exact composition of their gas sales. The price in zone a, linked to the GO price is substantially higher than prices in the zones with HFO related prices (as an indication, the average purchase price in zone a in 1995 was around 30 cents per cubic metre, whereas the price for the lowest volume in zone b was around 23 cents). When negotiating the prices for the different zones the differing distribution costs are taken into account. This shows clearly in the price zone a which is much higher than prices in the other zones since distribution costs are higher for smaller consumers. The relationship between GO-related and HFO-related prices will, however, typically vary over time.

Industrial Tariffs

As the above table describing the zones indicates, the LDCs are supposed to serve customers taking up to 10 Million cubic metres a year. There are, however, lots of examples of exceptions. This is reflected in the fact that Gasunie's tariffs for sales to the industrial sector also contain prices for lower volume tranches. As shown in table 5, the structure of the industrial prices are very similar to the wholesale prices applied for the LDCs.

Table 5 Gasunie Prices to Industrial Customers (1996)

Zone	Cubic metres/year	Commodity charge (NLG cent./cubic metre)	Standing charge (NL G/year)	Price 4th quarter 1996 (ex.taxes)
a	0-170.000	G<550:[(0.8*G+0.2*550)/500]*37.2+1.70 550<G<750:(G/500)*37.2+1.70 G>750:[(0.8*G+0.2*750)/500]*37.2+1.70	82	41.039
b	170.000-3 Million	(P/500)*38.2+7.35		24.944
c	3 Million-10 Million	(P/500)*38.2+3.60		21.194
d	10 Mio.-50 Million	(P/500)*38.2+1.80		19.394
e	above 50 Million	(P/500)*36.3+1.75		

G: the value of gasoil including excise duty, COVA surcharge and the surcharges for the costs of trade and transport.
P: corresponding value for HFO
Source: Gasunie

In zone a, the price is linked to the GO price. Price fluctuations are limited to 3 NLG cent/cubic metre semiannually. When the G value is outside a band between 550 and 750 the price is tempered by the formula in that the gas price does not follow the gas price completely.

For large customers the price is linked to the HFO price (in zone e it is linked to the HFO price at 95%). The price declines with increasing volumes. Although differing load factors are not directly taken into account through the formulae, there is an additional price element (albeit very small) to be paid by customers with low load factor having an annual consumption above 1 Million cubic metres. These customers have to pay a surcharge that can have a maximum value of a little less than 1 cent per cubic metre.

In principle, Gasunie does not offer interruptible contracts because the seasonal and daily flexibility in the delivery system is sufficient.

In addition to the tariffs in the table, Gasunie also offers a special feedstock tariff to large industrial customers, especially fertiliser producers. This tariff implies a rebate to such customers compared to tariffs paid by other large consumers. Customers in the provinces of Groningen, Friesland, Drenthe and Overijsel benefit from a rebate of approx. 0.75 NLG cents per cubic metre. This rebate is not allowed to surpass 5% of the total net price.

As from 1 January 1999, Gasunie will apply a new tariff structure to industrial customers and LDCs taking over 50 million cm per year (from 1 January 2000 over 10 million cm per year) with distinct price elements: commodity, transport, services.

Price to Electricity Producers

Gasunie has only one contract for sales of gas to power generation which means that all power stations pay the same price for their gas. The price formula used is similar to the zone e price for industrial deliveries but the price is somewhat lower and the lag structure is different. In 1995, a rough indication of the gas price paid by the electricity industry would be about 18 cents per cubic metre (over the last few years this price has been as follows: 18.3 cents in 1992; 18.6 cents in 1993; 17.6 cents in 1994).

Under a contract between SEP and Norwegian gas sellers to supply gas into the combined cycle facilities in Eemshaven, the gas price has a completely different structure. It is based on the so-called "indifference" principle: in the price formula gas fired and coal fired facilities in terms of capital costs, operation cost and fuel costs are compared. A gas price is paid which would make the buyer indifferent between a gas fired and a coal fired facility. The base price arrived at is escalated with an international coal price and inflation. Through this mechanism the buyer is guaranteed to remain competitive vis a vis the coal alternative. The base price in this contract is substantially higher than the contract with Gasunie, one of the reasons being that coal prices are supposed to develop more evenly over time.

End User Prices in the Distribution Sector

The end user prices charged by LDCs consists of the supply price paid to Gasunie (described above) plus an element to cover distribution cost plus a premium to cover risk and profits. Technically the selling price has three components:

■ Connection charge, which is a one-off contribution the cost of connection to the gas system. Enegiened sets a recommended charge annually. The charge for small consumers reflects the actual cost of connecting the property to the mains and installing a meter. The charge paid by large consumers is based on a cost-effectiveness calculation and theoretically applies for the duration of the contract.

■ Standing charge, which is a flat rate annual charge paid by all consumers to cover the cost of maintaining their connection to the mains. EnergieNed sets annual maximum and minimum recommended charges.

■ Consumption charge.

This part of the gas price is divided into graduated series of tariff zones. Under this system, consumption up to the first zone boundary is charged at the appropriate rate even if the total amount consumed is greater. Any consumption within the second zone is then charged at the rate for that zone, and so on. An exception, however, is that greenhouse growers pay the greenhouse tariff for the whole volume once they take more than 30 000 cubic metres. Apparently the gas price decreases dramatically when consumption goes beyond 170.000 cubic meters since the price is then linked to HFO and no longer to GO. This fall is, however, mitigated by the fact that the b-tariff has a considerable premium element. The consumption charge covers the purchase price of the gas and the distribution system cost. The zones are as follows:

Zone	Cubic metres per year	Category
a	0-170.000	small consumer
b	170.000-3.000.000	
c	3.000.000-10.000.000	large consumer
d	10.000.000-50.000.000	
e	50.000.000 or more	

The LDCs are in principle free to set their prices but will normally stay within the tariff intervals recommended by EnergieNed. Enegiened consults Gasunie before setting recommended prices for zones a and b. The intervals defined by EnergieNed, however, give some liberty in terms of price levels: for deliveries in zone a during the first half of 1996, the minimum price per cubic metre (the commodity charge) was 36.2 NLG (ex. Taxes) cents whereas the maximum price was 42.6 NLG cents, i.e. A difference of 17.6%. The standing charge in zone a could vary between 84 and 148 NLG per year, i.e. A difference of 76%.

Tariff for customers in this category varied from about 42.5 cents to about 51.5 cents with an average of some 47 cents (VAT excl., other taxes incl.). This means that the LDCs have significant possibilities to influence their economic result by changing their tariffs.

EnergieNed and Gasunie also agree a recommended price for sales in zone c; the LDCs are obliged to keep their zone c prices within a band around the recommended tariff. Mandatory uniform prices are set by Gasunie for zones d and e, in consultation with EnergieNed, which means there is room for negotiation only on the margin. The end user prices charged by the LDCs basically have the same structure as the industrial prices charged by Gasunie. Table 6 shows these prices as they were in early 1997:

Table 6   Tariff Recommendation from EnergieNed for Users Taking Less than 10 million Cubic Metres (first quarter 1997)

ZONES	Prop. Element A:0-170 000 M3	Prop. Element B:170 000 3 million cm	Prop. Element C:3 10 million cm	Fixed Element NLG/YEAR
Max. E.ned	44.9	27.1	23.4	154

Source: EnergieNed

Max. E.Ned: Maximum tariff as determined by EnergieNed.

Some observations linked to the table:

■ The commodity charge for small users follow the GO-price according to a 6-2-6 system.

■ The LDC prices to customers taking more than 170 000 cubic metres, are, as the Gasunie prices, linked to the Rotterdam 1%HFO price through a 6-0-3 system. EnergieNed's tariff recommendation varies within a narrow band for all industrial customers. The Gasunie prices normally are between the minimum and maximum prices recommended by EnergieNed.

■ Even in the small consumer category there is no distinction between different gas uses - all uses pay the same price. There are, however some exceptions to this rule in that there are special tariff for collective heating, market gardeners, co-generation plants and ammonia producers:

■ Collective heating complexes are charged at the small consumer rate, irrespective of the volume involved. This is to prevent a discrepancy in the price of domestic heating developing between collectively heated properties and individually heated properties. Consumers in collectively heated homes do, however, pay lower standing charges: these are reduced by 75-80% in relation to other small consumers.

■ Greenhouse growers benefit from a preferential tariff once they use more than 30.000 cubic metres a year. At a consumption of less than this volume, the ordinary small consumer tariff has to be paid. Once consumption surpasses 30.000 cubic metres, the special tariff is paid on the whole volume.

Since the alternative fuel in most greenhouses would be HFO, the special tariff is linked to this fuel (the exact formula being: $P/500*38.2+3.95+0.4$). In the second quarter of 1996, small consumers paid a consumption charge of 37.7 cents per cubic metre, whereas a market gardener has to pay 20.6 cents per cubic metre. The fact that under the special tariff a higher standing charge is paid, only marginally change this relationship since it accounts for only fractions of a cent per cubic meter.

■ Co-generation plants pay a HFO-related price that has been negotiated between EnergieNed and Gasunie. Since the tariff zoning system tends to discriminate against co-generation plants, particularly the small ones, it was found necessary to invent a special tariff for such installations in order for the electricity produced to be competitive. The price paid is somewhat (about 1 cent per cubic metre) lower than the market gardener tariff. No standing charge is paid.

■ As indicated above, ammonia producers and some other industrial customers using gas as feedstock benefit from a special tariff.

Cost Allocation and Reflection in Tariffs

The Dutch gas tariff system is primarily based on the market value principle and have few if any direct links to cost. There is an element of cost plus thinking in the Gasunie philosophy of pricing gas to the LDCs in the sense that their margin should cover distribution cost plus profits. In addition to the margin between average sales price and average gas acquisition cost, there is a direct contribution from Gasunie of 80 NLG per connected customer. This contribution is supposed to cover part of the fixed costs in the LDCs.

The tariff structure as such makes a distinction between fixed and variable elements, but the fixed elements are very low and are not at all geared at recovering the full share of fixed costs in transmission or distribution.

Cross-Subsidies

Similar to other countries, natural gas in the Netherlands is priced until now on the basis of market value, Prices, therefore, do not necessarily reflect the cost of serving the customer. With the low production and distribution cost in the Netherlands this has never been an issue, and all customer categories contribute positively to the coverage of fixed cost. However, with further market opening, there will be a need for clear cost allocation. The structure of the Dutch gas sector makes the calculation of the exact cost for each customer category difficult. E.g., how should the cost of using the flexibility in the Groningen field be calculated? Is the only cost that is attributable to load factor for the various customers overdimensioning of the transmission and distribution system in relation to a situation where deliveries are flat over the day and the year?

The contribution from Gasunie to the LDCs of 80 NLG per connection (which means a transfer of more than 500 million NLG) in 1995, might, if it is not a subsidy, be considered as a distribution of rent from one part of the gas chain to another. It is claimed, however, that the contribution is neither a subsidy or a redistribution of rent since it is cost related: it is in fact part of the margin earned by the LDCs and is a way of dividing the total margin into a fixed component and a variable component.

Customers in the four provinces of Groningen, Friesland, Drenthe and Overijsel benefit from a rebate of 0.75 NLG cents per cubic metre, which should not surpass 5% of the total net price. This is a rupture with the general principle of equal treatment otherwise applied in the

Netherlands, but may be defended by the fact that it reflects the lower cost of transporting gas to these provinces because they are located close to the Groningen field.

Large industrial users taking gas as feedstock, for instance for fertiliser production, benefit from a gas tariff which gives a lower gas price than the normal industrial gas price for the same volume. This special gas price reflects the absence of excise levy on oil feedstocks. For the period March- October 1992 an extra discount amounted to 0.88 cent per cubic metre and for the period May-July 1993 it amounted to 0.58 cent per cubic metre. The European Commission has accepted this tariff because it allows Dutch fertiliser producers to compete when market conditions for such producers are difficult. The same may be said about the preferential treatment of greenhouse growers: they both benefit from a special tariff and a lower rate of VAT.

Taxes and Prices in Relation to Other Fuels

Four different types of taxes are levied on gas in the Netherlands:

■ EAP (Environmental Action Plan) surcharge varies between 0.5% and 2.5% of the sales price ex. VAT and applies to small supplies tariffs, block/district heating tariff zone a,b,c; industrial advice tariff for CHP (0-10.000.000 min cubic metres and market gardeners tariff (category I)).

■ Fuel tax – applies to all consumers. The rate is 0.02155 NLG per cubic metre (excl. VAT) for consumption up to 10 Mio cubic metres and 0.01410 NLG per cubic metre beyond this volume.

■ Eco tax – applies to small supplies tariff, block/district heating tariff and zone a industrial advice tariff. There is no ecotax on the first 800 cubic metres consumed. The rate is 0.03200 NLG per cubic metre. The rate will increase to 0.0640 NLG in 1997 and to 0.0953 NLG in 1998.

■ Value Added Tax (VAT) – the rate is the same for all consumers (17.5%) except market gardeners who pay only 6%.

Taxes account for a considerable share of end user prices. The following figures show the approximate percentage of taxes in end user prices in some of the customer categories (1st quarter 1996):

Customer category	Total taxes as percentage of end user price* ex.taxes
Zone a customers	36%
Zone b customers	30.4%
Zone c customers	32.3%
CHP customers	33.2%
Market gardeners	13.7%
Market gardeners special tariff	17.1%

*Calculated as the average of EnergieNed recommended prices.

One may in fact say that these figures are an understatement because the gas price already before including the above mentioned taxes has a considerable tax element in them since the excise taxes on GO and HFO are taken into account when the market value of the gas is determined.

Application of the market value principle implies that the price of natural gas tracks the price of the oil products with a lag.

Table 7 shows that gas is treated relatively favourably in terms of taxes when compared to other fuels.

Table 7　Energy Taxes by Sector and Fuel in the Netherlands as of 1 January 1996 (ECU/GJ (NCV))

Energy	Residential sector	Commercial sector	Industrial sector	Power generation
HFO	0.87	0.87	0.87	0.87
GO	2.27	2.27	1.89	1.89
LPG	0.69	0.34	0.34	0.34
Natural gas	0.80	0.80	0.21	0.21
Coal	0.38	0.38	0.38	0.37
Electricity	3.88	3.88	–	–
Non-deductible VAT	17.5%	0	0	0

Source: Eurogas

For natural gas, the ecotax was increased from 0.032 NLG per cubic metre to 0.064 NLG in 1997 and to 0.0953 in 1998. In percentage terms about the same increases occurred for heating gasoil.

Price Transparency

In general, gas prices to end consumers in the Netherlands are quite transparent. It seems to be easy for consumers both to get a clear understanding for what prices have to be paid and to compare prices between companies. For outsiders, however, it is difficult to get precise information on some of the special tariffs applied by Gasunie, like the tariff to electricity producers and the feedstock gas price. Dutch LDCs in general have a fairly transparent way of accounting for their activities. The major difficulty seen from an external analyst's point of view is that of disentangling the economics of the various activities of the LDC although the accounting in the Netherlands is much better suited for this than in some other countries.

Costs, Value Added and Profits

Table 8 is an attempt at looking into average sales prices, gas acquisition costs and gross margins in the Dutch gas chain.

Table 8　Average Prices and Margins in the Dutch Gas Chain in 1995 (cents/cm ex taxes)

Gasunie average purchase price	19.3
Gasunie average sales price	20.5
Gasunie average export sales price	18.1
Gasunie average domestic market sales price	22.5
Gasunie total gross margin	1.2
Gasunie gross margin domestic market	3.2
Average purchase price EnergieNed companies	26.9
Average sales price EnergieNed companies	38
Average sales price to small customers	45
Average sales price to block heating/DHP	43
Average price to greenhouses	22
Average price large users	29
Gross average margin in LDCs	7.5

IEA calculations

The gross margin in the LDCs varies from one company to another because of differences in market composition; as it appears from the table the margins vary considerably from one market segment to another. In addition, the gross margin shown in the table the LDC will receive a contribution of 80 NLG per customer connection. The importance of this contribution will vary according to the market structure of each company, but is not unusual that total gross margin is around 10 cents/m³ when this contribution is included. Historically, this is confirmed by figures published by the Central Bureau of Statistics. The following table shows the gross margins for all the LDCs as a whole for the period 1992-1994.

Table 9   Gross Margins in Dutch LDCs 1992-1994 (cents/m³)

	1992	1993	1994
Total average retail sales price	40.5	37.5	38.1
Average gas purchase price from Gasunie	31.6	28.7	28.9
Average gross margin	8.9	8.8	9.2

Source: Centraal Bureau voor de Statistiek

Since only a minority of the LDCs are pure gas distribution companies, it is difficult to get a precise, aggregated picture of the economics of gas distribution in the Netherlands. The consolidated profit and loss account for the whole energy distribution sector, however, gives some indications. Table 10 gives this for the latest available years.

The 34 LDCs in the Netherlands vary a lot in terms of size, degree of horizontal integration, cost structure, tariff policy, etc. This make it difficult to talk about typical or average companies. The fact that the majority of the companies also have other activities than gas distribution often makes it difficult to disentangle the economics of each individual activity.

Table 10   Profit and Loss Account for the LDCs in the Netherlands (million NLG)

	1992	1993	1994
Total production value	**24.088**	**24.538**	**25.394**
Electricity sales	11.786	12.197	13.308
Gas sales	9.601	9.475	9.179
Heat sales	372	376	411
Water sales	333	321	328
Activated costs	542	513	532
Revenues from other activities	1.455	1.656	1.638
Total input	**18.036**	**18.310**	**19.064**
Electricity purchase	8.387	8.173	9.546
Gas purchase	7.545	7.415	7.171
Heat purchase	107	51	80
Water purchase	45	37	40
Purchase other energy	5	7	5
Own use energy	47	54	56
Other operating costs	1.901	2.033	2.167
Value added	**6.053**	**6.227**	**6.330**
Labour cost	2.278	2.341	2.390
Cost increasing taxes and charges	87	69	70
Gross operating result	**3.688**	**3.830**	**3.872**
Net interest charges	1.221	1.217	1.191
Saldo overige lasten en baten financiele vaste activa*	−244	−179	−213
Saldo buitengewone lasten en baten*	21	11	35
Stock variations	186	175	52
Depreciation	1.817	1.909	2.009
Net result for distribution	**686**	**697**	**798**
Dividends to shareholders	264	184	181
Municipalities	171	181	148
Addition to reserves	322	319	482
Other	−70	13	−12
Net result for distribution	**686**	**697**	**798**

Source: Centraal Bureau voor de Statistiek

UNITED KINGDOM

Statistical information

Natural Gas Supply/Demand Balance (Mtoe, 1996):

Indigenous production	75.8
Imports	1.5
Exports	- 1.2
Stock change	- 0.3
Total natural gas supply (primary energy)	75.8
Electricity and heat production	14.8
Other transformation and energy use	5.1
Total industry	14.4
Residential	29.1
Commercial	4.0
Other	5.3
Statistical difference	3.1

Natural Gas in the Energy Balance (1996):

Share of TPES	32.3%
Share of electricity output	23.6%
Share in industry	34.5%
Share in residential/commercial sector	56.5%

MARKET AND INDUSTRY STRUCTURE

The United Kingdom accounts for 6.5 percent of total gas consumption in OECD Europe and is basically self-sufficient in terms of supplies. The UK imports about 5% of her total supplies from Norway but is expected to become a net exporter over the next few years. The country already exports modest volumes of gas to the Netherlands and Germany from the Markham field, but the potential for exports will grow significantly after the UK-Continent Interconnector will have come on stream in 1998. In a European context, the UK is the only country which has introduced competition in gas markets to any significant extent by establishing a third party access regime. Since 1986, the British gas market has developed from a situation where a state owned monopoly through a wholly integrated transmission and distribution network supplied all customers to a situation where there is no state participation anymore and where all segments of the gas market feature competition between a multitude of suppliers. Another salient feature of the UK market is the recent breakthrough of gas in large scale power generation.

Table 1 shows the trend in UK gas demand in recent years.

Table 1   Size of the Gas Market in the UK 1993-1995 (million therms)

	1993	1994	1995
Domestic market	**11.700**	**11.400**	**11.400**
Non-domestic market			
– small firm			
– large firm	1.900	1.900	2.000
– interruptible	3.000	3.100	3.400
– power generation	2.700	2.900	3.000
– other*	2.800	3.900	5.000
	900	1.100	1.100
Total non-domestic market	**11.300**	**12.900**	**14.500**
Total market	**23.000**	**24.300**	**25.900**

*chemical feedstock, gas industry use and road transport
Source: British Gas

Over the past few years, power generation has by far been the most dynamic segment of the market in terms of growth. The coverage of the transmission and distribution system is very extensive, indicating that the UK is a relatively mature gas market: 89% of households are within 25 yards of a gas main, and 92% of those use gas. The high gas penetration in the domestic sector means that the potential for further growth in demand is limited. In its latest gas demand projections, the Department of Trade and Industry, however, still sees a sizeable growth in total final gas demand, increasing by 25% by the year 2000 on 1990 figures and by 57% by 2020. In its low price, central growth scenario it projects the following development in final demand (which means that gas for power generation is excluded):

Table 2   Long-Term Gas Demand Forecasts, UK (billion therms)

	1995	2000	2020	Growth*
Domestic	11.9	12.7	14.1	0.7
Iron and steel	0.6	0.8	1.1	2.5
Other industry	4.1	4.8	6.6	1.9
Service sector	3.2	3.9	5.7	2.3
Total	**19.8**	**22.2**	**27.5**	**1.3**

*Average annual growth from 1995 to 2020
Source: DTI

Structure of the Industry

Figure 1 is an attempt at describing the structure of the British gas industry.

In spite of the numerous recent changes in the British gas industry, British Gas plc. (as indicated in graph 1 the name of the mother company as well as the transportation and marketing affiliates have recently changed as a consequence of the demerger of the company -see below), up to 1986 a wholly state owned monopoly, is still the most important actor in the industry: in addition to being a sizeable producer of gas (about 17% of total UK gas production in 1995) and the major buyer of gas from the British continental shelf (taking about 64% of total gas supply in 1995), it still holds a de facto monopoly on transportation and storage at the same time as it holds a considerable market share (64% in 1995). Up to 1 April 1996 it had a monopoly in the

Figure 1   UK Gas Industry

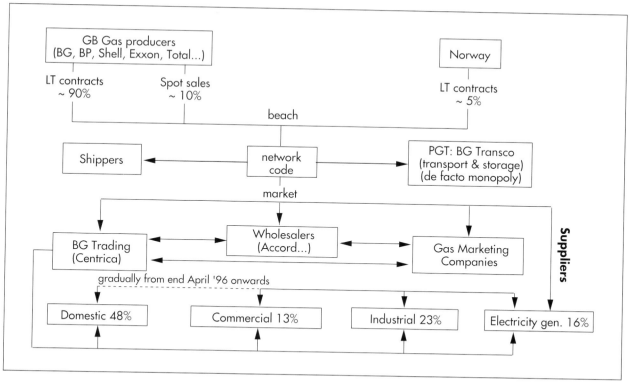

domestic market, whereas in the other segments of the market its market share has decreased with the introduction of competition in recent years.

The upstream part of the British gas chain is not dominated by any single company; there are more than 30 gas producers selling gas from more than 80 fields. Among the producers, oil and gas majors like Shell, Exxon, BP, Mobil, Texaco, Total, Elf, etc. are present. Traditionally, the upstream companies have sold all their gas to British Gas "at the beach", i.e., delivered at the various gas terminals at the coast. In recent years, a large number of the producers have extended their activity to the downstream sector by selling gas to power producers and large industrial users. From 1 April 1996, the marketing subsidiaries of a number of the producers also sell gas to residential consumers in the first pilot area.

The transportation part of the British gas industry is simple in terms of the number of actors: in addition to BG Transco, the transportation part of BG, there are few actors: Kinetica, a joint venture between Conoco (upstream oil and gas company) and PowerGen (one of the two major power producers in the UK) operates a 52-kilometre pipeline from the Theddlethorpe terminal to two power stations in Killingholme but is not a licenced public gas transporter under the Gas Act 1986. As of January 1997, there were three such licence holders in addition to BG (Transco). These three companies own and operate small distribution networks. Basically, however, BG Transco takes care of gas transportation, including storage, from the terminals to the final consumers.

Historically, BG has been the only gas distribution company in the UK : it used to sell gas to all customers through its own integrated transmission and distribution network. This situation started to change only in the beginning of the 1990s when other suppliers were allowed to sell gas to customers of a certain size (see below). Up to 1 April 1996 it was the only company selling and distributing gas to domestic customers, but after this date other suppliers are also allowed to sell gas into this part of the market. It may still be argued, however, that BG is the only gas distributor in the UK (neglecting the three exceptions mentioned above) since it is the only company physically distributing gas to end consumers through its own pipes. There has, however, never existed distribution companies in the UK in the form known in continental Europe, i.e., companies that buy their supplies at the city gate and distribute the gas through its own gas distribution network. In spite of the fact that BG's supply activity was divided into 12 regions and 90 districts, gas distribution as such has never been undertaken by subsidiaries taking care of both the physical and the commercial aspects of this activity.

A large number of companies in addition to BG are now selling gas into all segments of the market. Rather than gas distribution companies, however, they are gas marketing companies either producing gas themselves or buying it at the gas landing terminals and selling it to end consumers, using BG Transco's transmission and distribution facilities to bring the gas physically to the final consumer. In recent years, some middlemen like wholesalers and brokers have also entered the scene. These companies would typically buy gas from producers and from other marketing companies.

The BG marketing activity is taken care of by a separate subsidiary called BG Trading (after February 1997 BG Trading is a part of Centrica, one of the two new companies formed as a result of the BG demerger). The way BG is organised to perform its activity is intimately intertwined with the way the British market is regulated. A more detailed description of how BG is organised will be given below.

In addition to the usual distinction of market segments indicated in figure 1 (which also indicates the importance of the domestic market in the UK), one also makes the distinction between the tariff market, where customers are served according to a published price schedule, and the contract market, where each customer has a more or less standardised contract with the supplier. The contracts may be publicly available but there is no obligation to disclose individual contracts. The border line between tariff customers and contract customers is a function of volume. Originally the border line was drawn at 25000 therms, but in 1992 this was lowered to 2500 therms (the average consumption of a domestic customer is around 660 therms, equal to about 1742 cubic metres). This means that the distinction between tariff and contract customers cuts across the classification into domestic, commercial and industrial customers. As will be pointed out below, the tariff market is regulated in terms of pricing whereas this is not the case for the contract market. As indicated in figure 1, three categories of actors may be distinguished in the UK gas industry after the adoption of the Gas Bill in 1995 (see below) which introduced a three-tier licence structure covering transportation, supply and shipping. BG Transco, the transportation subsidiary of BG is, for all practical purposes, presently the only gas transporter. Suppliers are the companies selling gas to end consumers. As of January 1997, 64 companies held a licence to supply industrial and commercial taking more than 2.500 therms a year. 15 of these companies are also licenced domestic gas suppliers, which means they are authorised to supply domestic customers in the areas opened up to competition in this sector. The companies buying transportation services on the transportation network from licenced transporters are called shippers. As of January 1997, there were 52 licenced shippers. Shippers can be suppliers and vice versa. Transporters, however, cannot hold a gas supplier's licence or a shipper's licence. To regulate the relationships between the different actors, a Network Code has been introduced (see below).

Market Structure and Market Shares

The tariff market, basically composed by residential/commercial consumers, is by far the largest market segment in the UK. Until 1 April 1996, BG had a monopoly in this market. At that date, however, a region in the Southwest of England was opened up for competition in that other gas suppliers were also allowed to serve customers in this area. Over time, competition will be extended to other areas as well, and the intention is to make the entire tariff market competitive from no later than 31 December 1998. As of January 1997, other companies have obtained about 90.000 customers in the Southwest, i.e., about 18% of the total number of customers.

The contract market is divided into three different segments, firm contracts, interruptible contracts and power generation contracts. Since the opening of these markets for competition in the early 1990s, BG's market shares has been gradually decreasing. In 1995, BG held only a third of this market. Up to very recently, BG held a high share of the interruptible contracts, but over the last couple of years, its competitors have captured about two thirds of this market. Since liberalisation of the contract market to a large extent coincided with the "dash for gas" in power generation, BG never held as dominating position in this market as in the industrial markets except that it was the first company to sell gas into this market.

Table 3 gives an overview of BG's market shares in the various parts of the contract market over time.

Table 3

BG Market Shares in the Contract Market as Estimated by OFGAS

Date	Small firm	Large firm	Interruptible	Power stations
Oct-89	100%	100%	100%	0%
Oct-90	100%	93%	100%	0%
Oct-91	100%	80%	100%	100%
Oct-92	100%	57%	100%	26%
Oct-93	77%	32%	100%	12%
Jan-94	71%	22%	100%	12%
Mar-94	67%	20%	99%	12%
Dec-94	52%	9%	93%	17%
Apr-95	45%	10%	57%	32%
Jun-96	43%	19%	34%	24%

Source: OFGAS

Degree of Vertical and Horizontal Integration

BG before the demerger in February 1997 was still vertically integrated in that it was active in gas production, transportation and distribution. Its activities in the various areas are, however, taken care of by different subsidiaries. In recent years the company has diversified into power generation both in the UK and abroad. It is already a power producer in the UK and intend to become an electricity supplier. The company is also horizontally integrated in that it sells gas appliances and installation services.

Some of the power producers in the UK, like the two biggest ones, National Power and Power Gen, have invested upstream and now own shares in gas producing licences. The main objective is to supply their own power stations with gas, but some of them also trade gas and are licenced as gas suppliers. Some of them have joint ventures with oil and gas companies to sell gas to end consumers. All of the regional electricity companies in the UK now have interests in gas supply ventures.

As indicated above, a lot of the gas producing companies in the UK have gone downstream to supply gas to end consumers. All the ten biggest gas suppliers into the industrial and commercial market are ventures set up by gas producers like Shell, Esso, BP, Mobil, Conoco, Elf, Total, Texaco, Statoil, etc. Joint ventures with regional electricity companies are quite widespread, the objective being to benefit from synergies in gas and electricity distribution. Quite a few of the new gas suppliers also offer energy management services.

Contractual Arrangements

Up to around 1990 the situation in terms of contracts in the UK was fairly simple: BG bought all gas from the producers in the North Sea and from Norway at the terminals, mostly under long-term depletion contracts with take or pay clauses. BG then had sales contracts with the various type of customer categories; in the contract market often with a duration of ten to fifteen years. In general one may say that these contracts to the extent possible were back-to-back arrangements with the purchase contracts in terms of pricing and duration. What has happened since BG lost its monopoly in the contract market is that other buyers have also entered into long term purchase contract similar to the ones that BG used to enter into. In the beginning of the 90s the contracts used by the new suppliers to sell gas into the competitive market were not very different from the ones earlier used by BG. In recent years, however, contracting in the contract market have changed radically: instead of long term contracts with price escalation reflecting the beach contracts, one year contract with fixed prices have become widespread. This means that companies which have bought gas under long term contracts are squeezed when prices in their short term sales contracts decline in relation to the prices in their long term purchase contracts. A surplus of gas combined with competition in the contract market has led to the emergence of a spot market for gas, and prices in the contract market are influenced by the spot price. This is the explanation for the take or pay problems faced by BG and the losses made in the contract markets by other marketers recently.

Gas to the domestic sector has always been supplied by BG on a rolling contract basis, i.e. The contract goes on as long as it is not cancelled. This also seems to be the principle adopted by most of the companies now supplying gas in the areas that have been opened up to competition. Residential consumers can cancel their contract with 28 days written notice.

Organisation of British Gas

The way British Gas is organised is to some extent the result of regulation. After privatisation in 1986 it was still a strongly integrated company. The second MMC report, published in August 1993, recommended that BG should divest its supply business. This recommendation was never implemented, but from March 1994 Transco, the BG transportation business, was separated operationally from BG's gas supply business as part of the process designed to assure gas shippers that they would have equal access to the national gas transportation and distribution system. In 1996, it became clear that a demerger of the company 's transportation and supply business would be appropriate. This demerger took effect in February 1997. The following graph shows the structure of BG before the demerger:

Figure 2   Structure of British Gas

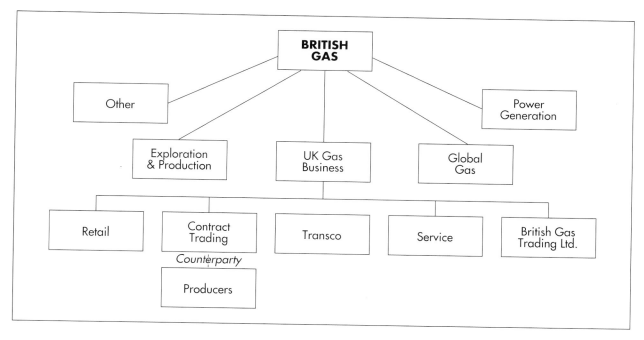

Through the demerger the company was split into two separate companies: BG plc and Centrica. The latter company is a holding company having several subsidiaries of which the most important are indicated in the following figure:

Figure 3   British Gas Demerger

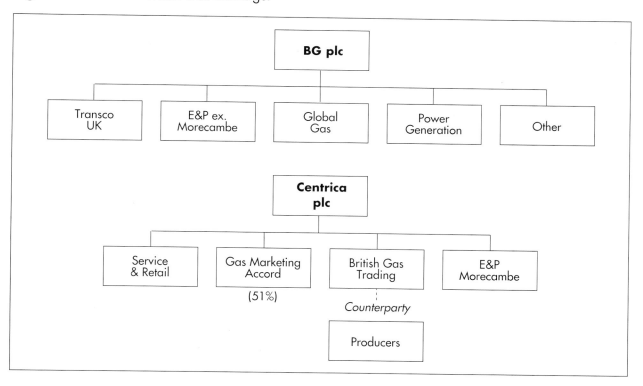

Centrica is responsible for the marketing of gas to end consumers and has taken over the responsibility for all the supply contracts previously entered into by BG. It has another link to the upstream sector through its ownership of the Morecambe fields, which play an important role in gas supply modulation in the UK. It also holds subsidiaries selling gas appliances and providing energy advice. A subsidiary called Accord is a company active in gas trading.

The rest of what was BG is now found in the other company called BG plc. The most important subsidiary in this holding company is Transco, the company responsible for transportation and storage. Other subsidiaries within the group are responsible for international E&P activity and international downstream ventures.

By this demerger the UK has probably gone further than any other country in the world in separating marketing and supply of gas from transportation (including distribution). The US and Canada have separated the merchant function and the transportation function on the interstate and interprovincial pipelines but the LDC still in many cases have an integrated merchant and transportation function. In Continental Europe these functions are still integrated.

REGULATION

The evolution from a gas market dominated by a state owned monopoly to a market intended to feature full competition in all segments in 1998, has to a large extent been the result of regulation and intervention by the Government since 1986. One of the main experiences made in the UK over this period is that competition does not come about by itself - regulation is needed. The regulatory system put in place in the UK is that of a relatively small number of framework laws which leaves considerable powers to institutions outside government to implement these laws. The most important of these laws is no doubt the Gas Act 1986. After privatisation of BG in 1986, a complex regulatory system largely based on licences was enacted through the Gas Act. It designated the Department of Trade and Industry (DTI) as the government supervisory body and established a regulator for the gas industry, the director general for gas supplies (DGSS).

In the integrated regulatory system established by the Gas Act, there are seven principal components:

■ the Gas Act itself which sets out the rules under which the gas supply industry is to operate;

■ the authorisation issued under the act on 28 July 1986 containing terms and conditions under which BG was authorised to act as the sole public gas supplier in Great Britain (but not Northern Ireland);

■ the Office of Gas Supply (Ofgas) as the central mechanism charged to regulate the gas industry;

■ the Gas Consumers Council (GCC) as the principal vehicle for consumers' grievances to be registered;

- the minister as the holder of certain ultimate powers of veto such as the prevention of referrals by the DGSS to the Monopolies and Mergers Commission (MMC) under specified circumstances;

- the Office of Fair Trading (OFT) to ensure that no unfair trading policies are adopted by BG as a monopolist, with the power to refer such practices to the MMC;

- the MMC to which both the OFT and Ofgas may refer BG: OFT may do so with respect to industrial and commercial matters, while Ofgas may do so with respect to domestic tariff matters affecting gas supply.

All the above elements form one system designed to balance constraint and freedom. Ofgas and the GCC are the specialist organisations which monitor aspects related specifically to the gas industry. By contrast, OFT is the generalist charged with assessing the fairness of business operations in all industries. The MMC is the final point of reference for all three agencies.

The Gas Act provided that BG should operate as a monopoly in the tariff sector subject to the regulator (it was licenced as the only public gas supplier in this sector at the time). It is important to note, however, the Gas Act abolished the exclusive privilege which BG had enjoyed with respect to gas supply in general. Instead, the minister was empowered to authorise any person to supply gas through pipes to any area designated in an authorisation. The act also contained provisions for third party access to BG's transportation network.

In 1996, a new Gas Bill replaced part of the provisions in the Gas Act, the most important being the parts dealing with licencing. As described more in detail below, there is now a three-tier licencing system in place. The Gas Act is supplemented by general competition law, essentially the Fair Trading Act 1973 and the competition Act 1980. Under the Gas Act, the DGGS may make reference to the MMC to require it to investigate and report on matters relating to gas supply to tariff customers, which operate or may operate against the public interest, and whether any adverse effects could be remedied or prevented by modifications to the conditions in the gas supplier's authorisation.

Under the Fair Trading Act 1973, references may be made to the MMC by the minister and the OFT. If the result of a reference is a report which recommends modifications to BG's authorisation and the recommendations are adopted by the minister, the DGGS must modify the conditions of the authorisation to give effect to the minister's requirement.

In the following, a brief overview of legislation and regulation in the various parts of the gas chain in the UK will be given. This treatment will, however, in no way be exhaustive, the aim being primarily to point out regulation relevant for gas distribution in a wide sense.

Gas Production and Contracting

Gas production in the UK basically takes place offshore, and before competition in end user markets was introduced, BG bought all gas produced "at the beach". After the first MMC report published in 1988, BG was disallowed to contract more than 90% of production in new fields, the remainder being intended for companies which wanted to compete with BG in the contract market. In 1992, BG reached an agreement with OFT on the release of gas to other suppliers.

The gas release scheme was intended to speed up access to gas supplies by the independent traders, so as to enable BG to meet its undertaking to secure the development of genuine self-sustaining competition in the contract market. BG agreed to release 500m therms for the gas years 1992/93, 1993/94 and 1994/95 and 250 m therms for the year 1995/96. More than thirty applicants have received gas under this scheme. Up to 1996 all of these volumes went into the contract market, i.e., power generation, industrial and commercial markets.

Imports/ Exports

There is no such thing as an export or import monopoly in the UK, but in practice exports and imports of gas have not been straightforward, as illustrated by the problems concerning imports of new volumes from Norway through the Frigg pipeline.

Formally, there is still a statutory requirement on UKCS producers to deliver the produced gas onshore in the UK unless the minister gives written authorisation to delivery elsewhere. In practice waivers to the requirement to land gas in the UK are granted as a matter of course, and the current statutory requirement is to be phased out.

The current rules on imports are contained in the Import of Goods (Control) order 1954, as amended. In 1986, the minister announced that BG would be able to import gas in the future subject to the normal consent and, in appropriate cases, to the conclusion of inter-governmental treaties. In light of later legislation it is assumed that any gas supplier in the UK is free to import gas provided consent can be obtained. The import contract concluded by National Power with the GFU in Norway has, however, been stalled several years for failure to reach agreement on a revision on the Frigg treaty governing pipelines between the UK and Norway. In the case of the Markham field, jointly developed with the Netherlands, the landing requirement has been waived, and the gas is exported to the Netherlands and Germany.

Transportation

Access to transportation is a condition for effective competition to function. Given the British government's desire to make all sectors of the market fully competitive there is no wonder why changes in this area compared the pre-1986 period have been so crucial and hotly debated. The main development over the last ten years has been easier access for third parties to both offshore and particularly onshore transportation facilities. It would, however, take too long to go through the development leading up to the recent changes in area; focus will therefore be on the present situation.

Offshore Transportation

The offshore pipelines in the UK are all owned by private oil and gas companies, often on a joint venture basis. Access to pipelines has historically only been obtained by commercial negotiations with the owners of the pipelines concerned. There is no legal obligation to transport gas or oil for third parties but the minister has regulatory powers through the Petroleum and Submarine Pipelines Act including the power to facilitate access to offshore pipelines by third parties if access is arbitrarily denied. These power have, however, never been invoked.

With increased liberalisation of the gas market, some of the gas marketing companies have expressed interest in buying gas offshore before it reaches the beach terminal. Feeling the pressure in favour of easier access to offshore transportation, the DTI took the initiative to elaborate a code of practice in this area. Such a code of practice, called "Offshore Infrastructure Code of Practice" was adopted in 1996. As its subtitle says, it is "an agreed industry document on the rules governing access to offshore infrastructure". The main elements in the code are as follows:

■ established rights of access where capacity is available, with owners having some accepted prior rights for their own future use. The owner is also expected to provide incremental capacity where present capacity is insufficient and the user is prepared to pay for the needed investment;

■ terms of access, agreed by commercial negotiations, should be non-discriminatory and set at fair market-related terms; all additional costs would be recovered either through a tariff charge or through some other capital contribution;

■ an obligation on owners to make available relevant information, with similar obligations on potential users to set out their prior requirements at the start of the negotiation;

■ a set time for completion for the various stages on the negotiation and in the period to final completion;

■ an agreed arbitration procedure.

With the adoption of this code of practice the regime for access to pipelines offshore has moved in the same direction as the onshore one.

Onshore Transportation

The Gas Act 1995 adopted by Parliament in November that year is now the central piece of legislation concerning onshore transportation, to a large extent replacing the Gas Act 1986. Two of the main driving forces behind the new act was the necessity to unbundle BG's activities, in particular that of transportation from other activities and the desire to introduce competition also in the tariff market.

The Gas Act 1986 contained a formal system of third party access. This has been retained in substantially the same form under the provisions of the new Gas Act. It is important to note that this affects transportation through high, medium and low pressure systems to end users. Two provisos for TPA are that the gas is similar to the kind that the pipeline is designed to convey, and that such conveyance would not prejudice the conveyance of gas by BG necessary to fulfil its responsibilities as public gas supplier and its contractual obligations.(It should be noted, however, that the new gas act gives the concept of public gas supplier a somewhat different meaning from what it had.)

Under the new gas act a transportation licence will authorise the holder to convey gas through pipes to any premises in an 'authorised' area specified in the licences or to the pipeline system of another holder of a transportation licence. The licensees are designated as 'public gas transporters'. They will in fact inherit many of the features of the current public gas supplier. They will have the following rights and duties:

- exclusive rights to convey gas to premises at a rate of 75.000 therms or less a year within their "authorised" areas;

- a duty to develop an efficient and economical pipeline system in these areas and to comply, insofar as it is economical to do so, with any reasonable request to connect premises to their pipeline systems and to convey gas to those premises;

- a duty to connect to their pipeline systems any premises which are located within 23 metres of one of their distribution mains, if those premises are expected to consume gas at an annual rate of 75.000 therms or less.

In principle, it is not possible for a holder of a transportation licence to hold a gas supplier's licence or a gas shipper's licence. The reason is to ensure separation of the transportation function from the functions of supply and shipping. Though in reality, there are limited exceptions to the ban on transporters acting as suppliers, e.g. companies such as AGAS which perform both roles for some housing estates.

Licences are granted by the DGGS, who has fairly wide discretionary powers in this matter. The DGGS is required to incorporate into each licence the standard conditions appropriate to each type of licence which are set down by the legislation. The standard conditions for holders of a transportation licence will include:

- a condition requiring that the licensee keep separate accounts for his transportation and storage business;

- a condition requiring that the licensee furnish the DGGS with details of his charges;

- a condition relating to standards of performance in respect of activities affecting consumers and third parties;

- a condition that requires the licensee to establish a network code covering the arrangement under which others may use his pipeline system;

- appropriate conditions which require the licensee to furnish information to the DGGS and the Gas Consumers Council;

- a condition that requires the licensee to make effective arrangements in respect of the provision of emergency services to the public;

- a condition that requires the licensee in certain defined circumstances, but no earlier than 1998, to raise a levy through his transportation charges to compensate a supplier who has incurred additional costs by virtue of supplying an undue proportion of premises occupied by pensioners or disabled people, or by people likely to default in the payment of charges. Similar provisions would apply to levies on the industry to compensate particular suppliers for excess costs incurred by virtue of making supplies of last resort available to consumers;

- a condition that requires the licensee to obtain the approval of the DGSS before disposing of key assets to operate his transportation system.

One of the obligations of the transportation licence holder is the development of a network code covering the arrangements under which others may use his pipeline system. The government's plans to extend competition the domestic market has required a radical rethink of the way the national transmission and distribution system is operated. Because third party access to the network has been relatively limited until recently, BG as the only holder of a transportation authorisation, has had the main responsibility for ensuring that inputs of gas to the system is equal to offtakes from it. Such daily balancing is essential for the safety of gas transmission. Now that more and more of the gas transported is owned by independents, this responsibility has been passed to the shippers by virtue of the Network Code developed through a consultation process between BG Transco, the shippers, OFGAS and other interested parties. Originally the code was going to become operative from October 1995, but this took place only from October 1996. One may say the Network Code is a compendium of contractual terms intended to enable competitors to use the onshore gas transportation and storage network in Great Britain on an equal and transparent basis. The key areas for which business rules have been developed are as follows:

■ operational balancing of gas put into and drawn from the network on a daily basis. The new rules for balancing has lead to the creation of a "flexibility market" for gas, used by companies to balance their daily needs with supply;

■ handling of supply points (for example the systems that deal with the transfer of customers from one shipper to another);

■ information systems supporting the Network Code;

■ charging for pipeline capacity and the treatment of interruptibles;

■ maintaining the security of the network;

■ liabilities on BG Transco within the Network Code.

The counterpart of the public gas transporter is the holder of a shipper licence as defined in the Gas Act 1995. This licence authorises the holder to arrange with a public gas transporter for gas to be conveyed in his pipeline system. Some licenced gas suppliers (see below) will purchase their gas direct from producers and arrange for it to be conveyed to their customers by a public gas transporter. Such suppliers also need to become licenced gas shippers. Other suppliers may prefer to purchase supplies from aggregators or wholesalers which may or may not wish to become licensed gas suppliers. This is the reason why a separate shipper licence has been introduced. The standard conditions applying to the holder of a licence are:

■ obligations as regards the use by the shipper of the pipeline system of a public gas transporter, whether in accordance with the transporter's network code or otherwise;

■ obligations to assist the public gas transporter in an emergency;

■ obligations to furnish the gas transporter with information relating to suppliers for whom the licensee is shipping gas;

■ an obligation to supply the DGGS with appropriate information;

■ obligations in relation to meters.

The third tier in the new licence system is that of a gas supplier. This licence authorises the holder to supply to premises gas which has been conveyed through pipes to those premises. It is intended that the holders of supply licences should compete with each other for the supply of owners or occupiers of premises which they are licensed to supply. A supply licence may be issued to supply gas at any annual rate of supply and for rates of more than 2,500 therms a year (the industrial and commercial market). A licence would normally contain no restriction as to premises within Great Britain which may be supplied, though some supply licences are regional, e.g. North Wales Gas. The licence conditions generally relate to conditions of supply, not to who can be supplied. An interesting condition is that the DGGS is empowered to withhold grant of a licence specifying premises by area or description, where he considers that the area or description has been so framed as artificially to exclude an undue proportion of premises likely to be occupied by disabled people or pensioners, or by people likely to default in the payment of charges. This is intended to discourage cherry picking of the market by new gas suppliers.

The standard conditions applying to holders of a supplier licence include:

■ an obligation to supply all prospective domestic customers (i.e. consumers of gas at an annual rate of 2,500 therms or less) at premises within the area or description in the licence, subject to technical exceptions and giving priority to existing customers;

■ an obligation to publish charges and supply gas in accordance with these charges, except for special cases such as "energy packages";

■ an obligation to supply gas to former customers of another supplier, where directed to do so by the DGGS as a last resort to ensure continuity of supply to those customers;

■ an obligation to take out a financial bond, or make other arrangements to the satisfaction of the DGGS, to meet certain additional costs arising from the continuation of gas supply to the licensee's domestic customers should the DGGS be obliged to direct any other supplier to supply these customers;

■ a condition requiring the licensee to offer a reasonable range of methods of payment to domestic customers;

■ a condition requiring the licensee to provide reasonable rights for consumers to terminate their contracts;

■ a condition requiring the licensee to adjust charges if the meter is found to have been registering erroneously;

■ a condition requiring the licensee, where he is in a dominant position in the market on a national or local basis, not to practice undue discrimination or undue preference in his pricing, subject to being able to respond fairly to competitive pressures;

■ a condition requiring the licensee to keep a register of qualifying domestic customers who are disabled or pensioners; and to make arrangements to provide these customers, on request, free of charge, a number of special services to ensure efficient and safe use of gas;

■ a condition requiring the licensee to adopt methods of dealing with domestic customers who, through misfortune or inability to cope with gas supplied for domestic use on credit terms, incur obligations to pay for gas so supplied which they find difficulty in discharging;

■ a condition requiring the licensee to make available to domestic customers advice on the efficient use of gas;

■ a condition requiring the licensee to keep appropriate records in regard to the discharge of his 'social obligations' as set out above;

■ conditions relating to meters;

■ conditions requiring the licensee to make arrangements to check the background of officers authorised to enter the premises of customers and to provide means which allow members of the public to confirm the identity of these officers;

■ appropriate conditions requiring the licensee to furnish information to the DGSS and the Gas Consumers Council.

The licence condition contain a remarkable number of measures intended to protect captive customers.

The above conditions apply to all holders of a supply licence. BG or any subsidiary of BG is also subject to a number of non-standard conditions. These standards are considered part of the price control regime for BG since most of them have cost implications. These standards require BG to reach a certain performance level in the areas of :

■ Customer contact (response times to customer calls, answering correspondence promptly, timely making and keeping of appointments, etc.).

■ New connections and alterations to connections.

■ Meter reading- specified frequency.

■ Elderly and disabled customers – obligation to keep a register and offer certain services.

■ Customers with payment difficulties-agreement of payment plans or offering of prepayment meter.

■ Energy efficiency-obligation to provide energy efficiency advice to domestic customers.

The additional standards laid down for BG are quite extensive. They could be seen as a substitute for competition since BG is supposed to have a de facto monopoly over most of the domestic market for still some time to come.

In the context of a study on gas distribution it is interesting to note that under the new competitive regime there will still not be distribution companies in the sense encountered in continental Europe: BG Transco is envisaged to remain the only gas transporter for a while, physically delivering all gas sold by licenced gas suppliers to end consumers. The Gas Act 1986, however, contains provisions for the appointment of other gas transporters. The only area in which a natural monopoly is recognised is in the local distribution network. Competitors will be forbidden to establish distribution mains within 23 metres of any existing one. All other activities will be potentially competitive.

Storage

Storage is regulated within the overall framework for transport in the Gas Act 1995 as summarised above. Under this law, it will be possible for others than BG to engage in storage provision, for which no licence is required. BG's storage is regulated under Transco's licence with respect to price, unbundling and separate accounting from other transportation and non-disposal except with the permission of the DGGS. The storage assets of a gas transporter which are connected, directly or indirectly, to its transport system are also to be regulated under the Gas Act 1995. The expectation is that independent storage will develop over time.

PRICE REGULATION AND TARIFFS

At the beginning of 1997, the situation in the British gas market in terms of price regulation was as follows:

■ Prices to end users in the tariff market taking less than 2.500 therms per year (accounting for about 44.% of the total market in 1995) are regulated through a price cap formula on BG. Prices for customers above 2500 therms per year (but less than 25000) are not regulated but gas in this volume tranche is sold according to a tariff schedule.

■ Prices to end users in the contract market (above 25.000 therms a year) are not regulated (and have in fact never been).

With the introduction of third party access to the transmission and distribution system, it has become possible for end users to buy their gas "at the beach". This mean for instance that a large industrial user somewhere in the UK can buy gas from a producer which is delivered at one of the landing terminals and that he can buy the necessary transportation services from a gas transporter (for the time being only BG Transco) to have the gas delivered at his own premises. This means that he has to include the transportation charges when looking at the total price of gas. The prices or charges paid for transportation and storage, since still a de facto monopoly, are regulated. The prices paid in the supply market or wholesale market for gas under short or long term contracts, that is in the transactions between producers and buyers of gas, are (and have never been) regulated. This means that they are the result of negotiations between the parties.

In terms of publishing tariffs, there is now only an obligation for suppliers in the market for deliveries below 2.500 therms to do this. In addition to this, BG Transco is obliged to publish tariff schedules for transportation and storage.

Compared to the situation initiated by the first MMC report published in 1988 where BG was forced to publish price schedules also for the contract markets, one might argue that price transparency has deteriorated. Since the obligation on BG to publish price schedules for the contract market was suspended in 1995, the only tariffs published are the ones for consumers below 25.000 therms.

Domestic Tariff Price Formula

In the market under 73.200 kWh (formerly 2.500 therms) British Gas' allowable revenues are governed by a tariff formula set out by the DGGS. The formula is currently under review by Ofgas and will probably be replaced by a revised formula from 1 April 1997.

Tariff Price Formula Applicable for the < 73200 kWh/Annum Market
(with effect from 1 April 1994)

$$M_t = 1.015 \times \left[1 + \frac{RPI_t - 4}{100}\right] P_{t-1} \left[1 + \frac{F_t - Z_t}{100}\right] G + E_t - K_t$$

Maximum average price	=	Rebasing factor	×	Non-gas component	+	Gas component	+	Energy efficiency component	-	Correction factor

where:

M_t = maximum average price per therm in tariff year t.

1.015 = rebasing factor to allow for reduction in threshold from 25,000 to 2,500 therms per annum.

RPI_t = percentage change in the Retail Price Index between that for October in tariff year t and that for the preceding October.

P_{t-1} = non-gas component of price per therm in the prior tariff year t-1.

F_t = gas cost index in tariff year t.

Z_t = gas purchasing efficiency factor defined as 100 [1.01N-1] where N is number of formula years between 1 April 1991 and end of tariff year t.

G = base gas unit cost.

E_t = allowable energy efficiency cost per therm in tariff year t.

K_t = correction factor per therm to reflect over/under-recovery against allowable miximum in prior tariff year t-1.

The maximum average price represents the maximum average price per therm allowed in the tariff year, which runs from 1 April to 31 March the following year. Any difference between the allowed maximum rate and the actual price achieved is carried forward to the following year and inflated in the correction factor (see below).

The rebasing factor represents an uplift of 1.5% which is applied to the gas and non-gas components of the formula as recommended by the MMC and approved by Ofgas to allow for the reduction in the monopoly threshold (and scope of the formula) from customers consuming 25.000 therms per annum to 2.500 therms per annum with effect from 1 April 1994. This was introduced to reflect the disproportionate effect of standing charges on the lower average consumption that arose from the reduced threshold.

The non-gas component is linked to the previous year's non-gas component and allows for an increase in line with the Retail Price Index (RPI) less a cost reduction X-factor - the current value of X has been set as 4. The change in the RPI is measured between that for October of the current tariff year and that for the preceding October.

The original component of the tariff formula allowed for pass through of gas costs. This was revised in the last review of the formula to a base gas cost to which is applied a gas cost index and gas purchasing efficiency factor, currently 1% per year compounded from 1991. The allowable gas cost is tracked by an index comprising a basket of indices – a weighted average of gas costs. This is to give BG an incentive to contract for the cheapest possible gas.

The energy efficiency component was designed initially to provide an incentive for BG to undertake energy efficiency measures. The allowable energy efficiency costs are only those approved by Ofgas. The operation of the tariff formula depends upon forecasts of certain components during the year, for example, expected PPI and other indices for the gas cost index; expected October to October RPI inflation for the non-gas component; the total volume of sales (highly weather dependent). Any difference between the allowed maximum and the actual achieved average price per term is dealt with via the correction factor. If the actual achieved average exceeds the allowed maximum, then the K-factor is positive and an interest rate of anticipated base rate in the formula plus 3% is applied to the over-recovery carried forward to reduce the allowable maximum the following year. On the other hand, if the allowed maximum exceeds the actual average achieved, the under-recovery is carried forward with interest at the anticipated base rate only via a negative K-factor.

The tariffs constructed based on the price cap formula are relatively simple. Table 4 shows the structure.

Table 4   BG Tariffs for Customers up to 732.678 kWh/year (25.000 therms) without Special Agreement

	Standard credit tariff	Direct pay tariff
Standing charge (p/day)	10.39	10.10
Unit charge (p/kWh)		
0-73.200 kWh/a	1.520	1.433
73.201-146.536 kWh/a	1.420	1.420
146.537-293.071 kWh/a	1.380	1.380
293.072-732.678 kWh/a	1.340	1.340

Source: British Gas

The tariffs cover all customers taking less than 732.678 kWh/year, that is less than the former threshold monopoly volume. These tariffs apply to customer categories for which BG had a monopoly up to 1 April 1996 (customers taking less than 73.200 kWh/year or 2500 therms) and to customers on which BG lost its monopoly on 1 April 1994 (these customer categories can also be supplied according to special agreements see below). There are two tariffs according to the payment procedure of the customer; the standard credit tariff being by far the most widespread. BG also offers a prepayment tariff, which allows for gas to be paid for as it is consumed, and as such has certain budgeting benefits as well as being an acceptable alternative to disconnection for people in debt.

The structure of the tariffs is very simple: the standing charge is the same for all volume zones and the unit price is digressive with the volume zones. This implies that the standing charge (which is quoted in pence per day) will decrease with the volume taken as volume increases.

Prices are the same all over the country, and it is clear that historically the structure of the tariffs has led to a cross subsidisation in favour of small customers. In 1993, BG came to the result that the average price to customers below 600 therms/a did not cover average costs . The standing charge did not cover the fixed costs incurred in serving these customers, and the unit charge was not sufficient to compensate for this. The under coverage on small customers was compensated by larger customers paying a price covering more than the cost of serving them. During recent price control review OFGAS has focussed on this issue and has put forward proposals to remedy this.

The domestic sector has only very recently been opened up to competition on an experimental basis. The regulation of BG since 1986 has, however, been a substitute for competition and it is interesting to note that prices in the domestic sector have come down over the period, although not as much as in the contract market. There are several reasons for prices coming down over the period - gas prices are influenced by the escalators in BG's supply contracts- but this does not explain all. Some of the decline in prices in recent year is due to price regulation in transportation and supply. Table 5 shows the development in the Standard Tariff (the most widely used in the domestic sector) for an average consumption in the domestic market, since 1986.

Table 5   The Standard Tariff since 1986 (January 1996 prices)

Date of tariff change	Commodity charge (p/th)	Standing charge (pounds/year)	Annual bill (650 therms consumed)	Cumulative change (%)
1 May 1986	58.26	54.58	433.23	
1 July 1987	53.49	48.92	396.59	-8.46
1 April 1988	54.58	49.34	404.14	-6.72
1 April 1989	52.23	45.67	385.17	-11.09
1 March 1990	52.88	46.46	390.20	-9.93
1 Novemb. 1990	51.12	43.38	375.63	-13.30
1 April 1991	51.73	42.37	378.61	-12.61
1 July 1992	47.73	40.63	350.87	-19.01
1 October 1992	46.41	39.53	341.20	-21.24
1 January 1995	45.77	38.96	336.45	-22.34
1 January 1996	44.55	37.92	327.48	-24.41

Source : British Gas Supply and Central Statistical Office.

From April 1996 consumers in Somerset, Devon and Cornwall could choose between 13 suppliers. Probably for the first time in European history, consumers in the domestic sector were offered gas from more than one supplier; 12 companies in addition to British Gas. Table 6 shows the total price to be paid by an average gas user (taking 20500 kWh a year) to each of the companies and the saving in relation to BG's offer. As can be seen, an average consumer could save as much as 21% on the BG tariff by choosing the cheapest supplier. BG has the opportunity to lower its price in response but only under specified conditions to avoid predatory pricing.

Table 6 Comparison of Bills between British Gas and Other Suppliers, March 1998 (based on consumption of 18,000 kWh)

Company	Standard credit	
	Annual Bill (£)[1]	**Savings compared with British Gas (%)**
National		
Amerada	270	15.5
British Fuels	270	15.6
British Gas Trading	320[2]	
Eastern Natural Gas	257	19.8
London Total Energy	273	14.6
Midlands	258	19.5
Northern Energy	282	12.0
Norweb	262	18.0
North Wales	266	16.9
Sterling	259	19.1
SWEB (unit price first 3660 kWh)	261	18.5
Southern	271	15.1
SWALEC	262	18.0
York Gas	260	18.8
Yorkshire	263	17.8
Regional		
Beacon Gas	267 - 273	16.4 - 14.6
Scottish Power	262 - 276	18.1 - 13.8
Northern Electric	255 - 262	20.2 - 18.1
Calortex	270 - 277	15.7 - 13.3

(1) annual bill includes VAT at 5%
(2) you get a prompt-payment discount of £7.5 a quarter (ex VAT) if you pay within 10 days of billing date
Source: The independent consumer guide.

Prices in the Tariff Market

As pointed out above, there is no longer any obligation to publish tariffs in the contract market. A lot of the contracts entered into around 1994 are, however, still running. A brief overview of the tariff schedules used and their main characteristics will therefore be given:

- the F15 schedule - applying to firm gas supplies from 25.000 therms a year with 12 volume bands. Contract duration was one year as a minimum; the price consisted of commodity charge and a standing monthly charge. In terms of pricing there were several options, varying between fixed price plus an annual fixed increase in per cent and a mix of a producer price index and the gasoil price;

■ the MT3 schedule – applying to volumes above 25.000 therms across 12 volume bands. Contract duration was between 3 and 10 years, and a 80% take or pay clause was applied. In terms of pricing options there was a choice between HFO price indexation and a mix of HFO, GO, coal and electricity.

Under similar schedules, interruptible supplies were also offered, stipulating interruptions of 7 to 90 days. There was also a separate price schedule for interruptible supplies int power stations. This schedule, the LTI3, proposed contracts of ten to fifteen years' duration with 7 to 55 days' interruptibility annually. There were three pricing options, all different mixes of GO, HFO, PPI and coal or electricity.

One of the reasons OFGAS instructed BG to publish price schedules was to facilitate market entry by competitors. When BG had lost about 45% of the contract market, tariff publication for this market was suspended. New entrants to the market gained market shares by lowering prices in relation to BG tariffs. BG responded by granting rebates on its published tariffs. After a while the type of contract entered into also changed: one year contracts with fixed prices without differentiation according to load factor or season became widespread. With a lot of new suppliers coming into the market and with an increasing surplus of gas, a market price arose which was no longer set by one dominating actor. During the course of 1994/95 this development led to a spot market for gas. This market has since then gone through several phases in terms of size, fluidity, transparency. The transition to daily balancing on BG TransCo' system in the autumn of 1996 led to the creation of a flexibility market which added another element to the spot market.

The following table shows quite clearly that prices in the contract market have come down over the period since the beginning of 1994.

Table 7 Prices to Industrial Customers 1994-1997 (pence per kWh)

Year (quarter)	Small	Medium	Large	Average	Firm	Interruptible
Q1 1994	1.221	0.952	0.752	0.805	0.941	0.647
Q2 1994	1.288	0.931	0.722	0.768	0.896	0.657
Q3 1994	1.264	0.960	0.736	0.759	0.853	0.684
Q4 1994	1.167	0.918	0.741	0.776	0.861	0.682
Q1 1995	1.143	0.930	0.739	0.784	0.889	0.668
Q2 1995	1.109	0.925	0.666	0.703	0.807	0.602
Q3 1995	1.146	0.821	0.584	0.613	0.740	0.505
Q4 1995	1.038	0.758	0.564	0.600	0.714	0.503
Q1 1996	0.960	0.673	0.451	0.494	0.546	0.433
Q2 1996	0.949	0.664	0.427	0.455	0.504	0.409
Q3 1996	0.960	0.639	0.420	0.437	0.480	0.402
Q4 1996	0.883	0.654	0.432	0.462	0.507	0.417
Q1 1997	0.886	0.687	0.455	0.496	0.567	0.428
Q2 1997	0.876	0.677	0.462	0.491	0.563	0.440
Q3 1997	0.896	0.685	0.467	0.492	0.552	0.452
Q4 1997	0.915	0.719	0.513	0.549	0.599	0.495

Source: DTI, Energy trends

Distribution Charges

The tariffs described above are all tariffs to end consumers and are prices paid for a bundled service, i.e. for gas delivered at the meter. In recent years it has become possible to buy gas and transportation and storage services separately. Given that transportation accounts for such a significant part of total gas price, there has in recent years been a lot of focus on the tariffs applied by BG Transco, the only public gas transporter so far. Important elements in the debate on transportation tariffs have been the potential for efficiency improvements and the correct reflection of costs in tariffs. It would take too long to look into this debate in this context, but since the UK is the only country in Europe having published tariffs specifically for distribution services, a brief overview of these will be given. Since the transmission and distribution system are heavily integrated, a few observations on the transmission system are necessary.

The National Transmission System (NTS) in the UK is a network of pipelines operating at pressures of up to 75 bar which transports gas from the entry points to local transmission and distribution systems, other connected systems and large volume consumers. NTS charges are split into capacity and commodity elements. At present each recovers approximately half of the NTS revenue.

The capacity charge for using the NTS is split into an entry charge, determined by the entry point, and an exit charge, determined by the exit zone which contains the offtake point. These capacity charges are based on the calculated long run marginal costs (LRMCs) of extending the system to meet a sustained increase in demand. Entry and exit charges are capacity charges. They reflect peak load requirements and so are levied in pence per peak day kilowatt hour. A uniform commodity charge is applied to all gas throughput on the NTS. In other words, each kilowatt hour of gas transported through the NTS incurs the same commodity charge.

Each Local Distribution Zone (LDZ), of which there are 12 and which take their gas from the NTZ, contains the local transmission system, a network of pipelines operating generally at pressures up to 38 bar, and the distribution system, a network of mains operating in three pressure tiers: intermediate (2 to 7 bar), medium (75m bar to 2 bar) and low (below 75m bar). The total length of the gas pipeline network in the UK is about 265.000 kilometres, out of these about 247.000 kilometres are distribution pipelines. Nearly 98% of end users are connected to the low pressure system. Charging for the use of the LDZs is separate from that on the NTS.

The LDZ charging functions – both capacity and commodity- use site peak capacity as the measure of customer size. The capacity and commodity charges each recover approximately half of the LDZ revenue. At daily metered sites shippers book a daily capacity which is monitored by use of dataloggers. It is not economic to install dataloggers at all 19 million supply points to obtain daily consumption figures. For non daily metered sites, the peak daily load is estimated through the use of a set of algorithms. These algorithms employ a range of end user categories, and are also used on a daily basis to estimate the consumption of non daily metered sites.

In 1994, a new methodology for the calculation of charges for the low pressure element of the LDZ system was introduced. This methodology is more cost reflective and rebalances the low pressure system charge between large and small customers.

The functions used to calculate the LDZ capacity and commodity charges are shown in the following tables.

Table 8   LDZ Capacity Charge

Load size	Pence per peak day therm per annum
Up to 2.500 therms per annum	427.9
2.500 to 25.000 therms per annum	827.8-242.3*ln{ln(PL)}
Above 25.000 therms per annum	1014.4-328.1*ln{ln(PL)}

Note: ln{ln(PL)} means the natural logarithm (to the base e) of the natural logarithm of the peak daily load, PL, in therms.
Source: BG TransCo

Table 9   LDZ Commodity Charges

Load size	Charge in pence per therm
Up to 2.500 therms per annum	7.650
2.500 to 25.000 therms per annum	7.635-2.499*ln{ln(PL)}
Above 25.000 therms per annum and below 327.497 therms per peak day	6.903-2.397*ln{ln(PL)}
Above 327.497 therms per peak day	2.317-0.595*ln{ln(PL)}

Source: BG TransCo.

The LDZ capacity charge is degressive with the peak daily load. The LDZ commodity charge is not uniform as on the NTS; it is degressive with annual offtake. In addition to the capacity and commodity charge the customer has to pay a customer charge designed to reflect customer related costs. At present these are service pipes, meters, emergency work and meter reading. Of these, the costs of service pipes, meters and emergency cover are related to customer size, while meter reading costs vary with the frequency of meter reading. For sites which consume less than 2.500 therms per annum, the customer charge is made up of a fixed charge plus a commodity charge. For sites which consume between 2.500 and 25.000 therms per annum, the customer charge is made up of a fixed charge which depends on the frequency of meter reading plus a fixed capacity charge. For sites which consume over 25.000 therms per annum, the customer charge is made up of a function related to the peak daily load and, for sites which contained datalogged meters, datalogger charges.

To give an impression of the relative magnitude of the various elements, an example will be given. A shipper brings gas onshore at St. Fergus to supply a customer in Glasgow taking an annual load of 20000 therms. The site has its meter read monthly. The total transportation charge in this case amounts to 11.86 p/therm and can be broken down as follows (figures in pounds per annum):

NTS entry capacity charge	321.03		
NTS exit capacity charge	17.23		
NTS commodity charge	219.80	558.06	23.6%
LDZ capacity charge	681.59		
LDZ commodity charge	733.46	1415.05	59.6%
Customer charge, variable element	230.05		
Customer charge, fixed element	169.50	399.55	16.8%
Total		**2372.66**	**100%**

Source : British Gas Transco.

Security of Supply and Modulation – Implications for Pricing

To understand the British thinking in terms of security of supply it is important to be aware of some fundamental facts:

■ although importing minor gas volumes from Norway, the UK is theoretically self-sufficient in gas. Considerations about building storage to counter politically motivated interruptions in supply have therefore never been on the agenda;

■ the former high degree of integration of the industry , BG's responsibility for security of supply and natural conditions on the fields made it natural to provide a high share of needed swing from the producing fields themselves. Storage in terms of number of days' demand that could be covered from storage is modest compared to most countries in continental Europe;

■ the high heating load results in violent daily and seasonal fluctuations in demand that require a lot of modulation capacity.

To face these fluctuations in demand, the following can be done:

■ provide swing capacity at the beach. Historically this has been done by BG entering into long-term depletion contracts with high flexibility in supply, which of course made development of these fields more costly than in a situation where supplies are level. Until recently producers were faced by a purchasing monopoly and had access to no other means of modulating supplies;

■ store gas in available depleted fields, salt caverns and onshore LNG peakshavers;

■ create interruptible contracts, which is an important element in the UK: in 1995 sales under interruptible contracts accounted for about 9% of total sales;

■ construct additional pipe capacity, gas holders and high pressure bullets to take care of diurnal swing in demand.

The UK has well defined criteria for security of supply (see UK annex in IEA gas security study) according to which BG historically has been supposed to operate and according to which licence holders will be supposed to operate in the future. In a monopolistic setting it is quite clear that BG did not have the same incentives to think explicitly about the cost of security of supply as it would have in a competitive situation.

In terms of costs and tariffs relating to security of supply, the recent development towards a competitive market has meant increased focus on these issues:

■ the unbundling of BG's activities into trading and transportation means that explicit tariffs have been developed for the commodity and for the services rendered. When there is pressure from competitors or from regulators to reduce costs, a more explicit thinking about where costs stem from and how they should be allocated is needed;

■ within the transportation system, separation of storage and transportation proper, presumably means a more explicit tradeoff between various ways of providing security of supply, for instance the tradeoff between building more storage and provision of flexibility from other sources on the transportation network;

■ the separate sales of and charging for storage services gives both BG and other actors an incentive to compare the cost of increased security of supply from building and using storage, using interruptible contracts and from flexibility in supply contracts. Producers now can compare the economics of providing flexibility through dimensioning of the infrastructure on the platforms with the alternative of buying storage services. To the extent that interruptible contracts and storage are substitutes, it is now possible to compare the economics explicitly. Competition also means that other suppliers than BG provide interruptible contracts, and that prices have come down.

Looking at the tariffs applied to end consumers, one will not find any explicit elements supposed to cover cost associated with security of supply (another aspect in this context is that there are fewer and fewer tariffs about). Generally, however, there has been a tendency in recent years to focus on cost drivers when constructing the tariff formulae. Peak day capacity has been found to be an important cost driver both on the transmission and distribution system. One requirement is that tariffs should reflect the extent to which the various categories of customers cause costs to arise because they need high flexibility and consequently investment in capacity. To the extent that this philosophy is prevalent, one might say that security cost is also reflected in the tariffs.

The cost of storage undertaken both for seasonal and daily modulation purposes may be considered as costs associated directly with security of supply. To the extent that this is admitted, one may say that it has now got an explicit price in the UK market. BG TransCo provides storage services as a regulated activity with well defined tariffs in the market. Transco operates seven storage facilities throughout the UK. Tariffs are site based and are supposed to reflect the long run marginal cost of providing additional storage capacity. For each site there is a capacity price (consisting of two elements, payment for reserved space in pence per kWh per annum, and a payment for reserved deliverability in pence per peak day kWh per annum) and a commodity price (consisting of a storage injection price in pence per kWh and a storage withdrawal price in pence per kWh). BG TransCo's obligation in terms of security of supply are, however, are reflected in the agreements with the customers buying services from the LNG storages: these customers are obliged to provide transmission support gas to TransCo on days with very high demand and to retain a minimum inventory level of gas so that transmission support gas is available all winter. The LNG storages are situated on the parts of the NTS most remote from the beach terminals. On days of high demand some of the gas required at the extremity of the network comes from the local LNG site. This means that pipelines feeding that locality do not have to provide gas for the full demand on a peak day. The required capacity of the pipeline has therefore been reduced and investment saved. This use of LNG storage to save on pipeline investment is known as transmission support.

Taxation

From a taxation point of view, natural gas is favourably treated in the UK: there are no taxes on this fuel apart from 5% VAT, the lowest rate within the European Union. The TVA is refundable for all consumers except in the residential sector. HFO and GO, the main competitors of natural gas are subject to excise taxes of 1.94 pence/litre and 2.5 pence/litre, respectively, in all sectors (corresponding to 0.71 ECU/GJ (NCV) and 0.91 ECU/GJ (NCV), respectively).

Gas Prices in Relation to Prices of Other Fuels

Figure 4 shows how gas prices in the residential and industrial sectors have developed on a quarterly basis since 1990 compared to the prices for gasoil and heavy fuel oil. Unsurprisingly, there is no close link between gas prices and the prices of oil products. In the residential sector, prices are set on a cost plus basis, recently to some extent also as a result of gas to gas competition. In the industrial sector, gas prices traditionally have been linked to oil product prices to some extent, but over the past few years, this link has been weakened by gas to gas competition.

Figure 4   Prices to Industrial and Residential Customers in the United Kingdom

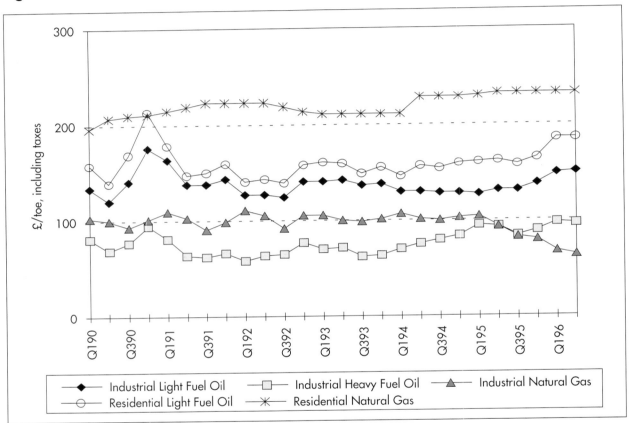

Value Added, Costs and Profitability

When speaking about value added and profitability in gas distribution in the UK, one is basically speaking about the performance of BG, at least up to the beginning of the 1990's. An attempt will therefore be made to look in to BG's performance and then to supplement this with some comments on the performance of the new marketers of gas.

Table 10 shows the development in prices, gas purchase cost and gross over the period 1983 to 1992.

Not surprisingly, the tariff market features the highest margins this is also where distribution costs are highest. In actual prices, total average gross margin increased over the period between 1983 and 1992 (from 17.4 p/th as the lowest in 1985 to 23.2 p/th as the highest in 1992). In real

Table 10   British Gas Average Revenue from Sales, Average Supply Price and Gross Margins, Actual Prices, over the Period 1983-1992,

Actual prices	1983	1984	1985	1986	1987	1988	1989	1990	1991	1992
Domestic Tariff	38.0	39.8	41.3	42.8	43.2	42.2	44.0	45.8	50.9	50.7
Contract	30.7	31.0	33.5	34.8	34.7	33.2	32.1	30.8	31.4	31.0
Total domestic	37.9	39.7	41.2	42.7	43.1	42.0	43.8	45.6	50.7	50.5
Commercial Tariff	**33.2**	**35.0**	**36.5**	**38.1**	**38.8**	**37.7**	**39.3**	**41.0**	**45.8**	**45.2**
Contract:										
Firm	31.1	31.6	33.2	34.9	33.9	33.1	32.1	30.6	32.8	33.3
Interruptible	26.0	25.9	27.5	29.2	21.6	21.3	17.6	19.4	20.2	21.0
Total	30.2	30.6	32.2	34.1	32.0	31.2	29.7	28.6	30.8	31.0
Total commercial	31.9	33.1	34.6	36.2	35.6	34.5	34.7	34.9	38.7	38.6
Industrial Tariff	33.0	34.7	36.2	37.7	38.6	37.4	39.1	40.8	45.7	44.8
Contract:										
Firm	24.6	27.4	30.1	32.4	30.4	29.6	28.8	28.0	28.2	28.7
Interruptible	21.4	21.5	24.1	26.0	17.3	18.0	15.9	16.9	17.6	18.3
Total	22.9	23.9	26.4	28.6	22.9	22.7	21.7	21.4	21.7	21.3
Total industrial	23.3	24.4	26.9	29.1	23.9	23.5	22.7	22.4	23.2	22.9
Actual prices	1983	1984	1985	1986	1987	1988	1989	1990	1991	1992
All Customers' Tariff	37.3	39.1	40.6	42.0	42.5	41.5	43.3	45.0	50.1	49.8
Contract:										
Firm	26.3	28.7	31.1	33.2	31.7	30.9	30.0	29.0	28.2	30.2
Interruptible	21.7	21.8	24.2	26.2	17.6	18.2	16.0	17.1	17.8	18.6
Total	24.1	25.1	27.5	29.7	25.0	24.6	23.6	23.1	24.1	24.0
Average total revenue	32.1	33.7	35.5	37.4	36.3	35.1	36.2	36.7	41.5	41.9
Annual Gas Supply – Average Price Paid by BG										
Pence per therm purchased - inc. levy	14.0	15.6	18.1	19.8	17.9	16.4	16.7	17.3	18.8	18.7
Gross margin	18.1	18.1	17.4	17.6	18.4	18.7	19.5	19.4	22.7	23.2

Source: MMC study of British Gas data

terms, the gross margin decreased somewhat over the period: the highest figure in 1991 prices was 29.4 p/th in 1983 and the lowest 22.0 p/th in 1990. It is interesting to note that the decline in the oil price from the middle of the 1980's is not clearly reflected neither in average sales prices nor in gross margins.

Taking the gross margin as the point of departure for each market segment, profits will depend upon the allocation of cost to each segment. Such cost allocation will never be objective in the sense that an element of judgement will have to come into play. Table 11 shows how BG perceived its profit in each segment over the period 1987 to 1992.

Table 11 Market Segment Profitability (pence per therm, at 1992 prices)

	1987/88	1988/89	1989/90	1990/91	1991	1992
Tariff: domestic						
Turnover	52.3	51.8	51.0	52.3	52.3	50.5
Costs	46.8	44.4	43.9	44.3	44.7	43.9
Profit	5.5	7.5	7.1	8.0	7.6	6.6
Tariff: non-domestic						
Turnover	46.9	46.4	46.0	47.3	46.9	45.1
Costs	34.6	33.1	33.1	33.2	34.7	35.0
Profit	12.4	13.3	12.9	14.1	12.1	10.1
Firm contract						
Turnover	39.8	36.7	33.4	32.2	31.6	31.9
Costs	28.5	27.2	26.6	26.8	26.8	28.1
Profit	11.4	9.5	6.8	5.4	4.7	3.8
Interruptible						
Turnover	22.6	19.0	19.3	19.2	18.2	18.4
Costs	18.1	17.0	16.6	16.2	16.2	15.9
Profit	4.5	2.0	2.7	3.0	2.0	2.5
Third party transportation						
Turnover				5.8	6.2	6.3
Costs				5.7	4.8	4.6
Profit				0.1	1.4	1.7

Source: British Gas

According to these figures, non-domestic tariff customers constitute the most profitable customer segment followed by domestic tariff customers. The table reflects the increasing competition in the contract market – profits are on a decreasing trend. From the beginning of the 1990s' some profits were also earned on third party gas transportation. As will be pointed out below, profits from BG's transportation business over time has become the most important element in the company's total profits in recent years.

Over the past few years, BG has undergone a number of changes in terms of organisation and accounting, especially as a result of changing regulatory requirements. The most important single factor has probably been the operational separation of Transco from the rest of the supply activity from March 1994. These changes make it difficult to compare margins and profitability in the various market segments on a consistent basis. In the context of the demerger of the company in early 1997, an attempt was made to analyse the historical profitabilty of the parts of BG that were supposed to constitute Centrica, the new holding company now responsible for gas supply, marketing and trading. (It should be noted that this company will buy all the transmission and distribution services it needs from Transco.) Over the period from 1993 to 1995, the new company would have realised the following margins:

	1993	**1994**	**1995**
Average sales revenue	41.91	41.99	42.03
Average purchase price incl. levy	16.03	16.37	16.62
Average gross margin	25.78	25.62	25.41

Source: Own calculations based on "Centrica - Introduction to the official list"

The recent discussions on the demerger of BG has of course raised questions about the relative profitability of BG's different activities. The analyses done by Ofgas shows that for 1994 and 1995, the major part of BG's profits were earned by Transco, the gas sales activity showing a much weaker result. In 1994, the total operating profit of the BG group on a current cost basis was 987 mill. GBP, of which Transco contributed about 64%. In 1995, operating profit was 977 mill. GBP with a Transco share of about 80%.

The analysis of Centrica's performance over the period since 1993 referred to above shows that gas marketing has remained profitable over the period, but when taking into account the appliance retailing activity and the sales of installation services, profits for the period were negative when exceptional charges are taken into account. In terms of relative profitability, there is no doubt that gas transportation is more profitable than gas marketing.

In terms of historical profitability for the BG group as a whole, it is probably fair to say that this has been relatively high over the last ten years, but declining. The second MMC report showed the following profitability figures (million pounds, per cent):

	1987/88	**1988/89**	**1989/90**	**1990/91**	**1991**	**1992**
Operating profit	976	1.017	879	1.248	1.268	989
Average capital employed	15.501	15.394	16.148	17.258	17.813	18.868
Profit on turnover	14.8	15.6	12.9	15.9	15.2	12.3
Profit on capital employed	6.3	6.6	5.4	7.2	7.1	5.2

Source: MMC

For the years 1993, 1994 and 1995, the profit margin on turnover was -3.1%, 10.18% and 6.78%, respectively. One of the reasons for declining profits over the past few years is the fact that BG's prices are regulated by price capping. In the 1997 Price Control Review for supply at or below 2.500 therms a year, Ofgas proposed that the profit margin should be limited to 1.5% on turnover.

In the 1997 Price Control Review of British Gas' Transportation and Storage, Ofgas proposed a regime intended to lower prices and profits for Transco. In the autumn of 1996 Ofgas made reference to the MMC asking to investigate whether Transco pricing is against the public interest. No matter what the outcome of the investigation may be, there is no doubt that Transco will remain under pressure to reduce its profits.

CANADA

Statistical information

Natural Gas Supply/Demand Balance (Mtoe, 1996):

Indigenous production	135.1
Imports	1.0
Exports	- 65.5
Stock change	- 0.2
Total natural gas supply (primary energy)	70.4
Electricity and heat production	3.7
Other transformation and energy use	11.1
Total industry	23.2
Residential	14.9
Commercial	9.7
Other	5.9
Statistical difference	1.9

Natural Gas in the Energy Balance (1996):

Share of TPES	29.8%
Share of electricity output	2.9%
Share in industry	36.1%
Share in residential/commercial sector	43.2%

OVERALL STRUCTURE OF THE GAS INDUSTRY

Canada is the third largest gas producer and consumer and the second largest exporter in the world. It probably has the most unfettered and most competitive gas market in the world which is the result of a series of deliberate policy actions stemming from a general economic policy of relying on market forces. The competitive regime, including also end user markets, is the result of significant deregulation by both federal and provincial governments.

The gas industry has gone through a number of radical changes from a situation in the 1970s where private contract carriage on the gas transmission system from western to eastern

Canada restricted consumer supply options to a single gas purchase monopoly, and prices all along the gas chain from wellhead to city-gate were regulated on an oil-equivalent basis. Today's industry is characterised by keen competition among producers and among suppliers of gas to final consumers. For instance, Canada is one of the few countries in the world where residential consumers can choose their supplier. The most important steps taken to deregulate the industry have been the scrapping of wellhead gas price regulation, evolution of the gas export requirements toward a more market-oriented approach, and the implementation of mandatory third party access to pipelines with unbundling of purchasing, transportation and sales activities. It should be noted that the transmission pipelines have kept their monopoly. They are regulated on a cost plus/rate of return basis.

The Canadian gas industry consists of gas producers, processors, transporters, marketers, and distributor. Unlike the oil industry, no companies in the gas industry are fully integrated from production to distribution. There are, however, linkages between the upstream and downstream sectors. Large producers often own facilities which process raw natural gas. As well, a few pipelines and local distribution companies have production and marketing subsidiaries, as there are major transmission pipelines that have ownership interests in LDCs.

Simplifying a bit, one might say that the Canadian gas industry consists of about 1000 producers whose gas production is transported on 8 major transmission pipelines to about 4.8 million customers (4.2 million residential customers, 0.47 million commercial customers, 18.000 industrial customers), most of them served by 16 distribution companies or their affiliates.

The production sector is composed a wide variety of companies, from large multinational oil and gas companies to small local firms. The smaller producers tend to sell their output through marketers and aggregators, while many large companies market their supplies directly. While there are around 1000 producers in Canada, the largest 100 firms account for more than 85% of production.

All transmission pipelines, both interprovincial and intraprovincial, are owned and operated by private sector companies, except the gas transmission system in Saskatchewan which is a Crown corporation. The major natural gas pipeline transmission systems are Westcoast Energy Inc. in British Columbia; NOVA Gas Transmission in Alberta and TransCanada Pipelines Ltd. east of Alberta. These systems carry gas both for domestic and export markets. In addition there are several export-oriented pipelines such as Alberta Natural Gas Company Ltd. and Foothills Pipelines Ltd.

Distribution is carried out by local, predominantly privately-owned utilities which have a monopoly as fas as the pure distribution function is concerned. For many years, large industrial customers and power generators have been able to buy their gas directly from producers. Recently, some provincial regulatory boards have permitted core market customers, i.e., end-use consumers in the residential/commercial sector, to buy gas directly from producers through aggregators, brokers or other middlemen. Canada was the first country in the world to do this. This arrangement does not mean, however, that the local distribution companies have lost their distribution monopoly: they still have a monopoly on physical distribution of gas.

MAIN ACTORS IN GAS DISTRIBUTION

Despite the emergence of new players such as aggregators, marketers and brokers on the Canadian gas scene in recent years, the LDC remain major actors because of their transportation monopoly from city gate or outlet of transmission pipelines to the end consumer and because they still sell gas to end users. In fact, some of the LDCs are also major transmission companies and transport gas for others both on transmission pipeline and distribution pipelines.

Table 1 shows an overview of most of the distribution companies in Canada. In total there are 16 LDCs, but the largest 8 account for about 95% of total LDC sales and about the same percentage of customers. The biggest utility, Consumers Gas has more than 1.2 million customers and sells about 10 bcm of gas a year whereas the smallest of the 16 LDCs has about 9000 customers and sells about 100 million cubic meters a year. In geographical terms, there is a concentration in the eastern part of the country in that the 4 companies here account for about 60% of sales and more than 50% of the customers.

With one exception, SaskEnergy, the gas distribution company in Saskatchewan which is a Crown corporation, all the Canadian LDCs are privately owned. There is some tendency towards concentration of ownership interests in that Westcoast Energy Inc., one of the leaders in the North American energy industry, having interests in gas transmission pipelines, gas processing and storage facilities, power generation and gas services business, owns 6 of the LDCs, the biggest of which are Union Gas Limited and Centra Gas Ontario Inc., two of the major three companies in Ontario. These two companies illustrate the variation in structure among the LDCs: Union Gas Limited is an integrated natural gas storage, transmission and distribution company. In addition to serving over 750,000 customers in southwestern Ontario, it also provides storage and transportation services for other utilities in Ontario, Quebec and the United States. Its throughput in 1996 was almost 27 bcm whereas sales to distribution customers were around 9 bcm. Centra Gas Ontario Inc., on the other hand, is a pure distribution company which sells the major part of the gas it transports through its distribution system to its own customers, but also serves some customers who buy directly. Union Gas and Centra Gas Ontario merged their operations effective January 1, 1998.

There are also examples of foreign ownership in some of the Canadian LDCs: until recently British Gas held a major participation in Consumers Gas. This participation has now been taken over by IPL Energy Inc., a company based in Calgary, which now holds 100% of the common share capital. Since 1994, Gaz de France holds interests in Noverco, the company owning Gaz Metropolitain in Quebec.

LDC SERVICES

Canadian LDCs are pure gas distribution companies in the sense that they are not involved for instance in electricity distribution. Some of them, however, have interests in co-generation. Sales

Table 1   Canadian Distribution Companies

Company	Total Sales (mcm)	Residential Sales (mcm)	Commercial Sales (mcm)	Industry Sales (mcm)	Volume Transported (mcm)	Total Customers ('000)	Residential Customers ('000)	Commercial Customers ('000)	Industrial Customers	Total Employees	Distribution Mains (km)
Consumers Gas	9,859	3,885	4,132	1,842	2,271	1,288	1,159	121	6,857	3,805	24,626
Union Gas/Centra Gas Ontario[1]	9,102	2,777	2,057	4,268	4,641	1,002	902	94	5,584	3,342	28,556
Gaz Métropolitain	6,211	756	1,681	3,774	N/A	175	133	40	2,028	1,466	7,433
BC Gas Utility Ltd.	3,744	2,078	1,430	236	821	716	641	74	594	1,779	16,775
Northwestern Utilities	3,275	1,531	1,568	175	9,293	382	342	40	167	1,162	15,063
Canadian Western Nat. Gas Co. Ltd.	2,754	1,457	1,183	114	3,283	356	326	29	154	1,068	16,113
Sask Energy	2,253	1,172	767	314	N/A	304	266	36	175	657	60,764
Centra Gas Manitoba	1,739	777	665	297	294	235	214	22	141	647	4,624
Fed. of Alberta Gas Co-ops. Ltd.	638	N/A	N/A	N/A	N/A	87	85	3	2,000	320	84,700
Centra Gas Alberta	376	224	152	N/A	99	50	45	5	N/A	130	16,237
Corp. of the City of Kitchener	284	N/A	N/A	N/A	N/A	43	N/A	N/A	N/A	41	789
Centra Gas BC	239	70	169	N/A	N/A	50	43	7	11	249	3,266
Pacific Northern Gas Vancouver	264	79	63	122	738	28	24	3	38	136	1,225
City of Medicine Hat Gas Utility	174	74	89	10	N/A	19	17	2	18	25	550
Gazifère	188	45	52	91	N/A	18	17	2	12	42	448
PUC City of Kingston	98	22	43	33	N/A	9	7	2	13	31	226
TOTAL	**41,197**					**4,765**				**17,792**	

(1) Effective January 1, 1998 Union Gas and Centra Ontario merged their operations

Sources: Canadian Gas Association, 1997 Natural Gas Utility Directory, September 1997; and Annual Reports

of gas appliances and provision of energy management services to customers are common LDC activities. The major utilities in Ontario are also engaged in gas based water heater rental business. Some of the major LDCs express an interest in diversifying into other activities like electricity production and distribution and water distribution, claiming that there are considerable synergies in metering and billing between these services. The advancement of technology in for instance electronic metering underpins this claim. One view expressed by some companies is that the typical distribution company in the future will be one-stop energy shops offering all types of energies and services that end consumers might require.

In addition to serving LDC system supply customers, i.e., customers who buy their gas from the LDC which delivers it at their site, the LDCs are also involved in cases where the customers buy their gas directly from suppliers. In the province of Ontario this has become a widespread phenomenon; in 1996 about two thirds of total volumes distributed were acquired through direct purchase. In these cases, the LDC offers transportation service. Technically there are two instruments for direct purchase: T-service and Buy/Sell arrangement In the first case the customer buys his own gas and delivers it to city gate (or upstream pipeline receipt point) and asks the LDC to transport it to the point of use. There is a daily link between delivery and use, but the agreement with the LDC usually includes load balancing. In the second case, the LDC buys the customer's gas at a receipt point on the upstream pipeline (or city gate) at a constant rate equal to its Weighted Average Cost Of Gas (WACOG). It then resells the gas to the customer at the point of use at regulated rates. The customer is saving the difference between the LDC's WACOG and the customer's acquisition cost. (This arrangement resembles the Single Buyer concept that has been discussed for electricity in Europe.) The Buy/Sell arrangement is most popular among all but the largest end users.

GAS SUPPLY

Over the last few years, the supply portfolios of the LDC have become increasingly short-term in terms of contract duration. There are still some long-term contracts (10 to 15 years in some cases), but some of the LDCs do not have a planning horizon stretching beyond two and a half year. The major part of the volumes supplied are delivered under one year contracts where prices are renegotiated every year. Most of such contracts would, however, have pricing mechanisms that reflect monthly pricing variations in the North American natural gas market, rather than fixed prices. These contracts could for instance be indexed to the New York Mercantile Exchange (NYMEX) or to price indices in the production regions like Alberta. Most companies have implemented risk management strategies to manage the price volatility of its long-term firm gas supply. Swap arrangements are frequently used to fix the NYMEX prices for the long-term supply volumes. Apart from the long-term contracts, gas is purchased on daily, weekly and monthly contracts.

REGULATION

LDCs in Canada, are exclusively regulated at the provincial level by Public Utility Commissions (PUC). While each province has an independent PUC, one common thread ties them together: they regulate the LDCs by determining the services that can be offered to various customer classes, by setting rates for those services, and by authorizing construction of transmission and distribution lines, including approving and recommending the granting of a franchise area. The PUCs endeavour to ensure that rates are fair, that gas supplies are secure, that environmental issues are addressed, and that the public interest is upheld. To protect core market customers, most provincial PUCs have imposed minimal supply conditions to Agents/Brokers/Marketers (ABMs). They are usually required to hold natural gas supply to cover all their direct sales for a number of years (1-5 years). In Ontario, when a consumer no longer purchases gas from an LDC, the consumer no longer has the security of supply that comes with the LDC. Moreover, ABMs do not have to meet any minimum supply requirements to serve residential consumers. In a case of supply disruption, the PUC relies on the other ABMs or the LDCs, which are supposed to use all "reasonable means" to mitigate any gas disruption. Table 2 shows by whom some of the major utilities in Canada are regulated.

Table 2   Regulators of Selected LDCs in Canada

LDC	Regulator
B.C. Gas Utility Ltd.	British Columbia Utilities Commission
Canadian Western Natural Gas	Alberta Energy and Utilities Board
Northwestern Natural Gas Ltd.	Alberta Energy and Utilities Board
SaskEnergy Inc.	Provincial Government
Centra Gas Manitoba	Manitoba Public Utilities Board
Union gas Limited	Ontario Energy Board
Centra Gas Ontario	Ontario Energy Board
Consumers' Gas	Ontario Energy Board
Gaz Metropolitain Inc.	Regie de l'energie

FRANCHISES

Gas distribution franchises are awarded by municipalities, most of the time in accordance with agreements containing standard rules which define acquisition of rights of way and specify who should pay for what. According to provisions in the Municipal Franchise Act, franchises are not exclusive from a strictly legal point of view, but in practice this is the case most of the time. It happens, however that two companies compete for the franchise in one municipality. The franchise period varies between 15 and 30 years. When a utility wants to sell its system or amalgate with another company, permission must by obtained from the PUC in the province in question. In some cases there are also restrictions on ownership: for instance in Ontario, approval is necessary when any person wishes to acquire or hold more than 20% of any class of shares.

The most conspicuous and maybe most important role of the PUC is to approve the rates for the services offered by the LDCs. The primary objective when setting rates is to guarantee that the public interest is served and protected. The PUC sets rates as low as possible while providing utility investors an opportunity to earn a fair return.

RATE MAKING

Rates are set according to a cost plus methodology (see the IEA gas transportation study for a discussion of this type of regulation). The resulting rates primarily reflect cost and in their pure form do not take competition with other fuels into account at all. Since LDCs in principle cannot discount prices, this means that for instance when oil product prices are low, gas may have problems in remaining competitive. For this reason, flexible rates that allow the utility to set a sales rate in response to competition have been allowed in some areas. This can be accomplished by creating a predetermined range within which a specific rate can be set. Although many LDCs in Canada use flexible rates in one form or another, they are usually limited and not globally defined. Practice varies a lot between provinces and LDCs which is illustrated by the following examples:

• For Centra Gas BC, 75% of its service area operates under rates that are decoupled from cost of service. Since its rates are indexed to the market price of oil, the rates do not necessarily recover the cost of service, so the shortfall are to be recovered from future customers.

• For BC Gas , a form of flexible rates is used for its interruptible/peaking service and off-system sales. Sales rates are either negotiated subject to regulatory approval, or indexed to reported prices, but should not be below the commodity cost.

• Centra Gas Ontario allows large volume interruptible customers to negotiate both their gas supply and delivery rates within pre-established ranges.

• Union gas views the use of flexible rates as a means to address a number of competitive situations including bypass (facilities competition), alternative fuel competition and competition in adjacent market areas.

• Gaz Metropolitain issues rebates, based on a set formula, on regular rates to defend against low oil prices. Currently this is a temporary program.

Regulators have not explicitly stated their views on flexible rates in all cases. In Ontario, The Ontario Energy Board has denied flexible rates for Consumers Gas but has approved them for Centra Gas Ontario and Union Gas. In general, utilities favour flexible rates and presently discuss this as one of the issues on which the industry could take a joint initiative.

INCENTIVE REGULATION

Another issue that has emerged in the wake of the recent criticism of cost plus regulation is that of incentive regulation, not only for interprovincial pipelines but also for distribution companies. Incentive regulation involves the establishment of criteria set for utilities aimed at increasing productivity without a decrease in quality of service, which would allow the opportunity for financial reward. This type of regulation is not widespread as yet, but there are some examples:

Centra Gas BC's present flexible rate structure has within it a form of incentive regulation based on meeting a set of targets.

■ BC Gas is currently operating under an incentive regulation program. This program was approved by the BC UC for a three year term, ending in 2000.

■ Gaz Metropolitain has an incentive regulation program and although the utility must absorb all deficiencies, the amount of surplus it is permitted to keep is capped. Its purpose it to demonstrate that excess returns are not a result of cost cutting which would negatively affect quality of service.

Incentives regulation is becoming prevalent in gas utility regulation in Canada and there is a general movement toward reduced administration in the regulatory process and self-regulated incentives are seen by the industry as being consistent with this development. The industry does not seem to be in favour of introducing incentive regulation that ignores the specific situation of each company.

GAS COST

The cost of gas is an important element in the price that LDCs are allowed to charge their customers. The LDCs, however, are not allowed to earn any money on gas supplies, i.e. even in the cases where they are still operating as merchant companies, gas costs are passed through without adding any margin. The WACOG forecast by the LDC is subject to approval by the PUC. All customers are charged the WACOG; "streaming" or discrimination is not allowed. The LDCs are generally required to organise tendering procedures for all gas purchases, and all purchases are in principle subject to prudency tests. The use of hedging instruments and other risk management techniques by the companies have made the regulatory task more demanding for the regulator.

OBLIGATION TO SUPPLY

For quite some time, gas distributors and industrial customers have been able to buy gas directly from producers. More, recently, the provinces have also permitted direct sales to core market customers which basically means customers that are captive gas customers. Over the last couple of years, a debate has arisen in Canada as to whether the LDCs should be allowed to keep their merchant function or not. A separation of the merchant function and the transport function would imply that the LDCs would no longer be allowed to sell bundled services Some gas users claim that the time has now come to complete the process of introducing competition by undertaking this type of unbundling. They maintain that gas end user markets can never become fully competitive as long as the LDCs control both activities since they would be inclined to favour their own transportation needs. Some LDCs seem to be of the opinion that the public interest would not be well-served if they were excluded from the merchant function. However, LDCs in Ontario are currently in the process to separate their merchant function from their distribution function.

One of the crucial issues in this debate concerning unbundling is the obligation to supply and the question of who should be the supplier of last resort. In general, all LDCs consider that commodity supply for system sales customers is the full responsibility of the LDC, which means that the LDCs take the full responsibility for security of supply to these customers. In British Columbia, Alberta, and Saskatchewan there is a firm obligation for the LDC to replace supply for system supply customers when a supplier fails to meet contractual supply obligations.

Any costs not recovered from the supplier are recovered from all customers. Manitoba and Ontario in essence have the same rules, but slightly different rules for how costs could be recuperated. Also in Quebec there is a legal obligation for the LDC to supply gas to every customer who so requests. When a supplier fails to meet its contractual obligations, the LDC will use its best reasonable efforts to get replacement supply. If the LDC succeeds, and unless the failure is related to force majeure, the cost of replacement gas will be the responsibility of the system supplier that has failed. For reason of force majeure, the cost of replacement gas will normally be borne by all customers (direct purchase and system sales). If the LDC does not succeed in replacing the gas, interruption of distribution service may be needed; but the level of interruption and the customer interrupted (direct purchase or system sales) will be no different than by reason of a cold snap.

All the provinces in Canada allow direct purchases by core customers. The province of Ontario has the most liberal system (and by far also the biggest number of customers). The rules concerning responsibility for gas supplies to direct purchase customers to some extent reflect the general attitude of the province in question to direct purchases. The provinces encouraging direct purchases also have stricter obligations on the LDCs to provide backup supplies. The main rules in the various provinces are as follows:

Ontario

The LDC will obtain replacement supply for a direct purchase customers when contractual obligations are not met. The LDC will charge the customer for the difference in gas cost between the contract price and the replacement price, and the direct cost of supply failure, if any. It may be questioned whether the LDCs legally have this obligation, but in practice they perceive it this way.

Alberta

The LDC must backstop core market direct purchase customers if the customer fails to meet contractual obligations to supply gas. Non-core direct transportation customers can be interrupted if their supplies fail, but the LDCs provide emergency backstop service to those customers if supply is available.

British Columbia

The LDC provides best efforts to replace supply for a direct purchase customer when he fails to meet contractual obligations. Any extra cost are paid by the customer.

Saskatchewan

Direct customers whose supplies fail may be provided backstopped service by SaskEnergy's transportation company TransGas under certain conditions.

Manitoba

The LDC will acquire replacement supplies to the extent that a direct purchase supplier fails to supply volumes which should be available under the gas purchase agreement between for instance a broker and the LDC. Additional costs will be paid for by the broker, or failing that, by the customer of that broker.

Quebec   The LDC may be exempt from its legal obligation to supply and does not have to act as a backstop supplier for a direct purchase customer when he fails to meet contractual obligations. In such case, however, the LDC will use its best reasonable efforts to get replacement supply.

DEVELOPMENT IN PRICES AFTER DEREGULATION

Table 3, showing Alberta average gas prices at the border, give a good indication of the development in Canadian gas prices after deregulation around 1985.

Table 3   Alberta Average Gas Price at the Border

Year	Total domestic price - $/GJ	Real $
1985	2.80	2.80
1986/87	1.91	1.83
1987/88	1.77	1.62
1988/89	1.66	1.46
1989/90	1.72	1.45
1990/91	1.65	1.32
1991/92	1.46	1.11
1992/93	1.50	1.12
1993/94	2.14	1.57
1994/95	1.55	1.14
1995/96	1.53	1.10
1996/67	1.85	1.30

In spite of the fact that the prices shown are export prices at the Alberta provincial border, they are a good indication of the development in end user prices since changes in gas prices paid by the LDCs are supposed to be passed on to the end user. As in the US, there has been a noticeable decrease in real prices since deregulation.

PROFITABILITY/RATE OF RETURN

In Canada, the rate of return earned by distribution companies is transparent since it is defined and fixed for each single company through hearings. In the rate-making process, the regulatory board determines the cost of providing service and approves a rate structure designed to collect that cost in total revenue appropriately from each company's various classes of customers. The cost of providing service includes the cost of gas purchases, depreciation, income taxes, operating and administration costs, and the cost of capital used to finance all assets used in the gas distribution and storage business (i.e., the rate base). The cost of capital, which is expressed as an allowable rate of return on rate base, is designed principally to meet the cost of interest on long and short term debt, satisfy the dividend requirements of preference shareholders and provide common shareholders with the opportunity to earn a reasonable rate on their investment. The determination of a reasonable return to the common shareholders involves a judgmental assessment by the regulatory board of many factors, including returns on alternative investment opportunities of comparable risk and the level of return which will enable the distribution company to attract the necessary capital to fund its operations. Typically, the distribution companies apply for a specified rate of return on common equity and get approval for a rate which is somewhat lower. Table 4 shows the applied for and approved rates for the major distribution utilities in Canada over the last few years.

The lower rates of return on equity allowed by PUC in recent years, reflect mainly the lower interest in Canada.

Table 4   Approved Rates of Return on Common Equity
Major Canadian Natural Gas Distribution Companies

Company	1992	1993	1994	1995	1996	1997
BC Gas Inc.	12.25	-	10.65	12.00	11.00	10.25
Pacific Northern Gas Ltd.	12.75 13.50	12.75 13.50	11.50	12.75	11.75	11.00
Centra Gas BC	13.00	13.00	11.12	12.50	11.50	10.85
Canadian Western Natural Gas	12.25	12.25	12.25	12.25	12.25	12.25
Northwestern Utilities	13.75	11.88	11.88	11.88	11.88	11.88
Centra Gas Alberta	13.25	13.25	13.25	12.00	11.75	11.75
Centra Gas Manitoba	12.60 13.10	12.13	11.25	12.12	11.28	10.58
Consumers Gas	13.13	12.30	11.60	11.65	11.87	10.30
Union Gas	13.50	13.00	12.50	11.75	11.75	11.00
Centra Gas Ontario	13.50	12.50	11.85	12.12	12.12	11.25
Gaz Metropolitain	14.00	12.50	12.00	12.00	11.50	10.75
Gazifere	14.00	12.50	12.25	12.25	11.75	11.75

Sources: Annual Reports

JAPAN

Statistical information

Natural Gas Supply/Demand Balance (Mtoe, 1996):

Indigenous production	2.0
Imports	54.3
Exports	0.0
Stock change	-0.2
Total natural gas supply (primary energy)	56.1
Electricity and heat production	38.5
Other transformation and energy use	-2.4
Total industry	7.7
Residential	8.3
Commercial	4.4
Other	0.1
Statistical difference	-0.5

Natural Gas in the Energy Balance (1996):

Share of TPES	11.0%
Share of electricity output	20.2%
Share in industry	5.8%
Share in residential/commercial sector	14.0%

MARKET AND INDUSTRY STRUCTURE

Japan produces less than 4% of the gas it consumes and relies on LNG imports for the remainder. The Japanese gas market is special in that about 70% of total gas supplies go into power generation. Gas distribution mainly takes place around the LNG import terminals and no country-wide pipeline system exists. Some Local Distribution Companies (LDCs) use other feedstocks than natural gas like LPG, kerosene and other coal and oil based gases for what is called city gas production. The number of LDCs using natural gas nearly equals the number of LDCs using LPG for this production. Different gas quality specifications are in fact one of the obstacles to creating larger gas distribution networks.

In 1995, the city gas companies (i.e., the LDCs) sold 20.2 bcm of gas of which 43.4% went to residential customers, 16.2% to commercial customers, 33.6% to industrial customers and 6.8% to other customer categories. Over the last ten years, demand for city gas in Japan has increased 1.7 times and the number of customers 1.3 times. In 1995, there were 23.58 million city gas customers, but only 5% of Japan's surface and 21% of urban areas are covered by gas distribution networks. The customers are served by 244 local gas utilities, of which 173 are private and 71 public. 20 of these companies have more than 100.000 customers. Only 6 of them have more than 500.000 customers. The three biggest ones, Tokyo Gas, Osaka Gas and Tokio Gas account for about 76% of total gas sales and about 65% of the total number of customers. Tokyo Gas, the largest one, has more than 8 million customers and sold 8.4 bcm of gas in 1995. It sells city gas wholesale to gas companies in neighbouring areas. It regasifies LNG used for power generation on behalf of Tokyo Electric Power Company (TEPCO), and supplies city gas for the generation of electricity.

The LDCs participating in the importation of LNG are vertically integrated in the sense that they also produce city gas and distribute it. Horizontal integration is not widespread in that the LDCs only distribute gas but they are involved in various services related to gas.

REGULATION

The Japanese gas industry is regulated by the Gas Utility Industry Law. This law gives MITI a wide-ranging authority to supervise and regulate the establishment of a gas utility business, the suspension and closure if its operations, the fixing of gas tariffs, the terms and conditions of gas supply, etc. Under this law, LDCs have historically been granted a monopoly position in their specified service area. The approach to tariff regulation is a cost plus approach where all the costs including feedstock, labour, operation and maintenance costs plus a rate of return are taken into account. When gas utility wants to change its tariffs it has to apply for authorization from MITI and to submit a detailed cost analysis. Tariffs are divided into two categories: the tariffs for gas customers in general and the tariffs applying to those customers whose consumption pattern contribute to efficient operation of facilities and hence to cost reductions. These special tariffs reflect the benefit from each customer group to the company's production costs. Examples are tariffs for air conditioning and tariffs for large industrial users.

In June 1994, the Gas Industry Law was revised with a view to introducing competition in the gas sector. From March 1995, regulations of prices to large-volume customers (defined as customers taking more than 2 million cubic metres a year) have been relaxed. Prices may now be set through negotiation between these large-scale customers and gas suppliers. Under certain conditions gas suppliers may supply gas to large-scale consumers other than those in their own service areas. New gas suppliers may also supply gas to large-scale customers. Regulation of gas supplies to small customers will continue as before, based on monopoly service areas and authorization of prices.

One of the effects expected from the modifications of the law is that large-scale customers outside defined supply areas may now be served. Gas supplies to customers in other service areas necessitates access to transportation. The revised law does not introduce mandatory third party access but rather a form of negotiated access.

As of March 1997, 645 contracts had been made through negotiation between large-scale customers and LDCs, which reaches about 30% of LDCs' gas sales.

In addition to the steps taken to introduce competition through the revision of the Gas Utility Industry Law, a more indirect form of competition between utilities has been introduced through a gas rate system imposed by MITI. The rate system, which came into effect in January 1996, is designed to promote further operational efficiency and competitiveness among city gas suppliers within the framework of the current full-cost method. Under the regulation, each company is expected to set efficiency goals and to incorporate these goals into the gas rate revision plan it submits to MITI. MITI then assesses the comparative operational efficiency of each company using a yardstick-based method. If a company is judged to be less efficient than others, MITI has the authority to cut the allowed cost of this company.

Because the full cost method is retained, gas companies are able to recover a "fair and proper" return, interest on bonds and loans, and dividends, in addition to operating costs.

Gas utilities shall prepare and announce efficiency improvement targets when applying for a rate revision. The utilities shall calculate the full cost of service incorporating the above-mentioned targets and apply for revision of rates accordingly. A regulatory authority (i.e. MITI) then makes a two-step yardstick-type assessment of the full-cost which the utility applied for. For this assessment, similar utilities are grouped together. The two steps are an individual assessment and a comparative assessment:

- The individual assessment basically follows the same principles as before. MITI shall investigate the appropriateness of the full costs by checking them against the past record of the utility. The conformity of the full costs with the plans prepared by the utility is to be checked.

- The comparative assessment is undertaken to compare the full cost of each single utility with those of analogous utilities to determine whether the efforts towards efficiency improvements which these costs reflect are at least as great as in other utilities. For this kind of comparisons 224 LDCs throughout Japan have been divided into 16 groups according to their management structure, raw material and production systems, and regional characteristics. According to each company's ranking within each group, MITI cuts a certain ratio of full costs, if necessary.

The utility itself shall make periodic evaluations of the extent to which efficiency improvement targets have been achieved, their revenue and expenditure situation, and other matters in order to assess its efforts of efficiency improvement and to judge the appropriateness of its rate levels. It shall also announce the results of this evaluation.

The regulatory authority is not to assess the efficiency improvement targets per se but the cost implications of the targets. The nature of the improvement targets is left to the responsibility and discretion of the utility. The substance of the targets should be meaningful and clear to the public, preferably including quantitative benchmarks to facilitate measurement of improvements. The time frame for attaining targets shall be three years.

The philosophy underpinning the yardstick-type of assessment is that a ranking of the companies will stimulate them to improve efficiency and reduce costs. This type of indirect competition is intended to bring rate levels down generally and to reduce rate disparities. The cost items on which the LDCs have no influence are excluded from the comparisons. The ranking of the companies in each single group is based on performance on a series of parameters, for instance cost per unit of sales volume. Each item is given a point-based rating used to calculate the total score. The companies within each group are ranked and these rankings are published.

The following table shows an example of the efficiency targets set by the companies.

The new gas tariff system came into effect in 1996 and under the new system, 80% of the LDCs have reduced their prices.

Table 1   An Example of Efficiency Targets Set by the Companies

Item	Nature of Rationalisation Efforts	Effect of Rationalisation						
I. Streamlining of work force	Organisational structures to be reviewed with aim of reducing work force by 1,300 between 1995-1999	Labour Productivity Targets	1994	1995	1996	1997	1998	1999
		Employees at year end	11,714	11,544	11,274	10,977	10,642	10,414
		Sales per employee ('000 m^3)	605	641	687	738	798	855
		Customers per employee	681	704	734	767	804	836
II. Rationalisation of capital investment	By reducing facility construction costs ¥28.7 billion annual capital investment from 1995 to 1997 will be kept below the 1995 level	Cost reduction (Yen billions)			1995 7.5	1996 7.9	1997 13.3	Total 28.7
III. Improvement in demand mix	Tokyo Gas will improve capacity utilisation efficiency by securing high load demand. Goals for period spanning fiscal 1995-1997 (3 years) – 5.1% increase in sales volume per customer – 0.7% improvement in load efficiency	A reduction in cost of ¥7 billion, resulting from the use of fewer storage tanks for LNG.						
IV. Rationalisation of coal-gas production system	The unit production cost of gas will be reduced by halting coal-gas production at the Tsurumi Works in fiscal 1997	Work force reduction: 180 people Cost reduction: ¥2.9 billion						

TURKEY

Statistical information

Natural Gas Supply/Demand Balance (Mtoe, 1996):

Indigenous production	0.2
Imports	6.7
Exports	—
Stock change	0.0
Total natural gas supply (primary energy)	6.9
Electricity and heat production	3.5
Other transformation and energy use	0.1
Total industry	2.0
Residential	1.4
Commercial	0.0
Other	0.1
Statistical difference	-0.1

Natural Gas in the Energy Balance (1996):

Share of TPES	10.4
Share of electricity output	18.1
Share in industry	13.0
Share in residential/commercial sector	7.9

MARKET AND INDUSTRY STRUCTURE

Turkey is a young gas market but one which has seen rapid growth since its inception in 1987. In 1995 it accounted for about 2% of total gas consumption in OECD Europe. Natural gas production started in 1982 and reached 0.2 Mtoe in 1995 (about 2.5% of total supply). Prospects for making further gas finds are considered as good, but domestic production is not expected to increase significantly. This means that Turkey will have to continue to import the overwhelming majority of its gas need. Imports in 1997 were 9.58 bcm of which 69% came from Russia, and 31% from Algeria. Imports from Algeria are LNG deliveries through a terminal which was inaugurated in 1994.

The build-up of the Turkish gas market started by introducing gas in power generation (1987) and in fertilizer production (1988). Already in 1988, the first volumes were delivered in the residential sector, one year before the first deliveries to industry (apart from fertilizer production). Table 1 shows how consumption in the various consumption sectors has developed over time:

Table 1　　　　　Breakdown of Natural Gas Consumption (1000 m³/year)

Year	Electricity	Fertilizer	Residential	Industry	Transport	Total
1987	671	64	–	–	–	735
1988	1 017	149	1	57	–	1,224
1989	2 712	376	7	67	–	3 162
1990	2 555	493	49	320	–	3 418
1991	2 859	477	187	673	–	4 205
1992	2 603	641	372	996	–	4 612
1993	2 530	798	553	1 207	–	5 088
1994	2 927	640	808	1 005	3	5 383
1995	3 602	718	993	1 623	1	6 937
1996	3 791	802	1 484	1 816	3	7 896
1997	4 900	734	1 955	1 827	3	9 419

Source: Botas

Initially, electricity production and feedstock use were the major markets before sales to industry got off the ground. In recent years consumption in the residential sector has grown quickly as the network has been extended to new cities.

The prospects for further increase in gas consumption in Turkey are bright. Table 2 shows how gas consumption is projected to increase over the period up to 2015.

Table 2　　　　　Natural Gas Demand Projection for Turkey, 1996-2015 (bcm)

	Electricity production	Fertilizer production	Other industry	Residential	Total
1996	3.8	0.8	1.8	1.5	7.9
1997	4.9	0.7	2.0	1.8	9.4
2000	12.2	0.8	4.1	3.7	20.8
2010	30.5	2.6	11.3	9.2	53.6
2015	40.5	2.6	12.1	9.6	64.8

Source: Botas

The projected demand growth for gas in power generation is remarkably strong: of the increase in demand of 55.4 bcm over the period, about 63% is assumed to come from this sector. In volume terms, the increases in industry and the residential sector are much more modest but are still quite vigorous. These projections imply future heavy investment in gas infrastructure.

The main transmission line runs from the Bulgarian border to Istanbul (the major industrial region in Turkey) and further to Izmit, Bursa, Eskisehir and Ankara. Its length is about 1000 km and it has a total capacity of 8.6 bcm/year and can be expanded to about 15 bcm/year. In 1994, an LNG terminal at Marmara Eriglisi was completed. The design capacity is 6 bcm of gas a year, and the storage capacity is 0.26 bcm of regasified gas.

As indicated on the map, the infrastructure is under expansion. The expansion of the grid and the capacity of the pipelines depend on future gas availability. For the time being, gas availability is a brake on the speed of gas development.

To increase gas supply several projects are planned:

■ In order to import gas from Iran a pipeline is being constructed between Tabriz and Erzurum(the Erzurum - Ankara pipe section is pending).

■ Several routes to bring gas from Turkmenistan to Turkey are under consideration, among which a pipeline through the Black Sea to supply Russian gas.

■ New LNG plants have been planned to receive increased volumes of LNG.

These new projects will increase the geographical basis for gas distribution considerably.

The structure of the Turkish gas industry is relatively simple and is similar to the structure found in some of the Western European countries: one company, Botas, is responsible for imports and transmission of gas to customers taking their gas from the high pressure transmission system, that is electricity producers, large industrial customers and local distribution companies (of which Botas owns two). The LDCs are responsible for retail distribution to residential/commercial customers and small industry. Botas has the right to sell gas directly to industrial customers taking more than 1 million m³ even when these customers are located in the area of the LDCs. It may, however, transfer this right to the LDCs.

Botas Petroleum Pipeline Corporation was set up in 1974 to transport Iraqi crude oil through Turkey. It was an affiliate of the Turkish Petroleum Corporation (TPAO) until the end of 1994. In 1995 Botas was separated from TPAO and restructured as a State Economic Enterprise. The scope of its activities is "to construct and have constructed all types of pipelines for petroleum, petroleum products and natural gas within and outside Turkey; to take over, purchase and lease those pipelines that have already been built; to transport petroleum, petroleum products and natural gas through pipelines and to purchase and sell the crude oil and natural gas transported in these pipelines". By virtue of Decree no. 397 on Natural Gas Usage, Botas holds a monopoly on gas imports, transportation and setting of gas prices except in the cities.

So far, there are five Local Distribution Companies (LDCs) in Turkey: EGO(Gas, Omnibus) in Ankara, IGDAS (Istanbul Gas Distribution Co.) in Istanbul, Izgaz in Izmit, Botas in Bursa and Botas in Eskisehir. Table 3 gives some more detailed information about the companies:

Table 3   Gas Distribution Companies in Turkey (1996 and 1997)

City	Ownership	Start-up year	Sales volume million m³		Number of customers	
			1996	**1997**	**1996**	**1997**
Ankara	municipal	1988	507	667	240,000	318,724
Istanbul	municipal	1992	1148	1446	832,000	794,607
Izmit	municipal	1996	637	6746	n.a.	10,048
Bursa	Botas	1992	137	188	55,000	56,329
Eskisehir	Botas	1996	8	42	22,191	26,904

Source: Turkish authorities

For the time being, there are no private interests in gas distribution companies. Botas has created a company called Turkish Gaz Ltd. To invest in gas distribution in co-operation with private partners in the future.

Through its direct sales of gas to large industrial users and power producers, Botas serves about 200 customers, 21 of which are taking less than 1 million m^3 of gas annually.

REGULATION

The principal governmental authority in the energy sector is the Ministry of Energy and Natural Resources (MENR) and the Ministry of State. The MENR is responsible for energy planning and for conservation issues. For licence applications, the competent authority is the Petroleum Directorate in Ankara, which reports to MENR. Licences are needed for exploration and production of natural gas. By virtue of Law no. 397, Botas has a monopoly on imports and high pressure distribution of gas. Local distribution companies need a permit from MENR. Since there is presently no underground storage in Turkey, the responsibility for this is under consideration. There is a presumption that Botas will own and operate such facilities.

Botas holds a monopoly on gas imports. The company does not need approval of its import contracts from Turkish authorities. Botas is responsible for construction and operation of gas pipelines. Investment decisions are taken by Botas and presented to the MENR and to the State Planning Organisation for approval. The criteria used for evaluation of such extensions are based on general energy policy guidelines and economic considerations.

By virtue of Law no. 397 Botas holds a monopoly on gas distribution to all customers except residential/commercial customers and industrial customers taking less than 1 million m^3 a year within municipal boundaries. As indicated above, three of the distribution companies are owned by municipalities, two by Botas. There is no regulation preventing private distribution companies from being set up. Gas distribution can be carried out by any private company but it would need a permit and the shareholding structure of the company has to be approved by the Council of Ministers.

The LDCs have a monopoly to supply gas in a specified geographical area, but do not have an obligation to supply in their area; supply can be refused based on economic criteria. There is regulation providing for investment incentives to LDCs to encourage development of gas distribution in the cities.

Gas price regulation in Turkey is relatively light handed: Botas sets the prices of all the gas it sells. Domestic producers set their own prices for gas sold to industry. These prices are not subject to approval by MENR. Gas prices charged by LDCs are set by the companies themselves but the prices are approved by the MENR. Up to February 1997 the prices applied by the LDCs were not allowed to exceed the prices paid by the LDCs to Botas by more than 70%. From August 1997 this limit has been lowered to 45%. Within this limit the LDCs are free to set their prices.

Decree No. 397 grants Botas, the natural gas transporter and trader, a monopoly in natural gas importation, distribution, sale and price determination. This Decree is currently under review.

CONTRACTUAL ARRANGEMENTS

Botas presently buys gas on long term contracts (20-25 years) from Russia and Algeria. Prices are linked to a basket of crude and petroleum products. Other contracts that have been signed (with Nigeria and Iran) have a similar structure.

Botas is selling gas on long term contracts only to power generation. Contract duration varies for the different power stations (which have individual contracts) but it is generally around 20 years. They all have take-or-pay clauses.

Botas sells gas on shorter term contracts to all other customers (industrial consumers, LDCs, fertilizer plants and autoproducers of electricity). These contracts generally have take-or-pay clauses based on the annual consumption of the customers. The contracts between Botas and the LDCs stipulate maximum hourly, daily, monthly and annual off-take.

DEMAND FLUCTUATIONS AND SECURITY OF SUPPLY

The following table shows gas demand month by month in the various gas sectors:

Table 4 Monthly Gas Demand – 1997 (million m^3)

Month	Electricity production	Fertilizer production	Residential sector	Industry	Total
January	400	44	315	137	896
February	315	32	314	119	780
March	381	68	315	158	922
April	381	72	226	134	813
May	430	76	46	149	701
June	430	73	26	139	668
July	417	73	17	142	649
August	435	73	19	140	667
September	458	48	25	156	687
October	461	71	110	189	831
November	405	75	213	185	878
December	407	31	327	182	947
Total	**4920**	**736**	**1953**	**1830**	**9439**

Source: Botas

Turkey has much lower fluctuations in demand over the year than all the other countries comprised by this study. The major reason for this is the high share of gas taken by power generation and industry, which in its turn reflects a typical demand pattern in an emerging gas market. Gas consumption in the month with the highest gas consumption in 1996 was only 46% higher than the month with the lowest gas consumption. Even on a daily basis the differences in sendouts were modest: the highest daily sendout of the year 1996 was only 2.24 times higher than the lowest daily sendout.

The structure of demand and its seasonality pattern imply a relatively modest requirement for load factor. Most of the time this need can be satisfied by using the flexibility under the supply contracts and variation in linepack on the system. Since the residential sector, the sector having the greatest need for load factor, accounts for a low share of total demand, no attempts have been made at allocating transportation cost (and thereby load factor cost since load factor takes place at the transmission level) according to the degree to which the various customers are responsible for costs. This means that all customer categories basically carry the same unit transportation cost. The extent to which the various demand sectors need load factor will have consequences for the cost of constructing the necessary infrastructure in the future. One question that will have to be discussed is how these costs are taken into account and allocated over the different customer categories.

Disruptions in gas supply from one particular source can be countered by having recourse to flexibility in supply under other gas contracts, drawing on storage and using interruptible contracts. The volume flexibility under the contracts with Russia and Algeria is only 10% of annual contract quantities, but the flexibility within a year is much higher. This allows higher imports during the winter than during the summer. Gas storage in Turkey is very limited (corresponding to some 8 days of consumption). Interruptible contracts exist; about 20% of consumption in industry is delivered under interruptible contracts. There is no legislation concerning the rationing of gas in the event of a disruption, but during recent disturbances in gas supplies caused by the Russian transit problems with Ukraine, priority was given to residential consumers whereas one of the gas-fired power stations switched to fuel oil and two fertilizer plants were put on standby.

Both the Government and Botas are well aware of the need for further diversification of gas supplies and the need to develop storage facilities. As in most Western European countries, however, little has been done so far to explicitly identify costs pertaining to security of supply, let alone try to allocate these costs over the various customer categories.

APPROACH TO PRICING

The general approach to pricing is of a cost-plus type tempered by considerations ensuring gas competitivity and penetration in the market. The prices are supposed to cover the import cost of gas, transmission and distribution costs. As indicated above the gas volumes to all sectors basically carry the same unit cost but prices to end users differ as a result of differing market values. These market values result from comparisons with other fuels. When such comparisons are made the full cost including investment costs are taken into account for the alternatives. If it is necessary to become competitive, prices are set below the levels of competing fuels. In some cases, however, natural gas imposes itself because the use of other fuels is prohibited for environmental reasons. This is the case in the centre of the major cities. The level of prices to end consumers in the market served by the LDCs is, however limited by regulation: the price to end consumers are not allowed to exceed the purchase price by more than 45% since August 1997. Within this limit the LDCs are free to set their prices. Until February 1997 this percentage was only 30% which implied that the LDCs were not able to cover their costs.

PRICES FROM BOTAS TO LDCs

Although Botas has the right to differentiate the prices to the LDCs, this price is now the same for all the companies. The price is composed of the import price for gas plus elements to cover transportation cost and profits. The price has an escalation similar to the average import price.

END-USER TARIFFS IN DISTRIBUTION

Prices to end consumers served by the LDC are set according to their own marketing strategy within the limitation mentioned above. As of April 1998, the prices to non-industrial customers served by the LDCs were as follows:

Table 5 Prices to Non-Industrial Customers Applied by the LDCs, April 1998 (US cents/m^3)

	Households	Commercial	Government offices
Ankara	20.85	20.85	27.11
Istanbul	26.01	26.01	26.01
Izmit	21.24	21.24	21.24
Bursa	19.36	19.36	21.29
Eskisehir	19.36	19.36	21.29

Source: Turkish authorities

Since each LDC is responsible for its own pricing, prices vary from one company to another. The tariff schedules available for the companies are reproduced and commented upon in the following:

Table 6 IGDAS[1] Domestic Gas Prices as of End 1996 (in TL/m^3)

Consumption (m^3)	IGDAS Price
0 – 50,000	17,350
50,001 – 100,000	17,177
100,001 – 500,000	17,005
500,001 – 1 000,000	16,835
1,000,000 – 999,999,999	16,667

Source: Turkish authorities

The domestic gas price is a flat rate without any distinction between fixed and variable elements. The rate is digressive with volume. As the next table shows, the industrial price is also a flat rate with no fixed element. Again, the rate is digressive with volume:

1. IGDAS = Istanbul Gas Distribution Company.

Table 7   IGDAS Industrial Gas Prices as of End 1996

Consumption in m³	Temporary	Production	Steam
0-300,000	14,369	16,397	15,965
300,001-1,000,000	14,420	16,108	15,600
1,000,001-5,000,000	13,874	16,012	15,415
5,000,001-15,000,000	13,682	15,570	15,203
15,000,001-50,000,000	13,600	15,402	15,111

Source: Turkish authorities

Table 8   Tariff Schedule for EGO, Ankara (US$ per m³)

	Group I 0-2500 m³	Group II 2501-5000 m³	Group III 5,001-50,000	Group IV more than 50,000
Households	0.1988	0.2485	0.311	0.3883
Industry	0.1988			
Commercial	0.1988	0.2485	0.311	0.3883
Public uses	0.2584	0.3231	0.404	0.5048
Employees	0.0994	0.1243	0.155	0.1942
Other	0.1988			

Source: Turkish authorities

This is not the latest schedule from EGO, but the structure is still the same. Gas used for public purposes has a higher price than gas sold to private users. EGO employees benefit from a preferential gas rate. The gas price is progressive with volume.

In addition to a price per cubic metre there is a fixed element of 1 US$ per month.

PRICES TO LARGE INDUSTRIAL CUSTOMERS SERVED BY BOTAS

Large customers are served by Botas unless its right to serve them has not been transferred to an LDC. Firm customers between 300,000 m³ and 100 million m³ are charged according to the following tariff schedule:

Table 9   Prices to Large Industrial Customers Served by Botas, as of April 1998 (US$/1000 m³)

Annual quantity in 1000 m³	Min. purchase obligation	Ceramics producer	Electricity producers	Process use	Steam raising
300-5.000	50	164.15	155.71	166.16	166.16
5.000-15.000	50	160.93	152.66	162.90	162.90
15.000-50.000	60	157.78	149.67	159.70	159.70
50.000-100.000	70	154.68	146.73	156.57	156.57

Source: Botas

Prices to interruptible customers have the same structure but they obtain a price reduction of 10 to 17% compared to firm deliveries. There is also a special scheme for enterprises located in organized industrial zones where prices are about 2% lower all over. In addition to a price per m³, industrial users in these categories pay a standing charge of 74 cents/m³ (fixed service fee) per contracted maximum hourly gas usage. The proportional gas price is decreasing with volumes. The minimum purchase obligation, however, is increasing with volumes. The prices in the various uses are supposed to reflect the value of gas in those uses. Electricity producers in this context only include autoproducers. The prices to fertilizer producers and industrial customers taking more than 100 million m³/year are negotiated with Botas. Large power producers have individual contracts with Botas. The contracts specify annual take-or-pay conditions. The prices are linked to escalators in the import contracts but also reflect the cost of transportation within Turkey.

PRICING IN RELATION TO OTHER FUELS

The following tables show a comparison of energy prices in the industrial, power generation and residential sector for 1996:

Table 10   Energy Prices in the Industrial Sector in 1996 ('000 TL/toe)

Heavy Fuel Oil		Natural Gas		Steam Coal	
Incl. Taxes	Excl. Taxes	Incl. Taxes	Excl. Taxes	Incl. Taxes	Excl. Taxes
15609	9209	16964	15707	7035	6118

Source: IEA Energy Prices and Taxes

Pre-tax gas prices in industry are higher than for HFO and steam coal. The full price of gas is also somewhat higher than the HFO price, but far lower than the coal price. HFO is the main competitor in this sector.

Table 11   Energy Prices in the Power Generation Sector in 1996 ('000 TL/toe)

Heavy Fuel Oil		Natural Gas		Steam Coal	
Incl. Taxes	Excl. Taxes	Incl. Taxes	Excl. Taxes	Incl. Taxes	Excl. Taxes
15609	9209	15386	14246	7147	6215

Source: IEA Energy Prices and Taxes

The average gas price to power generation including taxes is slightly lower than the HFO price but much higher than the coal price.

Table 12   Energy Prices in the Residential Sector in 1996 ('000 TL/toe)

Light Fuel Oil		Natural Gas		Electricity	
Incl. Taxes	Excl. Taxes	Incl. Taxes	Excl. Taxes	Incl. Taxes	Excl. Taxes
52865	19451	18910	17510	82697	68487

Source: IEA Energy Prices and Taxes

The natural gas price is much lower than the LFO price. The gas price in the residential sector, however, is only 11% higher than the industrial gas price- in most Western European countries it would easily be 100% higher. This means that the price of gas is kept artificially low. The main competitor in the residential`sector is lignite which has a much lower price. In central city areas, however, use of lignite is prohibited and thus gas has no competitor.

Generally the relative level of taxes on gas is lower than that for other energies. The VAT rate on gas is 8% whereas the rate for other energies is 15%. It could therefore be claimed that taxation policy is used to facilitate the penetration of gas. The relative price of gas is, however, considered to be a problem for further gas expansion. The Government is presently considering changes in energy taxation that would facilitate further gas penetration.

CROSS-SUBSIDIES

When Botas sells gas to its customers the element added to cover transportation cost is not differentiated according to distance from the import point. Implicitly, this is a kind of cross-subsidy from users close to the import point to customers further away. This may, however, be a necessary measure in a period of gas market expansion.

To ensure the penetration of natural gas into the market, natural gas prices are set below the level of competitive fuels. If the price arrived at is not sufficient to cover the cost of serving the customers in question, Botas compensates this price difference internally.

Preferential gas rates for LDC employees is a form of cross-subsidy in favour of one particular consumer category.

Only one of the Turkish LDCs is doing other business than gas distribution. Since its activities are not fully separated in accounting terms there is a risk that there are cross-subsidies between the different activities.

GROSS MARGINS AND PROFITABILITY

The following table gives an impression of gross margins in the Turkish gas chain:

Table 13 Gross Margins in the Turkish Gas Chain in 1997

Average sales price Botas	15.3 c/m^3
Average gas import price	11.2 c/m^3
Gross margin transportation (transmission and direct sales)	4.1 c/m^3
Average Botas sales price to LDCs	16.0 c/m^3
Average gas import price	11.2 c/m^3
Botas gross margin on sales to LDCs	4.8 c/m^3
Average sales price LDCs	18.98 c/m^3
Average purchase price LDCs	16.0 c/m^3
Gross margin LDCs	2.98 c/m^3

Source: Turkish authorities

In comparison with Western European countries, margins are low. This goes in particular for the distribution part of the chain. One reason for this is the limitation that a maximum margin of 30% can be added to the gas purchase price (this figure has now been increased to 70%). This limitation has contributed heavily to the fact that all the Turkish LDCs are running losses. Another important factor in this context is also the fact that gas distribution is a young business; in fact it is quite normal that distribution companies make losses over the first few years of operation.

Botas realises a profit on its gas activities. Its objective is to earn a rate of return on its investments of 15% before taxes, but historically it has not been able to achieve this. The fact that it does not distinguish between its gas and oil activities in its external accounts makes it difficult to analyse its performance.

UNITED STATES

Statistical information

Natural Gas Supply/Demand Balance (Mtoe, 1996):

Indigenous production	440.2
Imports	68.1
Exports	-3.5
Stock change	-0.6
Total natural gas supply (primary energy)	504.2
Electricity and heat production	119.3
Other transformation and energy use	47.1
Total industry	132.7
Residential	122.1
Commercial	73.5
Other	16.6
Statistical difference	-7.1

Natural Gas in the Energy Balance (1996):

Share of TPES	23.6%
Share of electricity output	13.2%
Share in industry	36.2%
Share in residential/commercial sector	44.5%

INDUSTRY OVERVIEW

The United States natural gas system consists of indigenous production from thousands of producing wells scattered over a vast area (from the Gulf of Mexico to the Canadian border) that are linked to consumers through a complex network of gathering, transmission and distribution lines. Domestic production is supplemented by imports from Canada that account for about 11 percent of US supply. During peak winter periods of high demand, consumers can be assured of a reliable supply of natural gas because of extensive use of underground storage. During 1996, the industry delivered to consumers 21.9 trillion cubic feet of gas which includes 2.7 trillion cubic feet imported from Canada. The US consumes about 25 percent of world wide marketed production of natural gas.

In 1997, the US produced 18839 trillion cubic feet of natural gas, about 1 percent over 1996. About 26 percent of indigenous production is produced offshore in the Gulf of Mexico. For the ten year period beginning in 1988 natural gas production has increased by about 10 percent. It should be recognized that this level of production was attained in spite of relatively low well head prices during that period. Even as late as 1996, prices were 18 percent below the levels of the previous year. The increase in the 1997 production level was accomplished through the combined impact of cost efficiency gains and improved technology. Improved productivity and lowering costs have helped US producers meet the challenge of relatively low prices since 1985.

The US has continued to experience a steady decline in the number of new gas well completions since the peak of drilling activity that occurred in 1981. Currently, new well completions are less than half the number experienced in 1981. Historically low natural gas prices have put a damper on drilling activity. Even though wellhead prices have improved significantly beginning in 1996, drilling activity has seen only a moderate increase. How long current prices will need to remain at this level or higher to stimulate major increases in drilling activity remains to be seen. The fact that wellhead prices for natural gas have not been subject to federal regulation for several years implies that market forces will send price signals to producers to increase drilling activity.

Since about the mid-eighties the natural gas industry has experienced a surplus in production deliverability. This condition has served as a strategic reserve by providing the industry with additional production capability that could be drawn upon to meet increased demand during prolonged cold spells, reduced production from some wells due to freeze offs or a combination of both. It is estimated that gas producers operated at a 94 to 95 percent of capacity in 1996. (The unusually mild 1997 winter across much of the US temporarily eased this pressure.) This is near the upper limits of deliverability because about 3 to 5 percent of producing wells are down for normal servicing or are experiencing operational problems. What this means is that the US surplus production capability is now gone. This fact has served as a stimulus to the entire industry to work closely together to insure that current production is available to meet consumer demand. In 1994, the Department of energy published a study that evaluated the ability of the gas distribution industry to supply its customers during a peak demand period that broke all previous records for daily demand. The study found that deregulation had in fact improved the ability of the gas industry to respond to the extraordinary demand. A benefit to opening the gas industry to more competition was the increase in communication resulting from all the negotiations required to complete gas transactions in a deregulated environment. The increase in communication was found to be a valuable asset during the January 1994 energy crisis because the increase in communication allowed gas industry participants to move gas quickly to where it was needed most.

The transportation of natural gas from producing areas to major centres of consumption is accomplished by a network of large mostly high compression pipelines often referred to as trunk lines. Further downstream these pipelines interconnect with other regional pipeline and distribution systems. The gas that serves the populous eastern third of the US is transported by about 25 interstate pipelines (those that cross state borders) however the system in its entirety consists of over 200 transmission companies with about 270,000 miles of pipe. Gas produced in the Gulf of Mexico can reach the most distant markets in the Northeast in seven to eight days. All of these pipelines are interconnected to form a grid that enable gas to move quickly from one system to another.

The distribution segment of the natural gas industry as a separate entity from the transmission and gathering systems begins at what is typically referred to as the "city gate". This is the point where gas enters the distribution lines of a local distribution company (LDC). Although large LDCs may operate transmission lines as part of their system, it is the fact that LDCs are entities that are primarily responsible for insuring gas is delivered to the consumer. Pipelines also make what are commonly called "off-system deliveries", deliveries made directly to consumers, (usually electric power plants and large industrial users) but pipelines are not considered part of the US distribution network. In 1996 only 17.6 percent of the natural gas deliveries classified as industrial deliveries were made by distribution companies. What further separates the transmission sector from the distribution firms is defined by regulatory oversight. The distribution segment is regulated by state and local governments. While most pipelines are regulated by the Federal Energy Regulatory Commission (FERC) but some pipelines do not cross state boundaries (intrastate) and are regulated by state governments. Ostensibly, the US gas industry is characterized by some overlap of pipeline and LDC transportation and sales functions. It is estimated that the US gas distribution system consists of approximately one million miles of pipe in the ground.

Natural gas consumption patterns in the US are characterized by wide variations in seasonal usage. On a national basis, consumption in January is about twice that in July but for residential consumption the demand can vary by a factor of seven or more. During cold spells the daily US demand (all classes of consumers) can exceed summer demand by a factor of three times or more. There has been a steady increase in summer demand for natural gas to meet the energy needs of electric utilities that has helped to improve load factors for the entire industry but, the US gas industry is still driven by winter demand. In order to meet seasonal demand surges the industry has increasingly turned to storage as a supplemental supply source. Total US storage capacity is 3.1 trillion cubic feet. On a typical cold day in January, withdrawals from storage can account for one third of total US supply.

IMPACT OF PIPELINE DEREGULATION ON GAS DISTRIBUTION

Beginning in 1985, FERC issued its first in a series regulatory policy changes to provide open access to the nations pipeline system. FERC's objective was to reduce the monopoly power of pipelines by allowing shippers equal access to pipeline space. The FERC orders created a level playing field such that a pipeline's own gas for resale could not have preferred access and delivery points over any other shipper. Since the full implementation of these orders in 1992, the US has seen a revolution in the way the natural gas is marketed in the US. However, transportation tariffs for interstate pipelines continue to be regulated by the Federal Energy Regulatory Commission (FERC), an agency of the United States government. Pipelines requesting tariff increases must receive FERC approval before they can be implemented.

In the early stages of pipeline deregulation one area of great concern for LDCs, particularly in growing gas markets, was the possibility that increased competition might thwart future or planned pipeline expansion projects. Since the mid-eighties, pipelines experienced the harsh realities of competition in the form of reduced profitability as a result of take-or-pay obligations. However, with the implementation of Order 636 in 1992 (with a resolution to the take-or-pay issue), natural gas pipelines have begun to earn rates of return that have lured investor interest in pipelines. In 1986 the rate of return for natural gas pipelines was a negative 3.93 percent and by 1993 it had recovered to a positive 10.97 percent. One of the key provisions of Order 636 was the methodology by which pipelines would be permitted by the FERC to recover their costs. Pipelines were allowed to switch from the modified fixed-variable (MFV) cost recovery methodology to the straight fixed-variable (SFV) cost recovery methodology. MFV fixed costs are recovered in a usage charge while for the SFV method pipeline customers are levied a reservation fee regardless of usage. This fee all but guarantees that fixed costs and return on equity will be recovered. The change in cost recovery methodology has been instrumental in encouraging investor interest in pipelines. The result has been a rapid expansion in US pipelines in the past two years: in 1996 a total of 26 major expansion projects were completed.

As noted above, since the mid eighties, the FERC has continued on a long steady path to create a competitive pipeline industry in the US (The process has been generally referred to as restructuring or open access). Ostensibly, the changes required by pipelines to meet new FERC guidelines have greatly impacted the transaction process by which distribution companies acquire and receive their gas supply. Consequently, it is essential that the current status of this on-going process be clearly delineated because many of the issues regarding reliability and supply risk for gas distributors are embedded within a restructured pipeline industry.

Prior to restructuring, pipelines were both a transporter and supplier of natural gas. They assumed the risk of assuring that on the coldest day of the year that gas would be available to all the distributors on their system. Because pipelines purchased large volumes of gas for resale to their LDC customers they were able to pool the production of many producers to reduce the risk of a supply disruption. Now, in a restructured market, pipelines provide a transportation function only, leaving the problem of finding a new supply source to each individual distribution company. The risk of a supply disruption is now assumed by the LDC; if a producer is unable to deliver contracted volumes it is incumbent upon the LDC to locate a new supply source to insure the integrity of service to their customers. In order to reduce this risk, LDCs have had to quickly integrate their operations into a complex and fast-paced environment (that is still evolving) to acquire gas supply, transportation and storage.

Changes brought about by deregulation have led to rapid growth of new participants in the US gas industry. Marketing companies have attempted to replace services previously offered by pipelines (including risk management). These new entrants into the gas industry have attempted to replicate the services formerly provided by their interstate pipeline supplier by dealing with many issues that are not very transparent such as balancing, access to storage, metering, tracking nominations and meeting unexpected demand surges. These service have been particularly helpful to small distribution companies that do not have the in-house expertise to arrange the same level of service formerly provided by pipelines. Recent experience has shown however, that some of these marketing companies have failed to deliver on their commitments, particularly during peak periods. LDCs must now seriously take into consideration the credit worthiness and reliability in selecting a producer, marketing company or other third party supplier. Furthermore, as a result of pipeline deregulation, LDCs

Figure 1   The Relationship Between Gas Distribution and the Other Segments of the Gas Industry Infrastructure

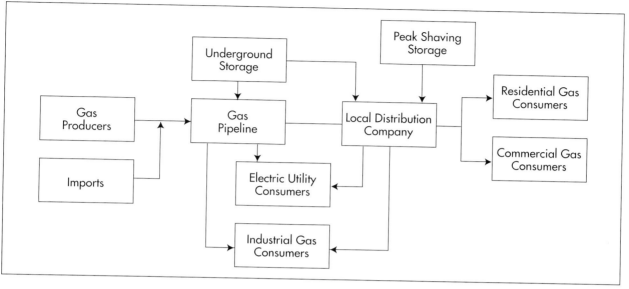

also provide a storage function or acquire access to storage. Typically, pipelines provide access to underground storage facilities but in areas where facilities are available, LDCs may acquire direct access to underground storage. During periods of very high demand, LDCs supplement their existing supply through peak shaving such as propane-air, LNG or above ground low pressure holders. The relationship between gas distribution and the other segments of the gas industry infrastructure are described in Figure 1.

ORGANISATION AND STRUCTURE OF GAS DISTRIBUTION

Natural gas distribution companies are both investor owned companies as well as public utilities. In a number of cases, gas distributors may be part of larger utility that also produces and distributes electric power, such as Baltimore Gas and Electric and San Diego Gas and Electric. There is also a very wide range in size from those that deal with a thousand or fewer customers to the largest that distribute gas to hundreds of thousands of customers. The largest gas utility in the US (in terms of number of customers) is Pacific Gas and Electric which distributes gas to approximately 7.5 million customers. There are at least another twelve gas distributors that serve over a million customers. On the other hand, there are hundreds or small (mostly publicly owned) gas distribution companies that serve less than a few thousand customers. LDC service territories are defined in part by state borders so that most companies have their operations confined to one state. There is also a wide variation in distribution company size by state. The state of California has eight gas distributors while the state of Georgia has 75, yet California has a population almost five times that of Georgia's. In a number of cases where an LDC may serve customers adjacent to its service territory in another state a separate subsidiary to serve those customers is often the provider. The fragmentation of the gas distribution industry is the result of each state having regulatory oversight of gas distribution within its borders.

The natural gas industry in the US is not vertically integrated to the degree as are other energy industries. There are many small firms in the US petroleum industry but, over 90 percent of all products distributed are done by firms that are usually involved in more than one operational phase. The electric power industry also has greater degree of vertical integration because of the close relationship between power production and distribution. There is very little upstream integration by gas distributors into production. Currently, the general trend for LDCs in the gas producing business is to divest themselves of such operations. In several producing states LDCs often own pipelines to nearby producing areas. There has recently been evidence of a trend towards mergers or acquisitions of gas distributors by pipelines or their holding companies and mergers or acquisitions of smaller companies by larger ones. Although these trends are evident, there is little data to quantify any major restructuring of gas distribution in the US with the exception of an increase in deliveries by pipelines directly to consumers. Recently, there has also been evidence of interest by small LDCs to form cooperative buying groups. The purpose of the groups is to reduce their purchased gas cost by increasing their buying power. Based on 1995 data, the percent share of total US gas deliveries by entity is as follows:

Combination electric and gas utilities (investor owned)	27%
Municipal - public ownership	7%
Vertically integrated - combined pipeline and LDC (investor owned)	19%
Investor owned LDC - gas only	33%
Direct deliveries by pipeline	14%

LDC SALES PROFILE

Since the mid-eighties, one of the most pronounced adjustments that LDCs have been required to makes has been the change in their sales profile since the mid-eighties as a result of the loss in industrial gas customers. Gas sales by local distribution companies have declined by about 30 percent since the mid-eighties. It is not a coincidence that this decline began at about the time that pipeline deregulation began. A closer examination of end-use profile data indicate that all of the sales loss were to industrial consumers. Table 1 shows this change rather dramatically.

In fact, sales to residential consumers have actually risen while sales to commercial consumers have remained level. In 1984 industrial gas sales accounted for 46 percent of LDC sales while ten years later they amounted to only 18.8 percent of sales. Some of this sales loss can be attributed to the general weakness in the manufacturing sector of the US economy but further evidence will likely show that most of the loss in industrial sales was captured by transmission pipelines or through bypass arrangements whereby large consumers purchased directly from producers or marketers using the LDC as a transporter only. LDCs have also fought successfully to retain a significant portion of this volume in transportation agreements.

Table 1 Gas Utility Sales by Class of Service (trillions of Btu)

Year	Total	Residential	Commercial	Industrial	Other
1984	13,161.7	4,628.3	2,396.0	5,991.0	136.4
1985	12,615.2	4,513.0	2,338.0	5,634.7	129.5
1986	11,125.1	4,380.8	2,238.9	4,338.2	167.7
1987	10,643.4	4,385.0	2,155.9	3,847.9	154.6
1988	10,705.3	4,695.0	2,306.0	3,544.1	160.2
1989	10,551.1	4,796.0	2,321.8	3,243.4	187.9
1990	9,841.8	4,468.4	2,192.0	3,010.4	171.0
1991	9,601.4	4,546.3	2,198.1	2,631.3	225.7
1992	9,906.5	4,694.4	2,208.7	2,772.2	231.2
1993	10,150.9	5,053.8	2,396.6	2,533.0	167.5
1994	9,584.3	4,972.1	2,353.8	2,109.7	147.7
1995	9,094.2	4,736.4	2,203.7	1,930.3	223.8
1996	9,531.9	5,198.4	2,395.1	1,790.6	147.8

The loss of the sales to pipelines (bypass) and direct purchase arrangements have been cause for great concern for gas distributors. Bypass arrangements whereby the LDC loses both the sales revenue and transportation charges has been a severe blow to many distributors. The decline in industrial and commercial customer participation in on system sales means that those customers who do remain on system are likely to be paying more of the fixed costs. If reductions in fixed costs are smaller than the decline in gas sales, consumers that are still full service customers of an LDC will experience price increases. If residential load does not expand rapidly enough or if the distribution costs cannot be reduced by efficiency improvements, the remaining on system customers end up paying higher prices.

REGULATION/RATES

Regulation of gas distribution companies in the US is vested in state administered regulatory agencies generally referred to as public utility commissions (PUCs). The fact that there are 51 state (50 states plus Washington, DC) PUCs is an indication of the possible variation in LDC regulation that can be found in the US Virtually all PUCs regulate rates or tariffs that allow LDCs to recover operating and commodity costs and earn a rate of return on their investment. It should be noted that gas distributors generally do not earn a profit but rather a return on their investment which is established by each PUC. Beyond rates and tariffs the extent of PUC regulatory oversight can include purchased acquisition cost, siting, use of spot and futures markets, marketing affiliates and most customer services such as storage, balancing and back-up.

It is the administration of rates and cost recovery allowances that is the primary regulatory focus of PUCs. Typically, rates are based on cost-of-service. The traditional design for establishing rates tends to be quite consistent for all states. Rates usually consist of two parts, a base rate and a volumetric rate that is based on the amount of gas purchased or transported. The base rate provides for fixed cost recovery and a return on capital investment. Base rates are usually set on a test-year representative levels, with a rate of return based on net investment in the base year. Future rate increases depend on the change in the level of investment from the base year. The second part of rate design allows for the recovery of the cost of the gas by the LDC or a transportation charge based on volume. Many states allow LDCs automatic commodity cost recovery in the purchased gas adjustment clause component of their rates. LDCs generate over half of their revenue from this source.

The traditional cost-of-service rates offered by all LDCs are now being supplemented by a variety of non-traditional rates such as market-based and incentive-based rates. Currently, about half of all states allow or are considering permitting LDCs to offer some form of a non-traditional rate. Several states now allow their LDCs to offer market-based or incentive rates. Market-based rates are designed to allow LDCs to compete in markets where alternate fuels or other gas suppliers have a price advantage. These rates are often established to meet the competition faced in suppling gas to an individual customer. In the case of New York state, LDCs are even permitted to sell gas to some customers below cost as long as the average sales price will exceed the commodity cost over the life of the contract.

Natural gas distributors in the US are not allowed to earn a return on the sale of gas but instead derive their revenue from transportation charges. Under traditional cost-of-service rates there is little incentive to reduce cost since any savings are passed on to the consumer. Incentive or performance-based rates (PBR) are designed to encourage utilities to pursue greater efficiency. The results are shared between the LDC and its customers. PBRs generally include a benchmark or standard for gaging the utilities performance based on cost and/or service measures. Additional earnings or cost savings are usually shared between the LDC and its customers. Regardless of the rate design, LDCs must recover their costs (and for investor-owned utilities earn a return on their investment) but it is the methodology by which LDCs are required to allocate their costs where we find less agreement by state regulators.

For tariff based rates about half of all PUCs apply a method of uniform cost allocation. This type of cost allocation uniformly projects the same cost of gas to all firm rate classes; it assumes it costs the same to supply an industrial consumer as it does a residential consumer. Uniform cost allocation subsidizes residential customers at the expense of higher volume commercial and industrial consumers. There are about ten states that adopted cost-based tariffs that take into account the cost differences in supplying different classes of consumers. The remaining state PUCs have adopted both uniform and cost-based allocation. Typically, for market based rates, the cost allocation method used by many states is based on weighted average cost of gas (WACOG). However, WACOG does not allow many LDCs to meet competitive prices. To overcome this limitation, an increasing number of PUCs are permitting LDCs to respond to market forces in assigning gas costs. Several states allow LDCs to allocate costs based on the cost of supplying a particular customer. This approach works well for large volume customers where the cost per mcf of service is less. At the other extreme, a few states even require incremental gas costs to market rates, which usually is the highest cost gas. However, the trend is definitely moving in the direction of market-based rates and cost allocation based on the cost of serving a particular customer. Table 2 shows the change in industrial LDC gas sales and revenues for the period 1984-94. For the same period, the revenue stream for retail consumers stayed about the same or increased slightly in the past year. When the industrial revenue data is compared to the sales dat in Table 2, it shows that for industrial class of service, revenues are declining more rapidly than gas sales.

Table 2 Gas Utility Revenues by Class of Service (thousands of US$)

Year	Total	Residential	Commercial	Industrial	Other
1984	67,496,298	27,484,877	13,205,106	26,093,776	712,539
1985	63,292,552	26,864,099	12,722,134	23,085,833	620,486
1986	51,200,810	24,759,281	11,273,843	14,494,725	672,961
1987	45,491,823	23,622,262	10,271,485	11,068,424	529,652
1988	46,161,619	24,828,402	10,681,487	10,113,425	528,485
1989	47,493,381	26,171,772	11,074,360	9,665,710	581,539
1990	45,152,716	25,000,275	10,603,513	8,995,989	552,939
1991	44,647,379	25,728,572	10,668,880	7,576,238	673,689
1992	46,178,067	26,701,927	10,684,830	7,913,202	698,108
1993	50,137,413	29,787,061	12,076,169	7,641,885	632,298
1994	49, 851,930	30,551,650	12,275,740	6427,610	596,940
1995	46,381,180	28,740,830	11,409,850	5,651,900	578,610
1996	51,114,500	32,021,470	12,727,150	5,039,393	545,450

REGULATORY REFORM

The process of natural gas regulatory reform began in 1978 with the passage of the Natural Gas Policy Act (NGPA). The NGPA provided for the gradual decontrol of wellhead prices: a process that was completed by 1989. As the process of wellhead decontrol came to an end, the process of pipeline regulatory reform began. The process of regulatory reform for the pipeline sector began in the mid-80s but the process was not fully institutionalized until 1992 with the issuance of Order 636. This order basically ended the pipeline merchant function and allowed all natural gas shippers equal access to pipeline space. Order 636 allows LDCs to purchase gas directly from producers using the pipeline as a transporter. The last link in the natural gas chain, the distribution sector, is now in the early stages of regulatory reform. However, unlike wellhead decontrol and pipeline restructuring that was accomplished primarily by one agency the FERC, LDC regulatory reform will be accomplished by 51 separate regulatory agencies.

The decontrol of wellhead prices and pipeline restructuring have opened up a plethora of competitive forces that have resulted in a growing number of natural gas customers to purchase gas service on an unbundled basis. Almost all large gas consumers now purchase their gas from a third-party supplier. It is estimated that about 75 percent of all industrial customers are buying their gas from the supplier of choice. The result has been a steady decline in gas costs to industrial consumers. The fact that industrial consumers have experienced a much grater decline in their gas costs as compared to residential consumers has not gone unnoticed by state regulators. This fact has encouraged many states to beginning experimenting with regulatory reform that will directly impact the way residential consumers will purchase their gas. Table 3 shows how prices to retail consumers have risen slightly while prices to industrial consumers have declined since 1984.

Throughout the US, state regulators are experimenting and in some cases implementing programs to engender natural gas competition for residential consumers. At the present time, over half of the states have looked at or begun pilot residential supplier of choice programs. The approach may vary for different states but the goal is the same, to reduce costs to residential

Table 3   Gas Utility Industry Average Prices by Class of Service, 1995 (US$/MBtu)

Year	Total	Residential	Commercial	Industrial	Other
1984	3.60	4.16	3.86	3.06	3.41
1985	3.68	4.36	3.99	3.01	3.51
1986	3.45	4.24	3.79	2.51	3.02
1987	3.31	4.18	3.70	2.23	2.66
1988	3.48	4.27	3.74	2.30	2.71
1989	3.70	4.57	4.0	2.50	2.60
1990	4.03	4.92	4.25	2.63	2.84
1991	4.22	5.13	4.4	2.61	2.70
1992	4.34	5.30	4.58	2.65	2.81
1993	4.71	5.63	4.81	2.88	3.61
1994	5.10	5.99	5.08	3.09	3.78
1995	5.10	6.06	5.18	3.00	2.59
1996	5.47	6.29	5.41	3.38	3.78

Note: Table 3. Was adjusted for inflation using the GNP implicit price deflator, 1995 base year

consumers by allowing them to chose their supplier. It is expected that providing freedom of choice to residential consumers to choose their own supplier will enable them to seek out the lowest cost gas supply. However, the LDC will still retain its requirement to transport and provide gas delivery service.

The regulatory reform changes now just beginning to take place at the distribution level mirror much of the change that has occurred for pipelines: the separation of the sales and transportation functions. Competition has forced pipelines to exit the merchant function which has left them to become transporters. But, allowing greater competition by third party suppliers for residential customers does face serious challenges that each state must deal with. The question at this stage of the regulatory reform process is how to deal with supply reliability if LDCs relinquish their merchant function and become transporters only.

Natural gas utilities are granted a franchise to serve a particular territory. This right to operate is usually coupled with an obligation to serve a particular area. One of the key elements of this obligatory service is to insure continued service during cold weather even to customers who have not paid their bill. A complete relinquishing of the merchant function to third party suppliers would transfer the risk of meeting peak day demand from the LDC to someone else. Residential consumers could not be sure that their chosen supplier was capable of meeting their needs on a peak demand day. There is no practical method by which residential consumers could evaluate the integrity and reliability of a third party supplier. This could put LDCs in the position of determining the reliability of suppliers but yet they would not have the control necessary to insure reliability. Therefore, any change in the merchant function for LDCs is likely to require that state PUCs, particularly in regions with cold climates, develop methods for insuring peak day reliability to residential consumers. The following is a sampling of the approach by different states in their attempt to deal with supply security and residential customer choice programs.

- Full risk is assumed by the consumer: The State of California (population over 30 million) has adopted the posture of not requiring an obligation to serve in the event the third party supplier fails to deliver.

- LDC makes emergency gas available at a higher price: Many states require the LDC to provide emergency backup service but allow them to asses a penalty to the consumer for providing this service.

- Severe penalties for supplier non-performance: The state PUC imposes severe penalties for suppliers who do not meet their obligations.

- LDC is fully obligated to serve if supplier of choice fails to deliver: Customers can revert back to their LDC without notice. No additional charges are permitted.

- PUCs approve suppliers: One state will only allow approved suppliers to compete for residential consumers. Suppliers are investigated for reliability before being added to the approved supplier list.

It is estimated that 70 percent of all gas consumed in the US could be purchased from non-utility sources under transportation agreements. The remaining 30 percent would represent primarily residential and small commercial consumers that are currently not able to choose their supplier. This dramatic shift has taken place over the past ten years or so as a result of regulatory reform but the full impact of this change on the financial viability of investor owned gas distribution companies in the US has yet to be fully assessed. The loss of gas sales revenue does not materially impact distribution companies because, as previously noted, they are not permitted to derive any returns on gas sales. However it is the loss of transportation customers that can significantly affect LDC income. This loss usually occurs when large industrial and electric power plants by-pass the LDC supplier by making a direct connection with a major pipeline. Small LDCs are particularly vulnerable to by-pass arrangements. The loss of a single major customer has forced many small LDCs to charge higher rates in order to meet their financial obligations and earn their approved rate of return. Of greater concern to the financial well being of gas distributors is how to deal with the transition costs to a competitive retail industry.

In the process of making the transition to a competitive market for natural gas services many LDCs are faced with the possibility of incurring "stranded costs". LDCs are in much the same situation with stranded costs incurred prior to unbundling as for pipelines that incurred stranded costs in take-or-pay contracts written prior to deregulation. The costs for LDCs include idled equipment, gas purchase contracts, storage agreements and unused pipeline capacity. Since these costs were approved by PUC regulators it is incumbent upon them to work with their LDCs to resolve this issue. Since gas marketing firms have not incurred these costs they often have an advantage in selling gas services. Some states have required gas markers to buy out LDC gas contracts. However, the issue of how to deal with stranded costs is currently unfolding but in the final analysis all parties including regulators, LDCs, consumers and all other stakeholder groups will need to reach an agreement that is fair and equitable to all parties. Current data indicate that for large gas distributors net income has not been adversely affected by regulatory initiatives to unbundle gas services. Table 4 is a list of major gas distributors net income from the four leading gas consuming regions in the US (Northeast, Southeast, Midwest

Table 4   Net Income for Selected Natural Gas Distributors (millions of US$)

Name	1996	1995	1994	1993	1992	1991
Brooklyn Union	122.8	91.8	87.4	76.6	59.9	61.8
Columbia Gas	221.6	(423.3) loss	246.2	152.2	90.1	(794.8)
Atlanta Gas Light	80.6	30.8	63.2	57.5	55.4	49.4
Conn. NG	17.4	19.0	17.6	16.8	15.2	12.2
Northwest	46.8	38.1	35.5	37.6	15.8	14.4
United Cities	17.2	9.9	12.1	12.2	10.2	not available
Enserch	19.1	13.1	102.1	59.4	(27.3)	19.2
ONEOK	52.8	42.8	36.2	38.4	32.6	35.9

and West). The data for the period 1992 through 1996 indicate that net income for most LDCs has grown significantly but some have had their problems. Columbia Natural Gas, for example, filed for bankruptcy because of its inability to resolve take-or-pay contracts with producers. In the final analysis, the issues relating to stranded costs and full residential competition have yet to be fully realized thus future net income may be adversely affected.

APPENDIX 1

PHYSICAL ASPECTS OF GAS DISTRIBUTION

This appendix, intended for non specialised readers interested in the gas industry, describes downstream operations of the gas chain with emphasis on distribution. The aim is to better understand the economic analyses described elsewhere in the study.

"Distribution" in this context includes gas supplies to power plants and to high volume industrial clients as well as gas supplies to the local distribution companies (LDC) that supply the low and average volume clients: i.e. The industrial and commercial companies, and the residential sector.

The gas chain generally consists of three main parts:

■ the upstream sector (national and foreign producers);

■ the transmission level, with transmission companies buying gas for resale to high volume customers such as LDCs, industrial clients and power plants, generally via their high pressure networks (most often from 60 to 80 bar at overregional level, and from 40 to 15 bar at regional level) which may extent to lower pressure links to individual customers;

■ the distribution level, with LDCs supplying gas to low and average volume consumers. Their networks are partly at pressures between 1,1 bar and 10/15 bar, and partly at below 1,1 bar.

The first part of this appendix is devoted to the operation principles to the physical supply of gas. The second part addresses the different equipment of the high pressure networks and those of the average pressure and low pressure networks. Finally, the last part is devoted to the influence of climatic conditions on forecasted demand both in the long term but also in the very short term.

Operational Principles to Physical Natural Gas Supply

In principle, a natural gas customer can be connected to either a natural gas distribution company, a natural gas transmission company or even directly to a natural gas producer (if in the vicinity). Depending on the offtake, his supply will require either some kind of a high or a lower pressure pipeline as well as some kind of flexibility provisions (e.g. storage). More often than not (although this is not a rule), a large gas customer will be connected to and supplied by a transmission company, and a smaller customer by a distribution company, the latter buying wholesale from one or several transmission companies.

Every gas supplying company aims at supplying its customers at all times with the quantity of gas that is required by or has been agreed with each individual customer at commonly agreed conditions and specifications. This requires the ability to cover the (aggregated) demand in gas at all times, even in times with extraordinary demand surges (e.g. demand peaks in an unusually cold winter) or in situations of own gas supply disruptions (e.g. breakdown in production, in deliveries/imports), or in times of a failure in the transmission/distribution network.

In Continental Europe, the main application of natural gas is for space and water heating. This among other things makes natural gas demand subject to short-term fluctuations (seasonal as well as daily or even hourly). Almost half of total gas consumption in north-west Europe fluctuates with weather and seasonal changes. This share could increase in the future if the bulk of the new gas-fired power plants that are expected to be constructed is going to be used in mid- or peak-load only.

However, providing supply flexibility at production is costly, and most upstream supply of gas in Continental Europe is characterised by long-term (10 - 25 years) high volume contracts. Even though these contracts may contain more or less developed flexibility elements (e.g. possibility for the buyer to defer a share of the obliged annual off-take by one or more years; sometimes reduction of gas flow during summer supply fluctuations; even hourly fluctuations in specific contracts) more is needed to balance supply and demand. Operational measures at transmission/distribution level therefore include:

- natural gas storage to level out supply in relation to fluctuating demand (by storing gas during times of low demand – e.g. summer –, and withdrawing it during times of high demand – e.g. winter);

- co-operation between gas utilities (swaps, back-up deliveries, mutual assistance in dispatching);

- back-up agreements with producers;

- back-up from line-pack (gas available in the grid);

- use of interruptible contracts during demand peak times (interruptible contracts provide the gas supplier the right to temporarily interrupt supply to the customer either partially or totally for a stipulated length of time and for an agreed number of times during a given period; the choice between an interruptible or a firm supply contract is with the customer; an interruptible contract will give him a lower price on his gas consumption; for the period of interruption he usually holds a second energy in reserve – e.g. fuel oil);

- back-up from locally installed synthetic natural gas.

High Pressure Networks

The **pipelines** constitute the major part of a gas transport network. Investment in pipelines is heavy and generally represents approximately 50% of total investment of a transmission company. The biggest cost factor is the civil engineering costs, the pipelines themselves representing only a minor portion of the total costs.

Compressor stations are used to compensate for the drops in pressure due to friction during the gas flow. They generally consist of centrifugal compressors driven, in most cases, by gas turbines. Investment in compression and pressure reduction stations represents approximately 20 to 30% of total investments.

Blending and odorisation stations are used to obtain the calorific value (CV) of natural gas required by the users. Depending on its sources, natural gas comes in different calorific values. Regions in north-west Continental Europe differ also by the calorific values of the gas consumed. Odorisation is a process by which a product is added to the gas causing it to have an odour. This facilitates the detection of gas leaks.

Transfer metering stations must be placed at all the entrances and exits of a high pressure transport network. These stations must provide continuous and real time information about the quantities brought in or drawn-off of the network and must control the operation of the network. Magnetic tape recorders provide an accurate measure of the activity on the system. Telemetering devices which provide real time information are increasingly replacing magnetic tape recorders. **Billing** is based on the energy consumed, which implies knowing the transited volume (described in normal cubic metres: at 0 C and 1,013 bars) and the calorific value of the gas. In order to determine the normal volume, each transfer metering station has

- a measuring instrument (turbine, orifice or rotating pistons) to measure the gross volume V_b, that is to say, the volume to the real pressures (P) and real temperatures (T),

- pressure and temperature sensors to give the real characteristics of the gas,

- a corrector to bring the gross volumes (V_b) to normal volumes (V_n) according to the formula:

$$V_n = V_b \frac{P}{P_n} \cdot \frac{T_n}{T} \cdot \frac{1}{K}$$

The factor K, called the compressibility factor, is the function of discrepancy between real gas and perfect gas. It is measured by chromatography or by direct determination (Z metre). Chromatographic measurements are also used to determine the calorific value of gas (by multiplying the concentrations of each component by its calorific value). The precision of the meter is better than 1%.

The transfer of gas from a high pressure network to high volume industrial consumers and LDCs calls for **delivery stations** that generally combine pressure reduction stations and transfer metering stations. The pressure reduction station is used to reduce gas pressure from 60-80 bars to less than 15 bars. A delivery station normally has

- an upstream insulating coupling,

- an inlet valve,

- a filter that stops the solid particles carried by the gas, particles that would damage the measuring instruments,

■ a main pressure reduction line equipped with:
  • a safety valve,
  • a regulator,

■ an emergency pressure reduction line equipped with a regulator,

■ a heating system used to avoid low temperatures which arise from gas pressure reduction,

■ a meter installed on the main line (less common, if it is a meter with rotating pistons) or downstream of two pressure reduction lines for other types of meters.

The meters based on orifice plates are used mostly for large outflows and high pressures. Turbine meters are used increasingly for heavy outflows and high pressure due to their high degree of precision. Such new techniques as vortex or ultrasonic meters, have been developed and are starting to be used. The cost of a delivery station varies considerably from one company to another, due in part to security reasons. On average, a delivery station will cost approximately 500,000 US$.

Calorific value measurements

Two methods are used:

■ calorimeter, a device which measures the energy given off by the combustion of a determined quantity of gas,

■ chromatography determines the exact composition of the gas, that is, the percentages in methane, ethane, propane, butane and pentane; these percentages vary from one gas field to another. Then, by calculating, the calorific value of this compound gas can be found, as the exact calorific values of each component are known (see next table).

Composition in Hydrocarbons of a "Characteristic" Natural Gas

Gas	Formula	Content in natural gas (%)	Boiling point (°C)	Calorific value (kJ/m³)
Methane	CH_4	94.00	– 161.60	37.780
Ethane	C_2H_6	4.00	– 88.50	66.630
Propane	C_3H_8	0.50	– 41.90	95.639
Butane	C_4H_{10}	0.15	– 0.38	126.191
Pentane	C_5H_{12}	0.07	36.20	159.701

Source : British Gas.

Storage facilities contribute to supply and demand balancing. Seasonal storage sites must be able to hold very large quantities while their injection or withdrawal rates are relatively low. On the contrary, storage sites for peak periods are characterised by high injection or withdrawal rates compared to their capacity. The investments allotted to storage are in the order of 5 to 10% of the total investments in the gas chain, but make up 15 to 20% of investments by transmission companies. For seasonal storage, the principal sites are depleted gas or oil fields, water zones in porous rocks surmounted by an impermeable layer, or abandoned mines. A part of the injected gas is not recoverable (so-called cushion gas) and represents a non-negligible part of the cost of storage. Possibilities for peak storage are provided by caverns dug in salt domes by injecting fresh water to dissolve the salt and by extracting the briny water produced, or by LNG. In most cases, it is necessary to maintain an elevated pressure in the cavern in order that its walls do not buckle. The volume of gas corresponding to this minimal pressure is equally a gas cushion which can only be recuperated at the risk of destabilising the cavern. LNG can be stocked in storage tanks (LNG is 600 times denser than natural gas) and re-gasified when necessary. LNG storage is most economic upon receival from an LNG tanker. The cost of gas liquefaction is much higher than the re-gasification cost, and would in most cases be too expensive (nevertheless, if the gas contains large quantities of propane or butane, liquefaction allows them to be separated, which can give off very often some interesting returns and can reduce the total storage cost).

Distribution Networks

The distribution networks are made up of so-called average or low pressure pipelines, pressure reduction stations (local stations or delivery stations for more important clients) and gas metres. They represent 70 to 80% of total investments in downstream gas supply.

Very approximately, pipeline investment is between 30 and 150 US$ per metre of pipe, the cost of the pipe itself being approximately between 5 and 6 US$ per metre. The wide variation in the cost is linked to the degree of difficulty in digging the trench in which the pipe is to be laid. The cost can be quite high in a city, and low in a rural zone. Polyethylene, is used more frequently because it is easier to lay and erosion-resistant. It systematically replaces cast iron pipes. Steel is used for larger diameter pipes. Connection costs between the main pipe and the client's meter are around 1,000 to 1,500 US$, varying with regard to the distance between the pipe and the meter (on average 10 metres) and, with regard to the degree of difficulty in digging the trench. There are generally between 30 and 100 connections per kilometre of pipe.

Regarding pressure reduction stations, one should distinguish local stations from client stations. The former supply, under low pressure, the low volume clients situated nearby to it, the latter are used to supply higher volume clients. They are used to reduce the pressure to approximately 1,03 bar. Their cost is generally between 20,000 and 50,000 US$. In general, there are 200 to 500 low volume clients per local station.

Metres are standardised across western Europe. The cost of the installed meter is around 150 to 200 US$ whereas the cost of the meter is in the order of 80 to 100 US$. In certain countries, one can find pre-payment metres, though they are not very common. For example, in Germany, pre-payment meters represent only 0,1% of the total number of meters.

Predictions for the Gas Supplies and Standardisation of the Earlier Consumption by Taking Climatic Conditions into Consideration

Efficient balancing requires forecasting of gas demand. This is difficult due to the extent of weather and climatic influence on gas demand. In order to circumvent this problem, forecasting is based on standardised, so-called temperature corrected consumption figures: gas consumption statistics are corrected in order to obtain the consumption during "normal" climatic conditions (e.g. the average of the past 25 years). The standardised values allow to predict consumption in the long term and to foresee investments to be made in order to cover them.

Two more reasons must be added to justify an in-depth study of the influence of climatic conditions:

- the exact calculation of a bill to be addressed to the clients for which the consumption is recorded only once a year while the price of gas fluctuates throughout the year,

- the calculation of the total probable consumption of all low volume clients of each supplier for the day(s) following and the day(s) preceding in order to be able to manage the network and to adjust supplies and draw-offs ascribed to each supplier.

The **exact calculation of the bill** necessitates – if the selling price of gas varies – knowing the drawn-off quantities during each of the periods determined by the modification in the prices.

The **calculation of the consumption** of the days preceding is necessary in the case of a competitive market, even for the residential consumers, as the system assumes a daily adjustment of supplies and draw-offs for each distributor. However, draw-offs of the low volume consumers can only be obtained by calculations which take climatic conditions into account. Moreover, each distributor must know, as precisely as possible, what the draw-offs will be for the day(s) following with regards to the climatic conditions so that supplies may be adjusted in the network.

The link between consumption and climatic conditions is not an easy operation and the different distributors have developed models which differ to some extent. The simplest take into account the degrees-days defined as being the difference between a threshold (18 °C in Italy and for Eurostat, 16,5 °C in other countries) and the average outside temperature. Others take into account temperatures of the preceding days, the wind and the sunlight. Still others retain only the part of consumption which is susceptible to climatic conditions or weight the degrees-days (the weight factor is less in summer and higher in winter) allowing the entirety of the consumption to be corrected. An important limiting factor remains the inherent uncertainty of meteorological forecasts.

APPENDIX 2

NATURAL GAS STORAGE - COST AND VALUE

Storage is a key link in the gas chain because of the range of services it provides to the gas network (security, reliability, flexibility, etc.). As European gas markets are about to change as a result of countries looking at ways to introduce competition in natural gas supply by way of third party access, storage could have new roles to play, enhancing its significance in overall gas supply. However, data on storage cost to a large extent remain confidential. At present in Continental Europe, storage is integrated in a chain of services (transport, load factor, distribution) usually invoiced by a single operator to the end-user. This provides little or no transparency on the costs of storage. Exact knowledge of these costs, however, is essential to assess its future role.

The first section of this appendix reviews the physical characteristics of the different types of natural gas storage reservoirs, relating them to the various roles storage can play in the physical gas chain, and keeping an eye on the costs involved. A particularity of storage is that unit investment and operating costs vary widely depending on specific site conditions. There are therefore no determinist rules regarding cost of storage. Nevertheless, and attempt is made here after to provide information about cost of storage. Also, the value of storage is discussed because a value approach allows considering load factor and security issues on the network as a whole (other means to balance supply/demand are mentioned and examined upon their the cost-effectiveness in relation to storage).

Natural Gas Storage: Why and How

Load balancing is the main function of gas storage, reflecting the need to keep gas supply as constant as possible to exploit production and transportation facilities at a maximum level, whereas gas demand is highly fluctuating (seasonal, daily and hourly variations), chiefly in the residential-commercial sector, which is especially sensitive to changes in temperature. Seasonal load balancing needs and daily or hourly load balancing needs are usually distinguished.

Balancing Gas Supply and Demand

The need for seasonal load balancing stems among other things from of the important share of the residential sector in total gas consumption. In the US, however, with the development of air-conditioning, another peak corresponding to increased electricity demand is apparent during the summer, mitigating the need for seasonal load balancing.

Daily and hourly fluctuations are partially absorbed through line-packing, i.e. The use of pressure differences in the network. Gas demand however can be twice the average daily demand during the winter season under extreme weather conditions, resulting in a need for short-duration, high-deliverability supply sources designed to provide service on the coldest 5 to 15 days of the year. The solutions can be: building sufficient pipeline or underground storage capacity (but this is often expensive for such a low annual utilization rate), interrupting supplies to end-users that have alternative or back-up forms of energy (but interruption alone, though heavily used and attractive, is generally insufficient to meet the total increase in demand on a peak day), using peakshaving facilities (e.g. LNG tanks). Peakshaving facilities may also be used more regularly to provide operational flexibility in meeting hourly demand variations.

Providing Security of Supply

This function (often called emergency back-up) consists in reducing the risk of supply shortages, either of a technical, generally short-term nature (failure of a compressor, unplanned shut-down of a gas field, etc.) or of a political, i.e. strategic nature which could have a longer duration.

Optimizing the Network

This function is a consequence of the first one: storage enables gas transportation companies to operate their networks more efficiently throughout the year by allowing a high load factor, thus reducing the final cost of gas.

Commercial Role

In Europe, some storage operators let out storage capacity to neighbouring gas suppliers on commercial terms. After the latest reforms of gas market regulation in North America, one saw the emergence of hubs (exchange markets consisting of a network intersection equipped with storage capacities to match supply and demand). New actors appeared, offering a wide range of value-adding sales services from storage to pipeline system balancing (making the producer less vulnerable to seasonal fluctuations in wellhead prices by building underground storage in production areas), price hedging (making users less vulnerable to seasonal price increases: they can buy gas when it is cheap, store it and use it when it is expensive) or no-notice service (providing the customers with storage capacities that have not necessarily been booked in advance, when needed). In addition, thanks to available storage capacities, futures markets have developed within those hubs: storage serves as the physical support for financial transactions between various actors.

Type of Natural Gas Storage

There are two natural gas storage reservoir families: underground reservoirs and liquefied natural gas (LNG) receiving terminals and peak-shaving units. Underground reservoirs can be: porous formations, i.e. oil or gas depleted fields and aquifers, salt caverns, mined rock caverns (as yet only in project development), disused mines. The choice between different storage methods is largely dictated by the geological formations available. LNG is stored in above-ground or in-ground cryogenic tanks (mostly in Japan).

The cost of developing an underground storage facility can be considered as a function of the inherent characteristics of the proposed storage site, its location in relation to the intended users, including its access to the transportation and distribution system, and the types of services it will offer. The design of a storage facility is influenced by the economic tradeoffs between the desired design requirements and the inherent characteristics of the storage site, including any existing infrastructure and geological constraints.

The following provides a review of the main storage facility types (porous formations, salt cavities, LNG storage) by summarizing their inherent characteristics and how they influence the costs of providing various types of services.

Porous Formations: Depleted Fields and Aquifers

Description:

Porous formations (usually sands, sandstones or carbonates at depths of about 1500 to 6000 ft) are the most common means of storing natural gas, in any quantity, especially very large volumes. A high percentage (typically 50%) of the gas stored cannot be withdrawn from the reservoir: it is the cushion (or base) gas. It maintains a sufficient pressure in the reservoir and prevents wells from being in contact with liquid. Base gas requirements are a function of design flow requirements, reservoir flow capacity, and compression economics. The maximum withdrawal rate will depend on the number of wells but also on the characteristics of the porous rock. Compression is used to inject the gas into the reservoir and in some cases for withdrawal as well. To prevent any leakage, observation wells are normally drilled, especially at the periphery of the reservoir. Reservoirs in porous formations have generally been designed for one injection/withdrawal cycle per year.

Reservoirs in depleted fields satisfy the permeability and porosity conditions required for storage as they formerly contained gas or oil. However, many depleted gas fields are not suitable to conversion: they are too large for effective storage operations. Two types of depleted fields are often distinguished: structural (traps) and reefs. Reefs are characterized by very thick reservoirs (350-700 ft) having a small area extent, very high effective permeability and great tightness.

Storage in aquifers consists in gas being injected into the pore spaces of a water-bearing sedimentary rock formation overlaid with an impermeable cap rock. The rock formation must be an anticline with a sufficient closure, the reservoir must be porous and permeable and the cap rock must be gas-tight.

Factors influencing design costs:

In porous formations, contrary to other reservoir types, total storage capacity and individual well flow rates are chiefly dependent on the reservoir characteristics.

Depleted fields – advantages:

■ The geology and producing characteristics of a depleted field, which influence strongly gas capacity and daily withdrawal rates, are well known. There is therefore no need for an exploration phase.

■ Access roads, control buildings, and existing surface developments can often be re-used, and much of the existing wells and production equipment can generally be converted to storage use.

■ Only minimum risk is associated with vertical migration of the gas through the cap rock, which offered a satisfactory seal since the formation originally contained gas.

Depleted fields – drawbacks:

■ The main determinants of reservoir size and flow capacity (native rock porosity, permeability, water saturation, pressure, and thickness) cannot be easily or significantly modified.

■ There may be oil or water leakage, and cementation being a delicate task, maximum pressure in the reservoir may be limited.

■ There may also be a large quantity of H_2S compounds to be removed from the gas withdrawn.

- Differences in quality between cushion and injected gases can cause complicated mixing problems.

- Existing wells may have to be cemented.

  Reefs – additional advantages:

- Very few wells are required, as reefs are thick formations.

- The deliverability per well is relatively high as compared to structural reservoirs.

- Very little cushion gas is necessary (working gas/total gas ratio can exceed 70%).

- They can be operated at very low or very high pressure.

As a consequence, reefs make for the ideal storage sites. Typically, among all the storage means, they have the lowest investment cost.

Aquifers – drawbacks as compared to depleted fields:

- Prior to gas injection, aquifer storage requires that geological and engineering studies be conducted to assure the proposed reservoir seal is adequate to prevent gas leakage.

- Greater monitoring of withdrawal and injection performance is needed to ensure the gas bubble remains contiguous so that all the gas remains recoverable. Consequently, aquifer reservoir storage may take several cycles to achieve the design volume, and withdrawal and injection rates are limited.

- There may be unexpected leakage.

- Constraining environmental aspects often have to be taken into account (e.g. when the aquifer is the ground water).

- Base gas requirement is high, averaging 60% of total gas in the reservoir, because it is important to maintain an interface between the gas being withdrawn and the aquifer liquid.

- Base gas can not be recovered.

- Somewhat greater dehydration and water handling capacity may also be necessary as native water will be withdrawn with the gas, at least during the first few cycles.

Salt Caverns

Description:

Salt beds of sufficient thickness and purity and of suitable geomechanical properties, typically found at depths of 2500 to 6500 ft can be leached out with fresh water to create underground caverns for gas storage. They serve to store relatively smaller quantities of gas than porous formations, the storage capacity for a given reservoir volume being proportional to the maximum operating pressure, which depends on the depth. The pressure must never go below a specified threshold, for the mechanical stability of the subsoil to be preserved, and the cavities must be located at a specified distance from one another (a facility generally consisting of several cavities). 7 to 10 cf of water are required to leach 1 cf of salt. Salt caverns are best fitted for multi-filling/draw usage(typically twice a year or more) and to provide random peaking flexibility.

Advantages, as compared to porous formations

Factors influencing design costs:

■ high flexibility (in switching from injection to withdrawal for example).

■ very high injection rate, hence short filling period.

■ very high withdrawal rate, hence high degree of availability.

■ high level of safety.

■ high working gas ratio (60-70%) because of the relatively deep location and high permeability of salt caverns, and total recovery of cushion gas (at the end).

■ no compression requirement during withdrawal generally.

■ contrary to porous formations, flexible choice of the depth of the caverns, allowing to optimize the tradeoff between compression power (high if deep) and base gas volume (high if not deep).

Drawbacks

■ cost of the leaching process (the largest single cost component of a salt cavern development): heavy equipment required, long lead time (several years), strong dependance of construction costs on the supply of fresh water, and on the possibilities that exist for the disposal of brine (ideally sold to the chemical industry, it can also be pumped into a deep geological formation or an old mine, or be transported through a pipeline to the sea, thus reducing costs); on the other hand, possibility of significant construction cost reductions (10 to 40%[1]) in expansions because leaching and water disposal facilities already exist;

■ limitation of the ultimate storage capacity by the size of the leached cavern and the strength of the surrounding salt;

■ relatively high minimum operating pressure to prevent cavity to creep and lose volume;

■ high risk of hydrate formation;

■ greater compression requirements during injection, because of the relatively deep location.

LNG Storage

Description:

The idea is to save space, as 1 cubic meter liquefied gas corresponds to 600 cubic metres of gas at atmospheric pressure. Gas is chilled at atmospheric pressure ($-163°C$) and injected into tanks which are not refrigerated, but merely insulated from the atmosphere. They can only store a relatively small volume but are able to eject it in a very short time. The majority of the LNG tanks currently in service are double-wall metal tanks, i.e. A gas and liquid tight inner tank made of 9% nickel steel, separated by perlite, which is an excellent, inexpensive and fire resistant insulation material, from a carbon steel (i.e. cheaper) outer tank acting only as a pressure barrier and additional protection, as it is not subjected to very low temperatures. To insulate the tank from the ground, the foundation may be built into the ground, in which case soil heaters are provided, or on concrete piles leaving an air space between the ground and the bottom of the base.

Factors influencing design costs:

1. Underground gas storage in the world, a new area of expansion, Cédigaz, 1995.

Selection of tank design and the method of construction will depend upon requirements in local regulations and national codes of practice, and capital and operating cost considerations. For example, the outer tank is 20% less costly if it is in steel rather than in concrete, but according to Technigaz it is less secure.

Drawbacks

For various reasons, LNG storage development costs are always high as compared to underground storage:

- Each equipment device (pipe, floodgate, etc) is required to be cryogenic.

- Energy requirements (generally the stored gas itself) are high (up to 10% of the stored gas).

- A high level of security is needed (gas, smoke and fire detectors, water spray cooling systems, high expansion foam and dry chemical powder extinguishers, etc.).

- Foundations (or in-ground tanks) are very expensive.

- Where seismic risks can occur, the development cost dramatically increases, because the tank has to be shorter, thus not being of optimal shape anymore (the optimal shape being height = 0,7*diameter).

Advantages

- LNG tanks are able to provide very high daily deliverability, close to the market (the site can be chosen).

- There is no cushion gas.

Optimizing Underground Natural Gas Storage

Optimizing underground natural gas storage means activities to improve its operation and economics. This can be achieved through either increasing site performance in terms of productivity and flexibility or reducing development and operating costs and the development period or else developing new techniques allowing siting reservoirs near the consumer zones, even when the geological conditions are unfavourable in the region.

Increasing Site Performance

For porous formations, the main source of possible improvement is the new technique of horizontal drilling, which increases the withdrawal and injection rates of wells. Optimization can also be achieved through the followings: increasing, if practicable, the pore volume of the reservoir layers (e.g. by hydraulic fracturing in a carbonate reservoir), increasing the reservoir pressure range (through better geomechanical knowledge and cementation technique), thus improving the working/cushion gas ratio, limiting water production.

For salt caverns, the main trends in optimizing natural gas storage performance are: extending cavity volume, improving the working gas/cushion gas ratio by increasing the reservoir pressure range, improving utilization of the salt formation available for cavern construction.

Reducing Costs

Performance optimization naturally brings about costs reduction. Actually all the techniques involved in gas storage (geology, geophysics, drilling, salt cavity construction and leaching) are likely to allow cost reduction. An important research area to reduce investment costs in aquifers consists in injecting inert gas to replace a portion of cushion gas (in France, a 20% replacement has been achieved in three aquifer reservoirs).

Developing New Storage Techniques

Traditional techniques are faced with geological limitations and are increasingly confronted with environmental constraints (for example, competition between storage and water resources for aquifers, brine discharge issues for salt cavities). As it appears more profitable to create storage sites as close as possible to consumer centres, research is done on new techniques, like mined rock caverns for example. Another new technique is storage in producing oil fields: the gas forms a bubble (gas cap) that is designed to remain static over the oil in the reservoir, and that can be converted into storage.

Different Types of Reservoirs According to Function

A natural gas reservoir is mainly characterized by two parameters: the working gas volume V (capacity) and the maximum withdrawal rate Q (limited by the number of wells and the surface equipment). According to these parameters, two reservoir families are distinguished: if $V = 50$ to $100\ Q$ (mostly porous formations), reservoirs are not very productive but allow to store large volumes, thus being suited for strategic purposes and seasonal load balancing; if $V = 10$ to $20\ Q$ (mostly salt cavities, abandoned mines, rock caverns and LNG reservoirs), reservoirs are a lot more productive, but their volume is limited: they are suited for daily/hourly load balancing and peakshaving. The ideal solution is a combination of these two types of storage, but countries often have only one type of storage at their disposal. In France, for example, aquifers are largely used for peakshaving, especially in the Paris region, where there is no peakshaving facility but several aquifers which have a daily withdrawal rate comparable to that of a salt cavity.

The way natural gas storage is used differs sometimes significantly between countries in function of storage availability, importance of indigenous production and market opening (access regulation).

E.g., in the US, where ample storage and transport capacity is available, storage functions are more diverse than in a country like France, and a wide range of storage services are supplied on the market, each one corresponding to an increased value added to the gas. In the US, storage is considered either as a seasonal source of gas supply reducing the need for peak pipeline capacity (storage close to the consumption areas, e.g. North East states) or as a means to guarantee a steady flow from the producing states (storage close to the production facilities, e.g. Louisiana and Texas). In France, storage fulfills the two main functions of balancing and security of supply: storage in aquifers ensures middle-term security and seasonal balancing, whereas interruptible contracts and salt cavities ensure short-term security and provide peakshaving.

As regards the security function, given that the US is almost self-sufficient in gas there is little political risk, hence no need for strategic storage. France on the contrary made the decision to have sufficient storage capacity to supply customers in case of a one-year interruption in supply from its biggest supplier (Russia, 33% of supplies).

Summary

The following table gives an overview of the different ways of storing gas and their characteristics.

	Storage medium	Means of emplacement	Operating principle	Advantages	Drawbacks	Uses
Depleted fields	porous and permeable geological formations originally impregnated with oil, gas and water	fluid in place is compressed and displaced by gas injection	gas compression and expansion combined with the effects of the liquid's compressibility and mobility	large capacities, reuse of existing installations, no exploration phase, fully gas-tight existing cap-rock	there may be oil or water leakage, limiting maximum pressure in the reservoir, hydrogen sulfide can be formed, not all the gas can be recovered	seasonal load factor, strategic reserves
Aquifers	porous and permeable geological formations originally impregnated with water	same as above	same as above	large capacities	prospecting risks and costs, non-recoverable cushion gas, withdrawal rate is limited, environmental constraints	seasonal load factor, strategic reserves
Salt cavities	salt formations where cavities are excavated by leaching with water	gas compresses and expels brine	compression and expansion, balancing by brine (especially to recover the cushion gas)	high ratio between withdrawal rate and working gas, flexibility, safety	cost of leaching and brine disposal, cavity can lose volume due to creep, hydrate formation	daily and weekly load factor, distribution security
LNG storage	double-wall insulated cryogenic tanks (above-ground or in-ground)	gas is chilled at −163°C to be stored in the liquid state	LNG is pumped into or out of the reservoir	high withdrawal rate, high flexibility, site can be chosen, no base gas	high cost compared to underground storage	daily load factor, peakshaving, emergency back-up

The Cost of Storage

About the Confidentiality of Storage Cost

Reservoir engineering companies make every effort to keep information on storage costs secret because they know that these will be highly valuable information with the near prospect of gas market deregulation in Europe, which is likely to bring, as it did in the US, an increase in construction and use of storage. On the other hand, big gas operators like Gaz de France (which develops its own storage capacity) or Distrigaz in Belgium (which uses the services of a reservoir engineering company to develop its storage capacity), which integrate a range of activities on the gas chain (e.g. Import, transmission, storage, distribution), are reluctant to talk about costs. One explanation could be that they use the net-back pricing principle: the price at which import contracts are negotiated is obtained by deducting from the maximum selling price of gas on the market (depending on the prices of substitutes on end markets) the downstream costs: distribution, transportation and load factor, including storage. This could give rise to midstream and downstream rents, by overvaluing distribution, transport or storage costs, or at least keeping them as opaque as possible. It goes without saying that it is not in these companies' interest to reveal their costs relating to distribution, transportation and storage.

In this respect, storage costs are particularly likely to be kept opaque, as transportation and distribution are inherently more transparent (it is actually rather easy to estimate the cost of bringing 1 cubic meter of gas from A to B through a given type of pipe, or to calculate how much it costs to distribute 1 cubic meter of gas to client Z), whereas storage cost, because of the intricate functions of storage, can never really be considered independently from the other costs of the transportation network and is therefore much more difficult to estimate.

Cost of Development

The cost of storage can be viewed as either the capital outlay or development cost of a project to the developer or the effective annualized cost of service to the customer. The former is discussed here.

Cost Components of Underground Storage Development

To study development cost, investment cost is split according to the various cost components. When comparing different projects of one storage type, one realizes that there are no general rules: the shares of one cost component differ widely between projects due to the inherent site characteristics. A typical range of cost components could be: 10% exploration, 20% well drilling and completion, 30% surface network and facilities, 40% cushion gas. Below, we focus on the cost components that account for the highest dispersion around average.

Exploration

Exploration cost, which represents a large portion of the initial investment, is extremely difficult to assess, because often the exploration stage took place much earlier, and for other purposes than finding a convenient place to store gas (for example, many oil companies have developed natural gas storage because they had found an appropriate site while looking for oil resources). One can wonder, for example, concerning depleted fields, whether a portion of the exploration cost that enabled to find the field should be allocated to the storage facility development cost.

Leases, Rights-of-way, Buildings

This cost component will vary dramatically from field to field, according to their location, the surface area needed for operating them, the number of inhabitants upon the site, etc. In addition, political and environmental concerns of local governments may interfere. For example Péronnes, a disused mine storage in Belgium, has been closed because a raise in local taxes suddenly made it non profitable. Landes-de-Siougos, an aquifer project in France, has been shelved because of environmental concerns. At Germigny-sous-Coulombs in France (aquifer), surface facilities have not been built at the optimal location because the village upon the site refused to receive them, resulting in increased piping, oblique and deeper drilling, etc.

Drilling and Wells

There is no simple rule to define drilling costs as a function of depth, especially at average depths (1000 to 6000 ft) like for most natural gas reservoirs, as they will largely vary also according to geological, geographical (difficulty of the access to the site, presence of water) and economic (prices can rise dramatically when the drilling activity in a region increases) parameters.

Production and Gathering System

The production system will be in compliance with the quality of gas withdrawn from the field. In a very thin aquifer for example, or during the first cycles, the quality of the withdrawn gas will be lower than later on. Gas may or may not need desulfurisation. At Germigny-sous-

Coulombs in France, the sulfur compounds rate in withdrawn gas is highly variable: there are years when desulfurisation is not needed. Dehydration may produce a lot of polluted water, and getting rid of it may raise costs significantly.

Compression

Compression requirements are generally driven by injection and withdrawal design. Several parameters influence this design: maximum reservoir pressure, duration of injection cycle, working gas volume, and pipeline delivery pressure. The choice of additional compression versus additional base gas to facilitate withdrawals is an economic tradeoff.

Pipeline Connections

Their cost is highly dependent on how far the reservoir is from the network. It also varies according to the terrain, the population density along the right-of-way and the pipeline diameter.

Base Gas

This is generally the highest investment cost component. Base gas requirements are an economic tradeoff between the cost of a pressure cushion provided by the base gas and the cost of compression. The cost of base gas is the delivered price of natural gas taken during the low cost season i.e. The summer (varying with pipeline charges and distance from a source of supply to the storage site), plus the cost of injection. In depleted fields, base gas requirements will vary dramatically with individual reservoirs and the particular design requirements (there is native base gas already in the field). To reduce the cost of base gas in aquifers, inert gas can be injected in replacement of a portion of base gas (up to 20%). In the US, some developers do not include the base gas cost in their estimates, because they plan to have storage customers purchase their own base gas.

Overhead and Financing Costs

These vary widely between companies.

The development cost of an underground storage facility may be looked at in terms of cost of working gas capacity (the simplest and most common measure of the development cost of storage) and in terms of design withdrawal capacity cost. The following table shows estimates of the unit cost of development of various types of storage, taken from publications from American organizations (upper half) and from interviews with European companies (lower half).

Unit: $/kcf working gas

Source	Depleted field		Aquifer	Salt cavity	LNG tank
	reef	trap			
US Study 1[2]	3		5	10	–
US Study 2[3]	4		5	7	–
US Study 3[4]	3 (average 4,5)	7	7	19	34
European Source 1	3 – 5		5 - 10	13 - 21	53 - 63
European Source 2	4 – 9		7 - 17	10 - 20	40 - 62
European Source 3		6 – 8		17 - 22	–
European Source 4	–		5 - 10	13 - 26	22 - 40
European Source 5	9		13	–	–
European Source 6	≥ 5		8 - 16	11 - 26	–

(2) *Development cost of new underground natural gas storage facilities in the lower-48 United States*, Energy Information Agency, February 1994.
(3) *Natural gas 1994, Issues and trends*, Energy Information Agency, March 1994.
(4) *Operational and cost parameters for natural gas storage*, Gas Research Institute, March 1993.

What is noticeable here is the difference between American and European estimates: American estimates are roughly at the lower end of European ranges, or even far below[5]. Some elements of explanation are as follows. First, regarding sources, American data are project costs, provided by developers, whose aim is to show that investment in their project will be profitable. Thus they have an incentive to underestimate costs, whereas in Europe the above displayed costs have been given by existing operators, who may overvalue their costs in order to dissuade potential entrants from entering the market. Then, regarding market structures, everything tends to be less expensive in the US, because most markets are more competitive. In addition, the storage activity is more mature in the US, and all activities linked to subsoil exploitation are more developed. Finally, in the US studies, costs refer to new projects as well as expansions of existing storage fields. As expansions cost less, these costs may be undervalued.

Thus roughly, to give a range, the development of a storage reservoir costs $3 (for large seasonal storage) to $30 (for small peaking facilities) per mcf of working gas (which makes $10 to 100 million total[6] for an underground storage project).

Capital cost per M(cf/d) of design withdrawal rate is as follows:

Unit: $/M(cf/d) withdrawal rate

Source	Porous formations	Salt cavity
US Study 1	60-1000 (average 300)	30-200 (average 100)

NB: These estimates represent only a fraction of total costs, and should be divided by 1000 to make them comparable with the development cost of storage.

We can conclude that, whereas on a volume capacity basis salt caverns are at the higher end of the underground storage cost scale, they prove more favourable on a withdrawal capacity basis. We do not have quantitative data as regards LNG, but as the design withdrawal rate of these facilities being very high and their working gas volume relatively low, they certainly prove a lot more favourable as well as regards the development of withdrawal capacity than as regards the development of working gas capacity. That is the reason why they are used as peakshaving facilities only (i.e. when a very high withdrawal rate is needed on a short period).

Annualized Cost of Service to the Customer

While development cost is a useful measure when comparing various storage facilities, it cannot capture how the operation of a field influences the effective cost of storage service to the customer. To do this, it is possible to calculate the annualized cost of service per MBtu of gas withdrawn from the reservoir. It will include cost of depreciation, cost of operation and financial cost.

Calculation

The annualized cost of service will be expressed per MBtu of gas withdrawn from the field (and not per Mcf of working gas), thus reflecting the actual use of the storage capacities. Actually, operation and maintenance costs are mostly fixed costs. Therefore, the more gas injected into/withdrawn from the facility (i.e. the more storage cycles performed in the year), the lower

---

5. On average, estimates for Europe are 50% higher than for the US)
6. *Underground gas storage in the world, A new era of expansion*, Cédigaz, 1995.

the annualized cost of service, because fixed costs are spread over a larger volume. Two examples will illustrate this.

Salt cavities, the development cost of which is typically twice that of a porous aquifer formation and three times that of depleted fields, can have a lower annualized cost of service, and consequently should be economically attractive in applications that require multiple cycles per year (6 to 10 cycles possible). It will then all depend on having markets/functions for storage that require multiple in/outs per year. The existence of such needs gives developers an incentive to offer multiple services to maximize their facilities' utilization. Markets that require multiple injection/withdrawal cycles in a year typically develop in deregulated gas markets, resulting in a diversification of the supply of storage services.

Another example in the opposite direction is that of strategic storage, resulting in a certain amount of the stored gas not being intended to be cycled (for instance, in France, the average number of storage cycles is 0.58). As a result, the annualized cost of service will be higher. This gives some indication of the cost of strategic storage, otherwise difficult to assess.

To calculate the annualized cost of service per MBtu of gas withdrawn from the field, various rules are used, more or less complicated. The simplest way is to add the annual operating cost, the annual depreciation of investment, and the annual financial cost. A more complete calculation should take into account taxes and means of financing. It is often calculated by expressing it as a percentage of the investment cost (hence the interest of calculating this first). We typically have: depreciation = 4% (typical linear depreciation on 25 years), financial cost = 6% and operating cost = 5 to 10% (fixed costs, including salaries, maintenance, property taxes, insurance, and variable costs, including compressor station operation, and gas treatment). Hence, the annual cost is roughly 15% of the total investment cost.

Comparison of Annualized Costs of Service

Sources are the same as above.

Unit: $/M(Btu) withdrawn

Source	Depleted field		Aquifer	Salt cavity	LNG tank
	reef	trap			
US Study 1[2]		0.3 – 0.8		0.5 – 2.5 (1 cycle) 0.3 – 1.2 (2 cycles)	–
US Study 2[3]		0.2 – 0.7		0.5 – 1.8 (1 cycle) 0.3 – 1 (2 cycles)	7
US Study 3[4]	0.4	0.5	0.47	5 (1 cycle)	6
European Source 1	3 – 5		5 - 10	13 - 21	53 - 63
European Source 2	0.9 – 2.3		1.7 – 4	2.3 – 4.5	8 – 13
European Source 3	0.6 – 1.3		0.8 – 1.5	–	–
European Source 4	–		1 – 1.6	1.8 – 3.1	–
European Source 6	≥ 0.9		1.3 – 3.2	1.8 – 5.2	–

The classification of storage types according to the annualized cost of service is the same as according to the development cost, except that salt cavities seem more favourable here, because of the number of cycles, which is generally higher than for the other types of storage (on average, $/M(Btu) 1.5 versus 0.5 for porous formations, 3 versus 1.5 for salt cavities, and 10 versus 6 for LNG tanks).

It is noticeable that also here European costs are significantly higher than American ones. Some elements of explanation: as investment cost is higher in Europe, depreciation is higher too; operation costs are higher for the same reasons; and there is the above mentioned issue about the number of cycles: in Europe, strategic reserves reduce the number of cycles, whereas in the US, the need for flexibility increases the number of cycles (even some porous formations have more than one cycle in a year, which has never been the case in Europe).

Pricing

What will be the storage cost to the end-user? The above calculated annualized cost of service is supposed to be spread over the total gas consumption of the customer in proportion to the need for load balancing. For example, industrial customers need little load balancing, as their consumption is almost constant over the year. Tariffs should reflect this situation.

For example, in the US as well as in the UK, rate design for a storage field typically reflects the annualized cost of service (plus a commercial margin) using the following components: demand charge ($/k(cf/d) of withdrawal capacity), capacity (or reservation) charge ($/k(cf/d) of working gas capacity) and injection and withdrawal charges ($/k(cf/d) of gas withdrawn from or injected into the field). Demand and capacity charges are typically calculated by taking stated portions of the annualized fixed cost of service (frequently a 50-50 split) and dividing that cost by the maximum daily withdrawal rate and single-cycle working gas capacity, respectively. Injection and withdrawal charges are annual variable operating costs divided by the annual working gas volume expected to be cycled. There can then be different rates according to the degree of interruptibility of the service offered to customers. Rates are generally set on a site by site basis.

Let us now take the example of France. How are storage costs distributed over the different types of clients? The load balancing cost (i.e. seasonal load factor and peak load factor) is distributed using the following rules:

■ If we call Cv the cost per kcf of working gas capacity, Cw the cost per k(cf/d) of withdrawal capacity, Dv the customer's demand for stored volume and Dw the customer's demand for daily withdrawal, the cost for load balancing through storage is:

$$C = Cv*Dv + Cw*Dw.$$

Cv and Cw can be calculated using the principles presented in the previous paragraphs, and Dv and Dw can be calculated as follows.

■ As regards Dv, if we call Q the yearly gas consumption of the client and Qw the total winter consumption (5 months withdrawal on average, between November 1st and March 31st) and Qm the maximum daily demand, we have:

$$Dv = Qm/w - 5/12 *Q.$$

■ As regards Dw, we have:

$$Dw = Qm - Q/360.$$

The cost of security is reflected in the load balancing cost in that, as outlined before, the annualized cost of service per MMBtu withdrawn gas is higher when gas is stored for security purposes, i.e. not being intended to be cycled.

Cost and Value of Storage

The value of gas storage can be defined as the maximum price that any consumer is ready to pay to get the service he is offered, i.e. The price he is ready to pay, all other solutions having been taken into consideration. Seasonal balancing and security is firstly provided by adapting the demand (e.g. making some large industrial customers switch from gas to another energy in winter - interruptibility - be it seasonal or short-term.

Let us first have a look at the following table, displaying storage capacity / consumed gas ratios for some European countries (data from 1993):

Country	Consumption C (10^9 m^3)	Storage capacity S (10^9 m^3)	Ratio S/C (%)
France	34	10.5	31
Italy	51	12.9	25
West Germany	70	8.5	12
Spain	7	0.5	7
United Kingdom	66	3.6	5
Belgium	12	0.5	4
Netherlands	48	0	0

Source: Underground storage in the world, Cédigaz, 1995

Storage requirements differ widely from one country to another (for a variety of reasons of which the most important ones are seasonality of demand and long term security of supply), which also means that storage has not the same value in every country. The ratio for France for example (31%) is one of the highest worldwide (20% on average in Europe).

This raises the question of the value of storage and to what extent use of storage can be optimised or be replaced by other (more) economical means. Hence the idea to compare storage cost with the cost of alternative solutions. Another reason for being interested in the value of storage is that storage projects are not always well accepted, in particular due to unfounded public fears of underground (as well as LNG) natural gas storage. One problem with this is that in traditional markets, storage developers and network operators (who are generally the same companies) have not clearly spelled out the value of storage.

Interruptibles vs. Storage

Interruptible contracts represent an instrument which use could lower the requirement for storage. Wider use of these contracts and of the possibilities they provide in interrupting supplies to interruptible customers could free some gas at demand peaks for supply to other buyers at potentially higher prices (most interruptible gas is sold at fuel oil parity - minus incentives - and can at times of interruption be sold at higher value to other buyers).

Competitive open markets typically offer higher potential for this. E.g., before the opening up of the British gas market there was a "cushion" of interruptible customers which were seldom interrupted. But with market opening and new suppliers entering the market, competition forced suppliers to interrupt interruptible supplies in order to optimize the management of balancing methods. At that time gas prices were at a low level, and there was not a significant difference between the rates for interruptible supplies and non-interruptible supplies. Thus gas users did not have an incentive anymore to contract interruptible supplies, and those sitting on older interruptible supply contracts pressed to return to firm supply contracts. Since then, however, operators have reintroduced a price difference in order to maintain this instrument of flexibility, and the market has then emphasized interruptibility.

The "network code", established in Britain in 1996, requests gas marketers to balance the gas they sell with the gas they introduce in the system on a daily basis. To measure the clients' consumption everyday, as gas delivered to an individual customer is not measured everyday, a computer system calculates theoretical levels of consumption, using standard clients models, etc. Since September 1996, if marketers do not comply with the network code, they face penalties (they have had to pay the marginal cost of putting gas into the system). Therefore, marketers need interruptibles to be able to match supply and demand. Hence, the network code gives a value to interruptibility.

More fundamentally, the need to supply gas to the captive customers creates an uncertainty for the suppliers because no-one is able to predict what the severity of the winter will be and, therefore, what the demand of the captive residential and commercial customers will be. In practice, suppliers are often requested to contract enough gas and deliverability to match the worst situation[7].

Seasonalised Production vs. Seasonal Storage

An additional instrument in balancing demand with supply is to seasonalise supply (as far as economic and possible): either by seasonalising its own production, if it is a producing country, or by purchasing seasonally balanced supplies. Though generally, storage is more cost-effective than production load factor, particularly in the case of remote production fields, and/or high development costs in production.

7. e.g., that of the '1 in 50 years'

GLOSSARY OF TERMS AND ABBREVIATIONS

Aggregator   party which couples more than one order of supplies to gain a price advantage.

Balancing   see '*load balancing*'.

Beach / border point, Beach / border price   (gas price at) point of delivery (entry from upstream into a downstream transport system, which in case of imports usually is the national border, and/or which is often conveniently called "beach point" since much gas is produced offshore by a producer or upstream shipper.

Bundled sales   sale of both fuel and transportation services to an LDC or end-user.

Bypass   delivery of natural gas to an end user by means other than the traditional local distribution company connected to the end user.

Capacity charge   price asked for reserving or usage of a particular part of the system (e.g. pipeline/s, storage).

CCGT   combined cycle gas turbine.

Combination utility   utility which distributes other fuels, like electricity, water.

Commodity charge   part of the gas price or price asked for the contracted gas volume as such (distinct from other charged costs such as customer charge, capacity charge or of other services).

Common equity   is an item listed on the balance sheet. All balance sheet items are listed at book value. Common stock is valued when it is purchased in the stock market. The fixed proceeds from these sales become part of common equity. For example, if 200,000 shares of common stock of company x were purchased at $21 per share, this would increase common equity by $4.2 million dollars. On the last day of the fiscal year, this total is listed as common equity. Once these purchases have been made and added to common equity, subsequent fluctuations in the stock market have no influence on existing common equity.

Customer charge   annual/monthly fixed price of connection (sometimes also called connection charge) charged to customers in addition to the commodity charge and in some cases also in addition to the capacity charge.

Economies of scope   cost savings due to synergies between different activities (e.g. combined meter reading and billing in mixed energy utilities supplying gas, electricity and water).

Economies of scale   the reduction in unit cost as a producer makes/conveys/sells larger quantities of one product. Such reductions result from a decreasing marginal cost due to increasing specialisation, use of capital equipment, benefits of quantity purchasing and other economies.

End-user (of natural gas)   is a natural gas consumer (can be a household, an SME, an industrial consumer or a power producer) that does not resell the gas (or part of it) he purchases.

EU	European Union.
EU Directive, EU Gas Directive	binding legislation for the Member States of the European Union (they need to implement its provisions in their national laws). The so-called EU Gas Directive was adopted on 11 May 1998 by the EU Member States upon proposal by the European Commission and in co-decision with the European Parliament.
Evergreen clause	renders a contract binding until one of the parties takes action to discontinue to the contract. The evergreen clause takes effect once the initial term of the contract terminates.
Feedstock	natural gas which is used as the essential component of a process for the production of a product. ie fertilizer or glass manufacture.
FERC	Federal Energy Regulatory Commission (USA).
GDF	Gaz de France.
Horizontal integration	a horizontally integrated utility is a combination utility.
LDC	stands for "local distribution company"; in the study, however, the acronym is used in the widest sense for distribution company (whether at local or regional level).
LDCs	plural of LDC.
Line pack	gas delivered from line pack is the volume of gas supplied by the net change in pressure in the regular system of mains, transmission and/or distribution (in other words, gas delivered from the grid system excluding storage and other infrastructures).
Load balancing (hourly/daily/seasonal)	to balance demand and supply (at any given point) in a grid/pipeline/supply chain.
Load factor	ratio of average to maximum requirement for a time period i.e. one day, one hour, etc. normally expressed over the year as a percentage.
MMC	Monopolies and Mergers Commission (UK).
Natural monopoly	is a monopoly in place because competition would be impractical or not deliver benefits (e.g. 30 water supply companies in one city with each having their own infrastructure. The extra cost of installing up to 30 set of pipes would be too high in terms of externalities, and offset any possible price reduction brought about by competition.).
Netback market value	delivered price of cheapest alternative fuel to gas to the customer (including any taxes) adjusted for any efficiency difference;
	minus cost of transporting gas from the beach/border to the customer;
	minus cost of storing gas to meeting seasonal or daily demand fluctuations;
	minus gas taxes.
Nomination	electing a certain quantity of gas.
OFGAS	Office of Gas Supply (UK gas regulator).
Off-take	actual amount of gas withdrawn.

Peak periods (e.g. peak day)	the period or 24h day of maximum system delivery of gas during a year.
Peak shaving	the use of fuels and equipment to generate gas (LNG stations) or manufacture synthetic gas to supplement normal gas supply from the system during periods of extremely high demand.
Peak storage	gas storage designed/used to supplement normal gas supply in periods of extremely high demand.
RoR	Rate of Return.
RWI	Rheinisch Westfälisches Institut, Essen (D), German Think tank.
SME	small and medium sized enterprise.
Take-or-pay (TOP)	by take-or-pay commitments gas buyers agree in principle to pay their supplier for the herewith contracted (yearly) gas volumes even when they do not use the full amount of these volumes. Most TOP contracts contain clauses which allow to report over time full off-take of contracted volumes. A TOP commitment from a buyer is his ultimate commitment to the seller to take the (yearly) contracted volumes. It is an important – some sources from industry and finance claim essential – instrument for setting-up new gas production and supply projects in that it guarantees the producer maximisation of the capital intensive production and upstream transport facilities.
Third party access (TPA)	principle by which the owner of infrastructure is required to allow other parties to access/use it (equivalent terms are 'open access' or 'common carriage').
TOP	see Take-or-pay.
TPA	see third party access.
TPES	Total Primary Energy Supply.
Unbundling	means splitting up integrated companies into their different activities as if these would be carried out by separate undertakings. In natural gas, this would in most cases mean splitting the supply/trading activity from the transport and/or storage activity and from non-gas activities. (The EU Gas Directive requires that integrated gas companies unbundle their internal accounting, i.e. to keep separate accounts for their natural gas transmission, distribution, storage and non-gas activities, and put these at the disposal of government authorities.)
VAT	value added tax.

INTERNATIONAL ENERGY AGENCY

IEA Publications, PO Box 2722, London, W1A 5BL, UK
Telephone:+44 171 896 2244 - **Fax:**+44 171 896 2245
e-mail: ieaorder@pearson-pro.com

▶ *Special Discount: Order three or more products to claim a 20% discount.*

I would like to order the following publications (please tick)

PUBLICATIONS	QTY	PRICE	TOTAL
☐ World Energy Outlook - 1998 Edition		$120.00	
☐ Caspian Oil and Gas - The Supply Potential of Central Asia and Transcaucasia		$100.00	
☐ Benign Energy? - The Environmental Implications of Renewables		$100.00	
☐ Mapping the Energy Future - Energy Modelling and Climate Change		$50.00	
☐ Nuclear Power - Sustainability, Climate Change, and Competition		$60.00	
☐ Natural Gas Pricing in Competitive Markets		$110.00	
☐ Energy Policies of IEA Countries - 1998 Review		$120.00	
☐ Projected Costs of Generating Electricity - Update 1998 (NEA/IEA)		$100.00	
☐ International Coal Trade - The Evolution of a Global Market		$35.00	
☐ Renewable Energy Policy in IEA Countries - Vol. II: Country Reports		$30.00	

Sub Total	
Postage and packing ($15.00)	
Less discount	
Total	

MONEY-BACK GUARANTEE:
*Refunds are given on one-off publications returned in resaleable condition
by registered post within 7 days of receipt.*

Delivery details (If your billing address differs from your delivery address please advise us.)

Name

Position   Organisation

Address

Country   Postcode

Telephone   Fax

As all products are delivered by express courier service, please enter your telephone number above.

Payment details

☐ I enclose a cheque payable to IEA Publications for the sum of $_____

☐ Please debit my credit card (tick choice).

☐ Access/Mastercard   ☐ Diners   ☐ VISA   ☐ AMEX

Card no:

Signature: Expiry date:

EU registered companies	Non-EU registered companies
☐ Please send me an invoice along with the publications. I enclose a copy of my company headed stationery with this order form. To avoid extra charges EU companies (except UK) must supply VAT/TVA/MOMS/MST/IVA/FPA numbers:	☐ Please send me a pro-forma invoice. The publications will be forwarded to me on receipt of payment.

Data Protection Act: The information you provide will be held on our database and may be used to keep you informed of our and our associated companies' products and for selected third party mailings.

FOR FAST PROCESSING OF YOUR ORDER, FAX THIS FORM TO CUSTOMER SERVICES ON +44 171 896 2245

IEA PUBLICATIONS, 9, rue de la Fédération, 75739 PARIS CEDEX 15
PRINTED IN FRANCE BY STEDI
(61 98 19 1 P) ISBN 92-64-16182-1 - 1998